Solid/Liquid Separation: Principles of Industrial Filtration

R. J. Wakeman
Professor, Department of Chemical
Engineering, University of Loughborough, UK

E. S. Tarleton
Senior Lecturer, Department of Chemical
Engineering, University of Loughborough, UK

ELSEVIER

UK   Elsevier Ltd, The Boulevard, Langford Lane, Kidlington, Oxford OX5 1GB, UK
USA   Elsevier Inc, 360 Park Avenue South, New York, NY 10010-1710, USA
JAPAN   Elsevier Japan, Tsunashima Building Annex, 3-20-12 Yushima, Bunkyo-ku, Tokyo 113, Japan

First edition 2005

British Library Cataloguing in Publication Data
Wakeman, Richard J.
  Solid/liquid separation : principles of industrial
  filtration
  1. Filters and filtration
  I. Title II. Tarleton, E. S.
  660/2´84´84245

ISBN 1 8561 74190

Published by
Elsevier Advanced Technology,
The Boulevard, Langford Lane, Kidlington, Oxford OX5 1GB, UK
Tel: +44(0) 1865 843000
Fax: +44(0) 1865 843971

Typeset by Land & Unwin (Data Sciences) Ltd, Towcester, Northants
Printed and bound in Great Britain by MPG Books Ltd, Bodmin.

Contents

Preface

Other texts have dealt with the theory and equations of filtration, sometimes with guiding examples. The link of theory to the design process is not developed in these texts, limiting their usefulness from a practical viewpoint. This seems to have been a contributory factor in promulgating the view amongst many that theoretical development and modelling of filtration processes is inappropriate and cannot be utilised in practice. These notions must be dispelled. A total theoretical description of filtration is not currently possible, although scientifically based data are available for many of the processes which can be modelled. The research literature is made confusing by researchers claims that some new models are better than older ones – and indeed this is in some senses the case when a model is developed that takes account of previously ignored parameters. Unfortunately, in the engineering context a more complex model that takes account of basic variables in the process such as the properties of the particles to be separated is not necessarily better than an earlier model where those properties do not appear explicitly but are grouped into a lumped parameter. Many of those more complex models have unknown parameters in them which can only be evaluated by comparison with experimental data, making them little if any better than earlier models that are more easily applied. Sometimes, however, the complex model has a role in providing an understanding in a general sense of what would happen if a variable were changed – exact values that come from models still cannot be relied upon without the backing of experimental data.

The focus of this book is the presentation of models and calculation methods relevant to industrial design and simulation, linking practical aspects of filter design to models that have been proven by industrial practice. The current state of knowledge is used to inter-relate the various processes that may be operated during a filter cycle (i.e. cake formation, compression, deliquoring and washing) to provide a frame-work for an integrated design strategy. This approach enables the engineer to take into account effects of upstream operations such as crystallisation or precipitation on the solid/liquid separation, and effects of the separation on downstream operations such as drying. Tried and tested models for each stage of the filter cycle can be related to the known operational and performance characteristics of equipment, to enable process calculations to be carried out with a minimum of prior testing – this does not mean that the need for testing is eliminated!

The models used in this book have a practical significance, and their origins have a fundamental basis. A thorough knowledge of the back-gound to the models enables their limitations to be understood, and can help an engineer to reach well argued decisions about the separation process and its integration into a flowsheet. This does not mean, of course, that he/she need memorise the differential equations – this would not lead to better decisions about the process, but a qualitative understanding probably will do. For example, knowledge of the general way that particle size affects cake formation, deliquoring and washing can be very helpful qualitatively, and the models in this book enable the effects to be quantified. Particle size is also important to the selection of both filter media and a filter, and although it is controlled largely by upstream processes it can affect the performance of downstream operations. Once this is realised, its importance becomes apparent.

Solid/liquid separation technology contains numerous heuristics that have evolved through experience. A majority of process engineers need to possess wide ranging knowledge covering many unit operations, but they rarely have the opportunity to gain in-depth specialist knowledge of filtration and separation technology. Consequently the large number of heuristics that have evolved in the technology can lead to confusion for the non-expert. For example, considerable confusion results when an engineer attempts to decide which type of filter or separator is most appropriate for his/her process. In attempting to select a separator a decision has to be made from a plethora of equipment types, with competing claims from manufacturers and suppliers about their equipment capabilities.

With these practical problems in mind, the authors have also developed and published a software package – Filter Design Software (2005) – that combines the calculation methods in this book in relation to specific aspects of equipment design and performance using expert knowledge of solid/liquid separation systems and computer simulations. The process of obtaining appropriate and useful experimental data, analysing the data in a correct manner, and then using the analysed data

Filter Design Software

The software is written for use by engineers, consultants and others concerned with solid/liquid separation equipment specification, design and operation, as well as for educational and training purposes. Designed to run on a desktop personal computer, the software offers features including:

- Analysis of filter leaf test results, jar sedimentation test data, and expression data
- Calculation of scale up parameters
- Direct comparison of data from different tests
- Selection of solid/liquid separation equipment
- Simulation of vacuum filter equipment (Nutsche, multi-element leaf filters, belt, drum, disc, table, and tilting pan filters)
- Simulation of pressure filter equipment (Nutsche, multi-element leaf filters, filter presses, diaphragm, and tube filters)
- Key features of over 70 types of solid/liquid separation equipment
- Web access to equipment suppliers

Filtration Solutions

www.filtrationsolutions.co.uk

for equipment selection, performance simulation and process modelling is brought together in the software. As well as enabling all these functions, the software is intended also as a guide to the non-expert, giving information about solid/liquid separation equipment characteristics and features, together with illustrative diagrams and web access to equipment suppliers.

The structure shown of this book is hierarchical. Chapter 2 on the fluid dynamic background to solid/liquid separation processes provides the main underpinning models for Chapters 3 to 8. It is not necessary to read Chapter 2 before making use of the other chapters, as these latter chapters have been written to present results from modelling work that has been proven in practice and each is complete in itself (although knowledge of Chapter 2 would help the non-expert to become more expert and understand the inter-relationships that exist between different parts of the whole solid/liquid separation process!). All of the information presented is usable for equipment design and process simulation. Some examples of such calculations are given in the text, but more sophisticated work is made possible through the Filter Design Software.

The chosen structure makes the text useful as a handbook for both researchers and practitioners, and hopefully underlines the importance of the knowledge that both types of expert may possess. The thorough knowledge required for process design and simulation and further innovation in equipment design are likely to arise only when researchers have a good understanding of the practical problems and practitioners possess a more 'in depth' theoretical background to their processes and equipment.

Richard Wakeman and Steve Tarleton

April 2005

1 Introduction

In the simplest sense solid/liquid separation involves the separation of two phases from a suspension. Although a filter is the kernel of most solid/liquid separations, a vast array of other techniques and equipment are available to effect a separation, ranging from highly versatile units capable of handling different filtration duties to those restricted in use by the fluid properties and process conditions. Filtration, or more generally solid/liquid separation, is used in many processes with one or more aims:

1. recovery of a valuable solid component (the liquid being discarded);

2. recovery of the liquid (the solids being discarded);

3. recovery of both the solids and liquid phases; or

4. recovery of neither phase (e.g. when a liquid is being cleaned prior to discharge, as in the prevention of water pollution).

Not only is the nature of the "product" important, but a further serious consideration in the selection of solid/liquid separation equipment is whether the separation is to be effected continuously or batchwise. In the latter case the separator is operated intermittently between filling and discharge stages of its operating cycle. Scale or throughput during operation also requires careful consideration, as scale not only affects equipment size but many separator designs make them suitable for limited ranges of throughput.

1.1 The separation process

Any system design must consider all stages of solid-liquid separation, including pre-treatment, solids concentration, solids separation and

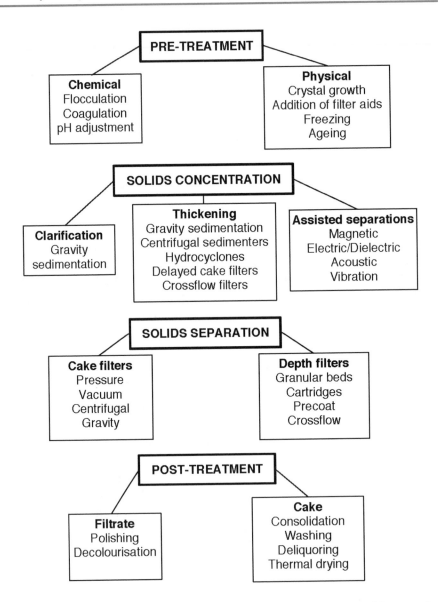

Figure 1.1 Some components of the solid/liquid separation process. A number of the separation processes above may be used alternatively as pre-treatment, solids concentration or solids separation techniques.

post-treatment. These factors encompass the wide range of equipment and processes summarised in Figure 1.1.

Pre-treatment is used primarily with "difficult to filter" slurries, enabling them to be filtered more easily. A slurry may be difficult to filter for a variety of reasons – for example, a proportion of the particles in the feed may be fine and tend to block the entrances to pores in a filter medium, or the concentration of particles in the feed may be low and cause surface or internal blocking of the filter medium, or the surface chemistry of the particles may cause them to gel or be attracted to the filter medium surfaces irreversibly, and so on. Pre-treatment usually involves changing the nature of the suspended solids by either chemical or physical means, or by adding a solid (filter aid) to the suspension to act as a bulking agent to increase the permeability of the cake that is formed during filtration. Some pre-treatment processes, notably chemical processes, simultaneously change the properties of the suspending liquid.

During concentration of the solids in a suspension part of the liquid is removed, for example by gravitational or centrifugal thickening or hydrocycloning. This reduces the liquid volume throughput load on a filter and generally enables a filter cake to be formed more readily. A number of "assisted separation" techniques are also gradually making their way into the list of technical alternatives available to the separations technologist – these utilise magnetic, electrical or sonic force fields either individually or in combination to provide for a more effective separation. Although still quite uncommon in practice, their existence should be recognised.

Solids separation usually involves a filter, the types of which can be classified in many different ways. For present purposes, the divisions in Figure 1.1 indicate those separations in which a cake is formed and those where the particles are captured in the depth of the medium is adequate. These general categories hide the complexities associated with the wide range of filter types available, which often causes confusion for people who do not need to have an in-depth knowledge of filtration problems and applications. The underlying fundamental models and approaches to the analysis of solid/liquid separation problems, which can also be used for the scale-up and sizing of filter systems, are amplified in this book. Cake filters can be further divided into pressure, vacuum, centrifugal and gravity operations.

Post-treatment processes involve making improvements to the quality of the solid or liquid products. In the case of the liquid product, the filtrate, these operations are often referred to as polishing processes. Polishing may involve, for example, fine filters to remove small contaminants from the filtrate. Further purification could involve removal of ionic and macromolecular species by reverse osmosis, ion exchange or adsorption. Cake post-treatment processes include washing soluble impurities from the cake voids, and the removal of excess liquid from the voids to reduce the cake moisture content.

Thermal drying is often the final stage of liquid removal from the solids that are discharged from a filter as a cake.

1.1.1 Solid/liquid separation in the process flowsheet

Solid/liquid separation is too often designed as though it is a 'stand-alone' process in a plant flowsheet – taking a traditional unit operations textbook approach. The performance of a solid/liquid separation device is sensitive to the history of the feed suspension, for example on the characteristics of the particles which result from conditions during precipitation or crystallization. Changes in the operating conditions of the particle production process often alter the mean size, the size distribution or the shape of the particles in the slurry fed to a filter, each of which can have a profound effect on the characteristics of separation and hence on the performance of a filter. *This is turn means that a change in particle production conditions can affect the best choice of filter for a separation!* All too often the choice of solid/liquid separation equipment is perceived to be between alternative separator types – for example between a centrifuge and a rotary vacuum filter. In fact the situation is considerably more complex as many equipment items up- and down- stream of a separator are frequently interdependent, and all ancillary equipment must be included in an analysis of the separation process. The choice of ancillary components varies according to the type of separator and the flowsheet, and on other units in the process. It is often possible to devise a multiplicity of solutions to any separation problem that lead to alternative process flowsheets.

The economics and viability of producing a product are often affected by the amount of liquid removed in post-treatment processes. For example, if the cake is to be transported, briquetted or pelletized the cake moisture content will need to be within a specified range, or if a bone dry product is required the thermal load on the dryers can be reduced by correct choice and operation of the filter.

Where interactions between process steps exist "local" and "global" trade-offs are identifiable – that is, the operation of one or more pieces of equipment may not be optimal if the overall flowsheet is optimised. Local trade-offs affect single items of equipment, whilst effects of global ones manifest themselves on the design of more than one item of equipment. Consequently the latter are of considerable importance when defining optimum design and operating conditions for a process as a whole. Consider the simple flowsheet shown in Figure 1.2 for the recovery of particles from an evaporative crystalliser. In this case particle size and temperature are major global variables and the effect of increasing either variable on the unit operation cost can be summarised as:

Variable ↓	Change in the operating costs of		
	Evaporative crystallizer	Solid/liquid separator	Dryer
	...... due to an increase in the variable.		
Particle size	increase	decrease	decrease
Temperature	decrease	decrease	decrease

Here two different effects of global variables are observed. Increasing the particle size usually enables easier separation and less moisture in the discharged cake, resulting in reduced costs for both separation and drying. However, the crystallizer size and cost are increased and an "optimum" particle size may be sought when size is not governed by market criteria or required properties of the particle. Increasing the temperature, on the other hand, leads to lower costs in all three unit operations – hence the "best" temperature will be one close to the temperature of an available utility such as steam. This is but one quite simple example, and each process flowsheet requires investigation in its own right using information pertinent to the particular problem.

On Figure 1.2 minor global variables can also be identified between two equipment items. Operating at higher slurry densities in the

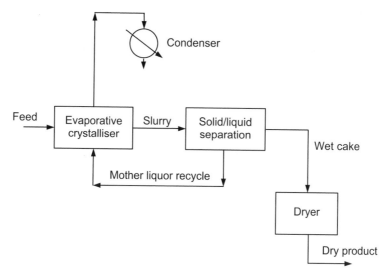

Figure 1.2 Possible simplified flowsheet structure for particle production.

crystallizer take-off can reduce the size of both the crystallizer and the separator. Producing a drier filter cake tends to increase mechanical separation costs, but reduce thermal dryer costs. Hence an optimum moisture content can be sought.

It is important, therefore, to consider simultaneously and to the same detail those unit operations that supply suspension to the solid/liquid separations plant, and the subsequent processing of the solid or liquid products.

1.1.2 Filter media

A filter medium is defined as *any material that, under the operating conditions of the filter, is permeable to one or more components of a mixture, solution or suspension, and is impermeable to the remaining components* (Purchas and Sutherland, 2002). The principal role of the filter medium is to cause a clear separation of particulates (or sometimes other components) from the fluid with the minimum consumption of energy. As such, the medium is a key component in the filtration system – a failure of the medium is tantamount to a failure of the filter and will often lead to product losses and process downtime.

The variety of filter media available is diverse and includes, amongst others, woven fabrics, non-woven media, fibrous materials, polymeric and ceramic sheets, sintered metals and perforated sheets. Woven (see Figure 1.3) and non-woven media (see, for example, Figure 1.4) are composed of single or multiple yarns or fibres that are structured to create the medium. Woven fabric media are the most commonly used type, with natural (for example cotton, wool) and synthetic (for example, polymers, metals carbon) fibres being used together with a variety of weave patterns, including modern three dimensional or double layer weaves. The combination of material and cloth structure (weave) gives a wide range of media properties, and a corresponding wide range of media from which to chose one that is most suitable for any particular filtration. This choice is further extended by layering the woven fabric with a microporous polymeric material to yield a composite medium (see Figure 1.5) which offers even greater variety of surface and filtering properties.

Non-woven media are accumulations of fibres that are held together by some form of bonding to create a flexible sheet of fabric. This includes felts which use the basic characteristics of the fibre to ensure mechanical integrity of the fabric, and bonded fabrics in which additional adhesive material or heat is used to fix the yarns in place. A common type of non-woven fabric are needlefelts (see Figure 1.4) in which layers of

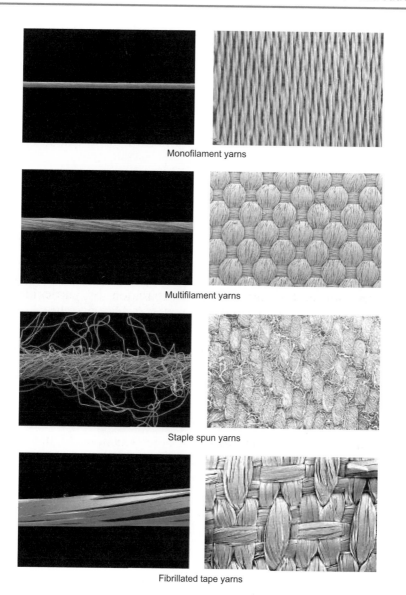

Figure 1.3 Common yarn types used in filter media and examples of the appearance of woven media manufactured from the yarns (Madison Filter).

carded fibre are stacked one upon another to form a 'batt', which is then formed into a denser structure of entangled yarns by needle-punching with barbed needles. Other non-woven types include resin or thermally bonded media and dry-laid spun media.

The industrial development of membrane filtration utilises thin sheets of permeable material, commonly made from polymers, ceramics and metals. Membrane filters also use tubular and hollow fibre media as well

Figure 1.4 Monofilament yarns made into nonwoven media – the picture on the right shows a needlefelt that is more commonly used in gas filtrations (Madison Filter).

Azurtex™ – Showing coating of the fabric and penetration between the fibres

Primapor™

Figure 1.5 Composite filter media in which woven fabrics are coated with a microporous polymer to lock the yarns into place (Madison Filter).

as sheets (which may be configured into pleated or tubular structures). Many membranes are of an asymmetric structure, composed of a thin skin which acts as the surface filter which is supported by a thicker layer designed to give mechanical integrity to the whole structure. The thickness of the membrane varies with the type of material from which it is made, but may be from <1 μm to several hundred μm. The aim of most membrane structures is to promote surface filtration whilst discouraging depth filtration; with this in mind, thinness is a desirable attribute which also leads to lower pressure losses. Polymeric membranes are probably the most widely used, and there are several methods for their production, including sintering of powders, stretching polymeric films, track-etching (exposing a polymer film to a beam of accelerated argon ions that pass through the film to form a more or less cylindrical pore that is then chemically etched), and solvent casting or phase inversion. Most commercially available membranes are obtained by phase inversion, which is a versatile technique allowing a range of different membrane morphologies to be produced. In summary, the technique is to transform a polymer in a controlled manner from a liquid to a solid state – the solidification process is often initiated by the transition from one liquid into two liquids (liquid-liquid demixing), and at some stage during the demixing a high polymer concentration phase solidifies to form a solid matrix. The process may use one of several methods – solvent evaporation, thermal precipitation, and immersion precipitation (which is the most common). Depending largely on the manufacturing methods employed, alternative membrane structures can be formed (see Figure 1.6).

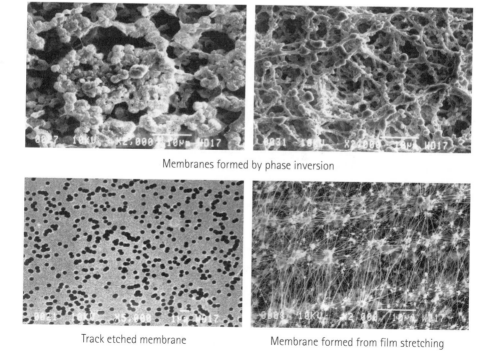

Membranes formed by phase inversion

Track etched membrane   Membrane formed from film stretching

Figure 1.6 Microfiltration membrane filter media formed by different techniques.

The above is not a comprehensive list of media types that are available, but even from these it is evident that many hundreds of media are available to meet the needs of filtration.

1.1.3 Cake and depth filtration

Filter media retain particles in two principal ways. When particles are predominantly larger than the sizes of the pores in the filter medium, they form a deposit on the medium surface known as a *filter cake* – for this reason this mode of filtration is referred to as *cake filtration*. When particles are generally smaller than the pores sizes in a filter medium, deposition occurs within the internal structure of the medium (of course, some very small particles may pass through the medium and be collected in the liquid product – the *filtrate*). This process of separation may be due to mechanical or surface chemical effects and is referred to as *depth filtration* and is the modus operandi of, for example, deep bed sand filters and some types of cartridge filters. In these applications the concentration of solids in the feed is usually very low, enabling particles to pass relatively unhindered into pores by following the trajectory of fluid streamlines. The principle of depth filtration is illustrated in Figure 1.7a.

At the microscopic scale cake filtration is achieved by a combination of two primary mechanisms, referred to as *complete blocking* shown in Figure 1.7b (a sieving process that occurs when the particles are larger than the pore sizes) and *bridging* shown in Figure 1.7c. Bridging occurs when particles smaller than the pore sizes in the filter medium form a cake; this occurs particularly when the particles are at a higher concentration in the feed. Several particles then attempt to pass simultaneously into a pore at the surface of the medium, but fail to do so and form a bridge over the pore entrance. This is essentially an arch stabilised by the flow environment around a pore entrance, and can become destabilised if flow velocities or directions are changed substantially.

1.1.4 "Laws" of filtration

The so-called laws of filtration inferred by the mechanisms described by Figure 1.7 have been studied by Hermans and Bredée (1935), Gonsalves (1950), Grace (1956) and Hermia (1982) amongst others. Their origins stem from a stochastic modelling of filtration and the behaviour of a particle arriving at the surface of the filter medium.

An extreme situation is described by the "complete blocking filtration law". Consider a filter medium similar to that used in liquid clarification: the medium may be a sheet of paper or similar porous material

Flow direction

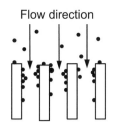

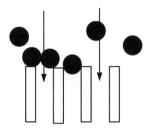

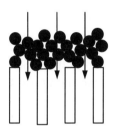

(a) Standard Blocking Filtration
- particle sizes < pore sizes
- low concentration in feed
- particle capture is predominantly inside the filter medium

(b) Complete blocking filtration
- particle size > pore size
- low/medium concentration in feed
- particle capture by sieving or screening process
- limited bridging is possible

(c) Bridging filtration
- particle size < pore size
- higher concentrations of particles in feed
- particle capture at surface of medium
- stable and permeable bridges formed

Figure 1.7 Mechanisms of filtration. Depth filtration occurs by the standard blocking law that describes gradual accumulation of particles in the medium, and most cake filtrations occur by a combination of blocking and bridging. An intermediate blocking mechanism allows for the particles to cause both pore blocking and cake formation.

or a thin layer of filter aid. Each particle that reaches the medium surface participates in a blocking process that results in sealing of a pore. This leads to the assumption that further particles are not superimposed on those already deposited. Hence, once a volume of slurry has been filtered a portion of the medium surface becomes blocked and is no longer available to support the flow of filtrate.

The "intermediate blocking filtration law" (this is not shown on Figure 1.7) considers that a particle arriving at the surface of the filter medium will block a pore, but particles arriving later may come to rest on already deposited particles. In other words, not every particle will block a pore but there exists is a probability that this event may occur.

A less severe blocking process is described by the "standard blocking filtration law". It is assumed here that the pore volume decreases proportionally with the filtrate volume produced, due to deposition of particles on the walls of a pore. In this case all pores are assumed to be the same diameter and length, and the decrease in pore volume is directly related to a decrease in the pore cross-sectional area. Hence, build up of a cake does not occur but particles are filtered by being trapped within the internal volume of the medium. Such entrapment may result from diffusional, inertial or electrostatic effects that cause a particle in the feed to contact the pores walls, and once contact has occurred it is assumed that the particle is filtered. The mass of deposit, or dirt holding capacity, is related to the amount of internal surface in the medium and the pore volume of the medium. When the limiting capacity has been reached, either particles will break through at the downstream surface of the medium and the filtrate will become contaminated or the filter will become blocked internally and no further useful flow of filtrate will be obtained.

When the solids are to be recovered from a filter, a relatively thick layer or cake is required. This is described by the "bridging filtration law" or "cake filtration law". In this case the particles may be slightly smaller than or larger than the pores in the medium, and the effect is enhanced by higher concentrations of solids in the feed. As several particles simultaneously approach a pore in the medium surface, there is a probability that they will form a bridge or arch over the surface of the pore. Such bridges or arches are quite stable provided the directions of flow do not change substantially, and they will support subsequent formation of a cake.

The characteristic form of the filtration laws is:

$$\frac{d^2t}{dV^2} = k_1 \left(\frac{dt}{dV} \right)^{k_2} \tag{1.1}$$

for constant pressure filtration, where k_1 and k_2 are constants. k_1 is dependent on the initial flow rate of slurry reaching the filter medium surface, and $k_2 = 2, 1.5, 1$ and 0 for complete blocking, standard blocking, intermediate blocking and cake filtration respectively.

Although these so-called "laws" are convenient for visualising and giving an understanding to the microscopic phenomena that may be happening at the filter medium surface, they do not describe the physics of particle deposition beyond the initial few moments of filtration. As a result their use for filter design purposes is limited.

1.2 Particle and liquid properties in filtration

Three parameter types may be identified to fully describe a solid/liquid system. These are the primary properties, the state of the system and the macroscopic properties.

The primary properties are those which can be measured independently of the other components of the system, and are the solid and liquid physical properties, the size, size distribution and shape of the particles, and the surface properties of the particles in their solution environment. The way the particle interacts with its surrounding fluid becomes important for smaller particles (notably when the particle size is smaller than about 20 μm), since the net attractive or repulsive forces between particles can become as significant as gravitationally or hydrodynamically induced forces. These factors decide whether the particles will, for example, settle slowly or quickly, whether they can be retained on some kind of filter medium, or whether the resulting cake will have a dry or sloppy consistency.

The description of the state of a system (porosity or concentration, and the homogeneity and extent of dispersion of the particles) combines with the primary properties to control the macroscopic properties that are measured to investigate the application of a particular separation method. Such measurements may be the permeability or specific resistance of a filter bed or filter cake, the terminal settling velocity of the particles, or the bulk settling rate of particles in a suspension.

1.2.1 Particle shape

Droplets almost always approximate a sphere due to the distribution of surface forces around their periphery. By contrast, solid particles are rarely either spherical or uniform. Certain classes of materials are essentially crystalline and may be made up from fairly uniform particles each of which is, for example, cubic or rhombohedral. But even crystalline materials may be a mixture of shapes, especially if, as often happens industrially, breakage of the crystals occurs due to the way they are handled. Indeed, breakage can be caused within the separator itself and is a common problem in, for example, pusher centrifuges. The great majority of particles are of irregular shape; fibrous particles are common, but they may possess a wide range of length to diameter ratios, they may have smooth surfaces or they may be fibrillated, and so on. *It is rare that the shape of the particles to be handled can be defined precisely.*

1.2.2 Particle size

Particles may vary in size from very fine or colloidal matter or molecular aggregates to coarse granular solids. Sometimes all the solids may be of the same material, that is, of homogeneous composition or, as is often the case with effluent suspensions, the individual particles may have very different compositions. In general terms particle size has a significant effect on the solid/liquid separation behaviour of a suspension. Knowledge of the techniques for measuring the size of particles is therefore important to the process technologist.

In solid/liquid separation four reasons for measuring particles size have been identified (Scarlett and Ward, 1986):

(a) To measure and specify the quality of a liquid which is the valuable product from a filtration process. In this case the particles remaining in suspension are dilute in concentration, and are therefore difficult to filter. Often only the total concentration of solids is required, as for example in water treatment processes. However, in operations such as the filtration of parental fluids or hydraulic fluids, the size (and occasionally shape) of the remaining solids is critical.

(b) An extension of this requirement is to specify the performance of a filter medium in terms of its ability to retain particles of different sizes. This type of evaluation is usually associated with fluid polishing operations and the specification of a nominal pore size for a polishing medium, or with the performance assessment of separating devices such as sedimenting centrifuges.

(c) In many operations the solid is the valuable product. It is rarely recovered in a completely dry state and is often processed further. The evaluation of the product is required for quality control and is not connected solely with the separation process. In this case the method of evaluation is often dictated by the customer or by the standards accepted by the particular industry.

(d) Occasionally the requirement is to evaluate the solids in order to predict their probable behaviour in a separation process. This may be to enable an initial choice between different separation methods, to select or test a suitable pre-treatment process or filter medium, to improve the efficiency of an existing machine, or to estimate the size of a new one. In any of these the objective is predictive, and the measuring technique must be selected more carefully than for a quality control application.

There are a large number of alternative methods for determining the size distribution of particles in suspensions. The results obtained are dependent on the measurement method used and frequently also on the skill of the analyst. To measure the size distribution of irregularly shaped particles it is usual to employ a measurement technique that represents the process under investigation. For example, techniques that give an equivalent surface diameter would be used to measure the size of particles involved in chemical reactions, but a projected area diameter might be used to represent the size of pigments in paints. The particle size most commonly required in cake formation, deliquoring or washing calculations is the surface-volume mean diameter, x_{sv}. If a liquid permeameter were used to measure the mean particle size the results might be analysed using the Kozeny-Carman equation, in which case the volume specific surface (S_0) would be calculated. For a sphere, the surface-volume mean diameter is related to the volume specific surface by:

$$S_0 = \frac{6}{x_{sv}} \tag{1.2}$$

There are often practical difficulties behind obtaining this size unless a well established particle sizing laboratory is available, and so the size measured is actually one that is related to x_{sv}. It is nowadays common to use laser diffraction based instruments to measure particle size distributions.

In sedimentation or thickening calculations where a particle size is required, one that is calculated by an application of Stokes' Law is most appropriate. This suggests the use of sedimentation measuring techniques. Incremental methods such as the Andreasen pipette and photosedimentometer fall into this category, as well as cumulative methods such as the sedimentation balance. The particle size range of gravity sedimentation techniques (1 to 60 μm) can be extended to smaller sizes (0.05 to 5 μm) by the use of centrifugal devices.

1.2.3 The solution environment

Interactions between a particle and the liquid in which it is suspended have greatest influence when the particles are smaller, particularly when their size is smaller than a few microns. Origins of interparticle repulsive forces lie in the distribution of solution ions around the charged surface of the particle, and the resultant electrical charge is dependent on the chemical species present at the surface. A potential energy of repulsion may extend appreciable distances from the particle surface, but its range may be compressed by increasing the electrolyte content of the solution. For practical purposes the magnitude of the net repulsive force between particles is represented by the zeta (ζ-) potential, and the following statements can be made about the influence of the ζ-potential in solid/liquid separation (Wakeman et al, 1989):

(a) The net repulsive force between the particles increases with the solids volume fraction in the solid/liquid mixture;

(b) The net repulsive force increases with increasing ζ-potential;

(c) Reducing the magnitude of the repulsive force causes the dispersion to become unstable, and generally more easily separated;

(d) Repulsive forces can be reduced by either (i) adding a non-adsorbing electrolyte to the liquid to change the distribution of solution ions around the particle, or (ii) altering the electrical charge on the surface of the particle by the specific adsorption of certain ions or charged polymers.

The link between the state of dispersion of a suspension, the rate of settling and filtration and some other cycle operations is shown succinctly on a plot of the ζ-potential against the solution pH for the suspension, as in Figure 1.8.

The filtration engineer often manipulates the ζ-potential unwittingly by using coagulants or flocculants to create the greatest separability of a suspension, and more often than not this is done using an *ad hoc* approach. Understanding the link between the ζ-potential and separation characteristics such as cake formation rates and settling rates can often shorten testing programmes when evaluating the separability of new suspensions.

In summary, around the isoelectric point of the suspension the process engineer can expect:

(i)   faster settling rates,

(ii)   more rapid filter cake formation, and

(iii) slightly higher moisture content cakes and sediments,

due to the aggregation of particles in the suspension where the interparticle repulsion forces are very small. The rates of deliquoring and wash liquid flow may also be slightly higher, provided the cake bulk volume does not reduce during these operations. Wash liquor feed can change the ion species distribution in the cake pores and cause the cake porosity to reduce, which actually makes wash liquor utilisation

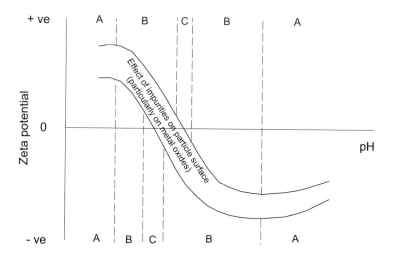

A: High ionic strength causes compression of the double layer.
   Intermediate settling and filtration rates and cake moisture contents.

B: Higher zeta potential and greater repulsion between particles.
   Well dispersed, stable suspension.
   Slower settling and filtration rates.
   Lower cake moisture contents.

C: Close to the isoelectric point (zero zeta potential).
   Little repulsion between particles.
   Particles tend to aggregate.
   Faster settling and filtration rates.
   Higher cake moisture contents.

Figure 1.8 The link between particle phenomena at the microscopic scale and process separations at the engineering scale of operation.

slightly more effective. At the maximum or minimum ζ-potential the engineer can expect:

(i)   slower settling rates,

(ii)   slower cake formation rates, and

(iii) slightly lower moisture content cakes and sediments,

due to the existence of greater repulsive forces maintaining the particles better dispersed in the suspension. Slower rates of cake formation are largely due to the formation of a lower porosity filter cake, which then takes longer to deliquor or wash. However, the lower porosity may also lead to improved wash liquor utilisation.

At pH's beyond those at which maximum or minimum ζ-potentials first occur the process engineer can expect intermediate settling, filtration and consolidation rates and intermediate cake and sediment moisture contents. This is due to compression of the electrical double layer around the particle caused by high ionic strengths in the solution.

1.2.4 The nature of the fluid

Apart from the effect of the fluid components (such as ions in solution) at the fluid/particle interface, the density and viscosity of the fluid are most important in industrial filtrations. Density is generally only significant where the separation depends on a difference in density between the fluid and the particles, for example in thickeners or centrifugal sedimenters. Whatever difference in density exists must usually be accepted and cannot often be controlled to any significant degree. Occasionally the influence of temperature may be important, or it may be possible to alter the density to a very limited extent by varying the amount of dissolved solids.

Viscosity is more widespread in its effects and at the same time is more amenable to control, since it is usually sensitive to changes of temperature. The rate of filtration of liquids can be greatly accelerated in many instances by a relatively small increase of the temperature, which causes a drop in the viscosity. (The same benefits are not obtainable with gases as their viscosity increases as the temperature is raised.) A hot feed may also leave hot liquid in the pores of the cake prior to deliquoring, which makes deliquoring more rapid and may also lead to lower cake moisture contents. Heating the wash liquor reduces the its viscosity and hence causes higher local velocities in the pores of the cake; this may not assist cake washing, but the increased rates of mass transfer will do so. The net effect of heating the wash liquor is not always apparent.

1.3 Conclusions – perfection in separation!

A perfect separation of the phases is never possible, as there is always some liquid which leaves with the solid product and some particles which leave with the liquid product. The imperfection of the separation is classified in a number of ways. For example, the mass fraction of solids recovered can be expressed as a separation efficiency (%) and the dryness of the solids recovered may be characterised by the moisture content (% by weight). Other definitions are used in various specific operations such as cake washing and cake deliquoring, and these are amplified in the appropriate place in this book.

2 Fluid dynamic background to filtration processes

There are many texts which give excellent accounts of the theory of flow in porous media, and which have relevance to filtration and cake post treatment processes. Design related aspects of the processes, however, use only a small amount of that knowledge. This chapter is concerned with setting out the principles which are used for the purposes of designing filter systems in subsequent chapters, and which give a better understanding of flow through filter media and the formation, deliquoring, washing and consolidation of filter cakes.

Basic laws governing the flow of liquids through uniform, incompressible beds serve as a basis for the development of formulae for the more complex, non-uniform, compressible cakes which occur in many filtration operations. Filter cakes are a class of porous medium, that is an assemblage of particles which might be visualised as a solid body through which a fluid may flow. Analogues between fluid flow through capillaries or ducts of regular cross-section and the pores in a porous medium are commonly accepted. In fact, the so-called pores are void spaces and do not have a regular cross-section, nor are they uniformly distributed throughout the volume of the medium. However, such analogies and models enhance our understanding of filtration problems and are presented in this chapter when their inclusion would enable an engineer to have an improved knowledge about the process. Much of the information in this chapter is also built upon in subsequent chapters, leading to details of design methodologies that have proved useful in practice.

2.1 Motion of particles in a fluid

Many methods for the separation of particles of differing sizes and/or shapes depend on variations in the behaviour of the particles when they are moving relative to a fluid. A particle immersed in a flowing fluid is acted upon by both pressure and viscous forces from the fluid. The sum of the forces which acts normal to the free stream direction is the lift, and the sum which acts parallel to the free stream direction is the drag. Buoyant or weight forces may also act on the particle, but are differentiated from lift and drag forces as the latter are limited by definition to those forces produced by the dynamic action of the flowing fluid.

The particle Reynolds number is used to describe the flow of a particle through a fluid, and is defined by:

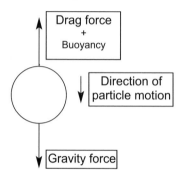

$$Re = \frac{\rho u x}{\mu} \qquad (2.1)$$

where ρ and μ are the density and viscosity of the fluid, u is the particle-fluid relative velocity, and x is the particle size. When Re < 0.2, flow around a spherical particle is completely laminar (creeping flow) and is amenable to theoretical analysis. Stokes (1851) solved the hydrodynamic equations of motion, the Navier-Stokes equations, to give

$$F_D = 3\pi\mu u x \qquad (2.2)$$

which is known as Stokes' law. F_D in this equation is made up from skin friction ($2\pi\mu u x$) and form drag ($\pi\mu u x$). Deviations from this equation become greater at higher Reynolds numbers when the skin friction becomes proportionately less than the form drag; at Re greater than about 20, flow separation occurs with the formation of vortices in the wake of the particle. The sizes of the vortices increases with Re until, at values of between 100 and 200, instabilities in the flow cause vortex shedding.

When a particle moves relative to a fluid in which it is suspended there exists a force opposing motion, known as the drag force. The drag force, F_D, is expressed by Newton's law as:

$$F_D = C_D A_p \frac{\rho u^2}{2} \tag{2.3}$$

where A_p is the area of the particle projected in the direction of motion and C_D is the proportionality constant known as the drag coefficient.

Elimination of F_D between equations (2.1), (2.2) and (2.3) gives an alternative form of Stokes' law:

$$C_D = \frac{24}{\text{Re}} \quad \left(\text{for Re} < 0.2\right) \tag{2.4}$$

This equation is shown on a graph of the drag coefficient plotted against the Reynolds number using logarithmic scales as a straight line. The complete relationship between C_D and Re is shown on Figure 2.1.

Several flow regions can be identified on Figure 2.1. In the region 0.2 < Re < 500–1000 the slope of the curve changes progressively to 0. In the region 500–1000 < Re < 2×10^5, Newton's law is applicable,

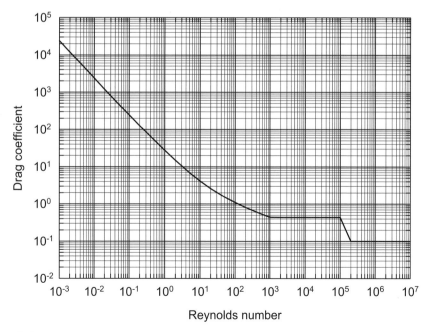

Figure 2.1 Drag coefficient versus particle Reynolds number for spherical particles.

$C_D \approx 0.44$ and the drag force on the particle is $F_D = 0.055\pi x^2 \rho u^2$. When $Re > 2\times10^5$ flow in the boundary layer around the particle changes from streamline to turbulent, flow separation takes place closer to the rear of the particle and $C_D \approx 0.1$.

In solid/fluid separation the greatest concern is with the fine particles that are most difficult to separate. Re is invariably small and often less than 0.2 due to low values of u and x.

2.1.1 Gravity settling at low concentrations

A single particle settling in a gravity field is subjected primarily to drag and gravity forces and buoyancy, and a force balance on the particle gives:

$$m\frac{du}{dt} = mg - m\frac{\rho}{\rho_s}g - F_D \tag{2.5}$$

$$\begin{Bmatrix} \text{inertial} \\ \text{force} \end{Bmatrix} = \begin{Bmatrix} \text{gravity} \\ \text{force} \end{Bmatrix} - \{\text{buoyancy}\} - \begin{Bmatrix} \text{drag} \\ \text{force} \end{Bmatrix}$$

where m is the particle mass, g is acceleration due to gravity, ρ_s is particle density, and t is time. In equation (2.5), the downward forces are taken as positive. Dividing through by m and introducing Stokes' law for F_D gives the following equation of motion:

$$\frac{du}{dt} = g\frac{\rho_s - \rho}{\rho_s} - \frac{u}{\tau} \tag{2.6}$$

where $\tau = \dfrac{m}{3\pi\mu x}$ is known as the particle relaxation time. For spherical

particles $m = \dfrac{\pi x^3}{6}\rho_s$, and $\tau = \dfrac{x^2 \rho_s}{18\mu}$. Equation (2.6) can be integrated to

give the velocity of the particle at time t as

$$u(t) = \tau\frac{\rho_s - \rho}{\rho_s}g\left(1 - e^{-t/\tau}\right) \tag{2.7}$$

when the particle is initially at rest. In this equation, as t becomes large $u(t)$ can be written as

$$u_t = \frac{x^2\left(\rho_s - \rho\right)g}{18\mu} \tag{2.8}$$

which is known as the terminal settling velocity in a gravitational field. For fine particles the terminal settling velocity is reached very rapidly and for most engineering calculations the acceleration period can be neglected. These considerations form the basis of simple scale-up calculations for non-flocculated systems, within the constraints of Stokes' law. There is a lower particle size (about 1 μm) below which Stokes' law cannot be applied. This is where Brownian diffusion becoming significant, which leads to settling rates lower than those predicted by Stokes' law.

Example 2.1 Calculation of the terminal settling velocity.

Calculate the terminal settling velocity of a 20 μm spherical particle with a density of 3000 kg m^{-3} settling in a liquid with a density of 900 kg m^{-3} and a viscosity of 0.001 Pa s. How long does it take to reach 99% of this velocity if it is initially stationary.

Solution

Using equation (2.8): $u_t = \dfrac{\left(20\times10^{-6}\right)^2 (3000-900)9.81}{18\times0.001} = 4.58\times10^{-4} \text{ m s}^{-1}$

Check that the Re < 0.2 criterion is not violated:

$$\text{Re} = \frac{900\times4.58\times10^{-4}\times20\times10^{-6}}{0.001} = 0.00824$$

Rearranging equation (2.7): $\quad t = \tau \exp\left(\dfrac{1}{1-\dfrac{u(t)}{\tau g}\dfrac{\rho_s-\rho}{\rho_s}}\right)$ (2.9)

At 99% of the terminal settling velocity,

$u(t) = 0.99 \times 4.58 \times 10^{-4} = 4.53 \times 10^{-4}$ m s^{-1}

$\tau = \dfrac{\left(20\times10^{-6}\right)^2 3000}{18\times0.001} = 6.67\times10^{-5}$ s

$\dfrac{\rho_s-\rho}{\rho_s} = 0.7$

Substituting these values into equation (2.9) gives the acceleration time:

$$t = 6.67 \times 10^{-5} \exp\left(\frac{1}{1 - \frac{4.53 \times 10^{-4}}{6.67 \times 10^{-5} \times 9.81} \cdot 0.7}\right) = 4.64 \times 10^{-4} \quad \text{s}$$

These calculations demonstrate that the acceleration period of small particles is very short and for most practical purposes can be neglected.

The Re < 0.2 criterion indicates that there is a critical size of particle, x_{cr}, above which Stokes' law not valid. This is estimated by setting Re = 0.2 and using equation (2.8) as follows:

$$Re = \frac{\rho u_t x_{cr}}{\mu} = \frac{x_{cr}^3 g}{18\mu}(\rho_s - \rho)\rho = 0.2$$

Rearranging gives:

$$x_{cr} = \sqrt[3]{\frac{3.6\mu^2}{g(\rho_s - \rho)\rho}} \tag{2.10}$$

2.1.2 Settling in a centrifugal field

For a spherical particle moving in a fluid in a centrifugal force field, the equation of motion in the Stokes' law region is:

$$\frac{\pi x^3}{6}(\rho_s - \rho)r\omega^2 - 3\pi\mu x\frac{dr}{dt} = \frac{\pi x^3}{6}\rho_s\frac{d^2 r}{dt^2} \tag{2.11}$$

where r is the radius of rotation and ω is the angular velocity. As the particle moves outwards from the centre of the radius of rotation the accelerating force increases, therefore the particle never reaches a terminal velocity. Neglecting the inertial terms on the right hand side of equation (2.11) gives:

$$\frac{dr}{dt} = \frac{x^2(\rho_s - \rho)r\omega^2}{18\mu} = u_t\frac{r\omega^2}{g} \tag{2.12}$$

That is, the instantaneous velocity of the particle is equal to the terminal velocity in a gravitational field increased by the factor $r\omega^2/g$.

Integration of equation (2.12) gives the time taken for a particle to move from radius r_1 to radius r_2:

$$t = \frac{18\mu}{x^2\omega^2\left(\rho_s - \rho\right)} \ln\frac{r_2}{r_1} \qquad (2.13)$$

For the Newton's law region the equation of motion is:

$$\frac{\pi x^3}{6}\left(\rho_s - \rho\right)r\omega^2 - 0.055\pi x^2\rho\left(\frac{dr}{dt}\right)^2 = \frac{\pi x^3}{6}\rho_s\frac{d^2r}{dt^2} \qquad (2.14)$$

If the acceleration term is neglected, the time taken for a particle to move from radius r_1 to radius r_2 is:

$$t = \left[\frac{\rho}{3x\omega^2\left(\rho_s - \rho\right)}\right]^{0.5} 2\left(r_2^{0.5} - r_1^{0.5}\right) \qquad (2.15)$$

2.1.3 Suspensions with high concentrations of solids – hindered settling

When the concentration of the suspension increases, the particles are closer together and the motion of any single particle is affected by the motion of neighbouring particles. The net effect is an increase of the settling velocity if the particles are not distributed uniformly, because the (upward) return flow due to volume displacement predominates in particle sparse regions – this is the so-called cluster formation effect, which is limited to near monosized particle suspensions. In most practical suspensions clusters do not survive for long enough to affect settling behaviour (Svarovsky, 1990), the return flow is more uniformly distributed, and the settling rate declines with increasing concentration. This process is referred to as hindered settling.

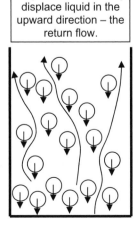

As particles settle they displace liquid in the upward direction – the return flow.

For non-flocculated systems Richardson and Zaki (1954) showed that the settling rate is linked to the terminal settling velocity (Re_t) of the particles by the function of the voidage of the suspension:

$$u_h = u_t \varepsilon^n \tag{2.16}$$

where u_h is the hindered settling velocity and ε is the voidage or porosity of the suspension. The index n was found to be related to the Reynolds number based on the terminal settling velocity and to the diameter of the vessel in which the sedimentation is taking place, D. The correlations for n are:

$$
\begin{aligned}
&\text{For } Re_t < 0.2, && n = 4.6 + 20\,x/D \\
&\quad 0.2 < Re_t < 1, && n = \left(4.4 + 18\,x/D\right)Re_t^{-0.03} \\
&\quad 1 < Re_t < 200, && n = \left(4.4 + 18\,x/D\right)Re_t^{-0.1} \\
&\quad 200 < Re_t < 500, && n = 4.4\,Re_t^{-0.1} \\
&\quad Re_t > 500, && n = 2.4
\end{aligned}
\tag{2.17}
$$

The correlations only apply when flocculation is absent, such as might be the case with coarse mineral suspensions. Suspensions of fine particles have a very high specific surface and tend to flocculate, and hence show a different behaviour. There are other similar correlations in the literature, but these have similar limitations attached to them.

2.2 Darcy's Law

Models for cake formation are based on Darcy's law which was developed originally to describe the flow of water through porous sand beds. Although the sand particles were quite large and fairly regularly shaped compared with many of the particles separated by filtration, the basis of flow through particle beds as set out by Darcy nonetheless provides a valid basis to describe flow through filter cakes.

Darcy (1856) carried out a series of experiments involving the flow of water through a bed of sand placed in a vertical iron pipe. During tests using two different sands Darcy found the flow rate to be proportional to the head, indicating that the flow regime was laminar. Although Darcy did not include viscosity in his original formula, the equation describing this observed relationship is customarily written as:

$$u = \frac{-k}{\mu}\frac{dp}{dz} \tag{2.18}$$

where dp is the dynamic (hydraulic) pressure difference across thickness dz of porous medium of permeability k, and u is the superficial velocity (volume flow rate per unit cross-sectional area of the bed, $m^3\ m^{-2}\ s^{-1}$) of liquid with a viscosity μ flowing through the bed. In packed beds it is customary to assume that the solids are stationary and that u is constant.

The permeability coefficient, k, has SI units of m^2, but other units are found, dependent on the user industry or the application of the coefficient. It is frequently measured in inconsistent units of darcies or millidarcies, notably by reservoir engineers where:

$$1\,(\text{darcy}) = \frac{1\,(\text{cm}^3\ \text{cm}^{-2}\ \text{s}^{-1}).1\,(\text{centipoise})}{1\,(\text{atmosphere cm}^{-1})}$$

while hydrologists and soil scientists prefer the term hydraulic conductivity, expressed as the velocity of the percolating water per unit hydraulic gradient. The conversions from m^2 to some other units are given in Table 2.1.

Table 2.1 Conversion factors for units of permeability

1 m²	≡	1.013 × 10¹² darcy
		9.8 × 10⁸ cm s⁻¹ (for water at 20°C)
		2.78 × 10¹² ft day⁻¹
		2.08 × 10¹³ US gal day⁻¹ ft⁻²

Darcy's Law can be obtained more formally as a force-momentum balance. The key assumption which underlies the law is that the Reynolds Number is considered to be sufficiently small so fluid inertia effectively plays no role in the dynamics of flow. A common misconception about porous media is that as the flow rate is increased Darcy's Law first breaks down upon the onset of turbulence in the pores. In fact, pore Reynolds numbers rarely become large enough for turbulent flow to exist in the pores. In practice, Darcy's Law is valid for a wide range of flow but breaks down at Reynolds numbers of the order 1 to 10 due to the onset of inertial forces in laminar flow. The Reynolds number (Re) used in packed beds is based on the particle size (x) usually taken as the mean surface-volume size, and Re is defined as:

$$\text{Re} = \frac{\rho u x}{\mu} \tag{2.19}$$

2.2.1 Darcy's Law in filtration

In filtration Darcy's Law is often used in a modified form, where the specific resistance (α) replaces the permeability (k) and the pressure gradient (dp/dz) is replaced by the pressure loss per unit mass of solid deposited (dp/dw) on the medium:

$$u = -\frac{1}{\mu\alpha}\frac{dp}{dw} \tag{2.20}$$

where w is the mass of dry cake per unit filter area deposited within distance z from the filter medium. dw and dz are related by:

$$dw = \rho_s(1-\varepsilon)dz \tag{2.21}$$

and the total mass of solids deposited is related to the cake thickness (which is most important to the design, specification and operation of a filter) by:

$$w = \rho_s(1-\varepsilon)L \tag{2.22}$$

k and α are related by:

$$k = \frac{1}{\rho_s(1-\varepsilon)\alpha} \tag{2.23}$$

In general, α varies less than k with pressure, and it is therefore used as a primary factor for the scale-up of filtration equipment. Its use is not universal, but most approaches to process design and scale-up use factors that can be shown to have some equivalence to α. The importance of α (and hence k) in filter cake formation is considered in detail in Chapter 4.

2.3 Permeability and specific resistance

The permeability of a porous medium may be regarded as a measure of the ease with which a fluid will flow through its voids. The magnitude of the permeability is determined by the degree of 'openness' of the medium, which would be more formally interpreted by the porosity of the medium and the sizes of the pores present in its internal structure. The complexity of the internal pore structure is illustrated by the

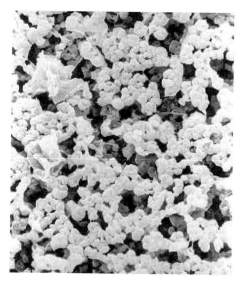

Figure 2.2 Scanning electron micrograph of a filter cake of tea solids formed on a polymer membrane, showing the complex pore geometry and the permeable structure in the cake (Scale: 1000:1).

scanning electron micrograph shown in Figure 2.2; such a complex geometry is currently impossible to describe mathematically, making it necessary to develop models to relate the properties of the porous medium one to another. The simplest, and probably most widely used, model to relate permeability to the porosity of a filter cake and to the mean size of the particles forming the cake is that developed through combining the porous media concepts of Kozeny (1927) and Carman (1938).

2.3.1 The Kozeny–Carman equation

Darcy's Law uses the single parameter k to account for the characteristics of a porous medium or filter cake in so far as they affect fluid flow. There have been many attempts to relate permeability to geometric considerations of a porous medium; in order to provide a basic understanding of these concepts, the theory attributed to Kozeny (1927) and Carman (1938) is developed in the following paragraphs. The model is based on the pores being represented by a bundle of capillary tubes whose orientation is at 45° to the fluid inflow face of the porous medium. In this case, energy is lost from the flowing fluid where it contacts the surfaces of the capillaries. In a similar manner, when the fluid is flowing through a porous body it loses energy where it is in contact with the internal surfaces of the body. The internal

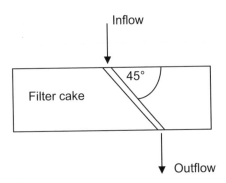

surfaces possess a very complex geometry which cannot be easily described mathematically, but for practical purposes can be interpreted through the porosity and the specific surface of the porous medium. Cake filtration involves laminar flow of the fluid and the Reynolds number is based on the particle size as defined in equation (2.19).

The average velocity of a fluid moving in laminar flow through a straight circular capillary of length L and diameter D is given by Poiseuille's equation:

$$u' = \frac{D^2}{32\mu}\frac{\Delta p}{L} \tag{2.24}$$

Assuming the porous medium to be a bundle of irregularly shaped, straight channels, the hydraulic diameter can be related to the porosity of the bed and the mean specific surface of the particles:

$$D_h = 4 \times \frac{\text{flow cross} - \text{sectional area}}{\text{wetted perimeter}} \times \frac{\text{length of the flow path}}{\text{length of the flow path}}$$

$$= 4 \times \frac{\text{volume of voids in bed}}{\text{total surface of particles forming the bed}} \tag{2.25}$$

$$= 4 \times \text{voids ratio} \times \frac{\text{volume of the particles}}{\text{surface of the particles}}$$

$$\text{i.e. } D_h = 4\frac{\varepsilon}{1-\varepsilon}\frac{1}{S_0} \tag{2.26}$$

where S_0 is the specific surface of the particles. For a circular pipe the actual (D) and hydraulic (D_h) diameters are identical:

$$D_h = 4\left(\frac{\pi D^2}{4}\right)\frac{1}{\pi D} = D \tag{2.27}$$

The effective pore length is somewhat greater than the bed depth because the fluid travels a tortuous path, thus the apparent length L in equation (2.24) should be replaced by the effective distance travelled L_e by an average fluid element i.e.:

$$\frac{L_e}{L} = \text{tortuosity factor} > 1 \tag{2.28}$$

However, because the tortuous fluid path is longer than L by the factor L_e/L, the velocity along that path must be correspondingly greater than that for travel through the bed in a non-deviating direction perpendicular to the fluid inlet and outlet faces, i.e.:

$$v = u'\frac{L_e}{L} \tag{2.29}$$

Setting $u = v$ and $L = L_e$ in equation (2.24) gives:

$$v = \frac{D_h^2}{32\mu}\frac{\Delta p}{L_e} \tag{2.30}$$

The interstitial velocity v, that is, the actual velocity of the fluid in the pores of the porous medium, is related to the superficial velocity u by

$$v = u'\frac{L_e}{L} = \frac{u}{\varepsilon}\frac{L_e}{L} \tag{2.31}$$

Substituting equations (2.26) and (2.31) into (2.30) gives:

$$u = \frac{\varepsilon^3}{k_0\,\mu\,S_0^2(1-\varepsilon)^2}\left(\frac{L}{L_e}\right)^2\frac{\Delta p}{L} \tag{2.32}$$

where

$$k_0\left(\frac{L_e}{L}\right)^2 = \text{Kozeny constant} = K_0 \tag{2.33}$$

If the average flow direction is at 45° to the inflow surface of the porous medium, $(L_e/L)^2 = 2$. k_0 is a factor which depends on the shape and size distribution of the cross-sectional areas of the capillaries (and hence is related to the particles which make up the bed). For circular capillaries $k_0 = 2$, but for rectangular, annular and elliptical shapes k_0 lies between 2 and 2.5. Putting $(L_e/L)^2 = 2$ and $k_0 = 2.5$ in equation (2.32) gives:

$$u = \frac{\varepsilon^3}{5\mu S_0^2 (1-\varepsilon)^2} \frac{\Delta p}{L} \qquad (2.34)$$

This is the Kozeny-Carman equation. It has important uses for determining the specific surface of powder samples from permeability data, and in the correlation of resistance data for fluid flow through porous media.

For particles that deviate strongly from a spherical shape, for wide particle size distributions and for most consolidated media the Kozeny-Carman equation is often not valid and therefore should only be applied with caution. There are a variety of reasons for poor agreement between the Kozeny-Carman equation and experimental values (Dullien, 1975):

(a) The equation is based on the assumption of flow through a circular duct. At higher porosities ($\varepsilon > 0.7$) this concept breaks down and experimental flow rate – pressure loss data are better described by a "flow around submerged objects" model.

(b) The existence of continuous flow channels of different effective diameters can invalidate the model. The pores in a real bed form a network interconnected in a random fashion, and at a juncture to enter a narrower or larger pore there is always a choice to be made by a fluid element. If at every juncture the largest available pore is entered, then the fluid element travels through a continuous flow channel of the largest possible effective diameter in the bed. The permeability is then seen to be greater than if the flow channels were selected in such a way as to contain all the pore sizes present in the bed.

(c) The effective cross-section of capillaries in porous media are known to vary along their length in a quasi-periodic manner, that is, wider and narrower sections alternate in a fairly random sense along the length of a capillary. This kind of sequential variation in the effective cross-section of flow channels results in a lower actual permeability.

(d) The Kozeny-Carman theory neglects any energy loss by viscous dissipation which arises from the sequential variations in cross-section.

2.3.2 Permeabilities calculated from cell models

The Kozeny-Carman theory has received much criticism, but it nevertheless correlates bed resistance data for a wider range of porous media types than any other permeability theory. More sophisticated mathematical models of particle removal by granular beds require a more detailed picture of the flow field near the collecting surfaces within the bed than is given by the Kozeny-Carman model. Progress to overcome some of the fundamental deficiencies has been made by adopting the Happel cell model to better describe the microscopic flow field (Happel, 1958; Happel and Brenner, 1965). Instead of viewing a packed bed as a bundle of tortuous channels, cell models view the bed grains as an assemblage of interacting, but essentially individual, spheres with the flow field about an average sphere being described in detail. Moreover, by summing the drag forces acting on the individual bed grains, cell models also permit self consistent predictions of permeabilities. These predictions agree with experimental data for porous media equally as well as the Kozeny-Carman equation.

Happel (1958) treated the problem of flow through an assemblage of spheres by assuming that a typical sphere was enclosed by a spherical envelope of radius b, whose volume corresponds to the porosity in the overall assemblage, i.e.

$$\frac{x}{2b} = \beta^{0.33} = (1-\varepsilon)^{0.33} \qquad (2.35)$$

where β is the packing density of the spheres. By writing the Navier-Stokes equation for isothermal incompressible fluid flow, simplifying it by introducing the stream function concept, assuming that no slip occurs at the particle-fluid interface and that the envelope at radius b is a free surface (i.e. no tangential stresses exist at the surface), Happel was able to derive an expression equivalent to equation (2.34) for the superficial velocity of the liquid:

$$u = \left(\frac{x^2}{36} \frac{6 - 9\beta^{0.33} + 9\beta^{1.67} - 6\beta^2}{\beta(3 + 2\beta^{1.67})} \right) \frac{\Delta p}{\mu L} \qquad (2.36)$$

The term in brackets corresponds to the Darcy permeability k. Equation (2.36) agrees closely with the Kozeny-Carman permeability in the porosity range $0.4 < \varepsilon < 0.7$. At higher porosities Happel's model is superior because it reduces to an assemblage of isolated spheres as $\varepsilon \to 1$.

2.3.3 Properties of porous media

Darcy's Law gives rise to a permeability coefficient k which embraces the bed properties affecting the nature of fluid flow through the voids. By equating a flow model equation, for example equation (2.34) or (2.36), to the Darcy equation (2.18) insight into the effects of the bed properties on the permeability are gained. Noting that $dp/dz = -\Delta p/L$, the result is:

$$k = \frac{\varepsilon^3}{5S_0^2(1-\varepsilon)^2} = \frac{\varepsilon^3 x^2}{180(1-\varepsilon)^2} \tag{2.37}$$

using the Kozeny-Carman model. The permeability is related to an equivalent spherical particle size through S_0, since:

$$S_0 = \frac{6}{x} \tag{2.38}$$

The equivalent expression for the permeability derived from the cell model is:

$$k = \frac{x^2}{36} \frac{6 - 9(1-\varepsilon)^{0.33} + 9(1-\varepsilon)^{1.67} - 6(1-\varepsilon)^2}{(1-\varepsilon)(3 + 2(1-\varepsilon)^{1.67})} \tag{2.39}$$

The fundamental importance of porosity to flow through porous media is clear when the influence of porosity (ε) on permeability (k) is examined using equations (2.37) and (2.39). The porosity of a packed bed is determined largely by the packing structure of the particles, that is, the position of the particles relative to one another in the bed. Consideration of the effect on ε of a variety of particle packings illustrates the sensitivity of k to ε; such data are shown on Table 2.2. Depending on the model used, there is between a 12 and 26 fold variation in the permeability coefficient when comparing the cubic regular packing of monosized spheres with the rhombohedral packing. Any real packed bed tends to contain a randomly distributed packing of the particles (except for some dense packings of colloidal sized particles), and point-to-point variations of k exist in the bed.

Particle size influences the permeability of a particulate bed in two ways. Firstly, there is a direct effect as shown by equations (2.37) and (2.39) where k decreases as the square of the particle size. This is to be expected as the drag force resisting fluid flow through the bed must be proportional to the surface area of the particles in the bed. If the particle distribution were monosized and >50 μm this would be the

Table 2.2 Effect of particle packing on the permeability function

Packing	Coordination number	Porosity ε	$\dfrac{\varepsilon^3}{180(1-\varepsilon)^2}$	$\dfrac{6-9\beta^{0.33}+9\beta^{1.67}-6\beta^2}{36\beta(3+2\beta^{1.67})}$
	3	0.7766	–	0.0373
	4	0.6599	0.0138	0.0119
	5	0.5969	0.00727	0.00662
Cubic	6	0.4764	0.00219	0.00202
	7	0.4388	0.00149	0.00134
Ortho-rhombic	8	0.3955	0.000941	0.000808
	9	0.3866	0.000853	0.000723
Tetragonal	10	0.3019	0.000314	0.000203
	11	0.2817	0.000241	0.000135
Rhombohedral	12	0.2595	0.000177	0.0000787

only significant influence of x on k, since the porosity of the bed would be independent of x. However, when a distribution of sizes are present the smaller sizes have the ability to occupy the voids between larger particles, thereby decreasing the bed porosity (and hence permeability) (Wakeman, 1975a).

Particle size causes other surface properties to become important for particles smaller than about 5 to 10 μm. The most important property to consider here is the surface charge on the particles, which is often represented by the zeta (ζ) potential. This is dependent also on the properties of the fluid surrounding the particle, such as the concentration and the valency of the ions in solution. Neither equation (2.37) nor (2.39) include ζ-potential effects, which limits their use to larger particles. Nevertheless, the equations have been applied to fine particle systems; when this is done it is necessary to measure the variation of permeability with solution pH to give an empirical interpretation of the effects of particle surface charge.

Example 2.2 Estimation of the specific surface of particles from fluid flow measurements.

A 4 cm internal diameter cylindrical tube is clamped with its axis vertical and both ends open to atmosphere. The lower end is covered by a wire gauze and the tube is filled to a depth of 5 cm with 88 g of particles with a density of 2450 kg m^{-3}. The tube is then filled by water (ρ_l = 998 kg m^{-3}, μ = 0.001 Pa s) to a height h above the gauze. The time taken for the liquid height to fall to $h/2$ was 360 s. Estimate the specific surface of the particles. What is the equivalent surface-volume size of the particles?

Solution

The pressure head decreases as the liquid height falls, i.e. $\Delta p = -\rho_l g z$ where z is the height of the fluid above the base of the particle bed. Darcy's law (equation (2.18)) is then expressed as:

$$u = -\frac{dz}{dt} = \frac{k\Delta p}{\mu L} = \frac{k\rho g}{\mu L} z$$

i.e. $\quad -\dfrac{dz}{dt} = \dfrac{k\rho g}{\mu L} z$

$\therefore \quad -\displaystyle\int_{h}^{h/2} \frac{dz}{z} = \frac{k\rho g}{\mu L}\int_{0}^{t} dt$

$\therefore \quad k = \dfrac{\mu L \ln 2}{\rho g t} = \dfrac{10^{-3}\times 0.05\times \ln 2}{998\times 9.81\times 360} = 9.83\times 10^{-12} \quad m^2$

The porosity of the bed is given by

$$\varepsilon = 1 - \frac{\text{mass of solids}}{\rho_s \times \text{volume of bed}}$$

$$= 1 - \frac{0.088}{2450\left(\dfrac{\pi\,0.04^2}{4}\right)0.05}$$

$$= 0.428$$

Using the Kozeny-Carman model (equation (2.37)):

$$S_0 = \sqrt{\frac{\varepsilon^3}{5k(1-\varepsilon)^2}} = \sqrt{\frac{0.428^3}{5\times 9.83\times 10^{-12}\times 0.572^2}} = 6.98\times 10^4 \quad m^2\ m^{-3}$$

i.e. the specific surface of the particles is $6.98\times 10^4\ m^2$.

The equivalent surface-volume size of the particles is $6/(6.98\times 10^4\times 10^{-6}) = 86\ \mu m$.

2.3.4 The friction factor for particulate beds

Although laminar flow is invariably involved in cake filtration problems, inertial or very occasionally turbulent flows can occur in pressure driven deep bed filters. Here flow is through a bed of granular media, and it is most important to be able to estimate the pressure loss across such a filter bed. This is most easily done by noting that in a turbulent flow the friction factor becomes constant and independent of the Reynolds number (see Figure 2.1); the value of this constant has been established by plotting experimental data of the friction factor against the Reynolds number.

After inserting equation (2.38) into equation (2.34) the Kozeny-Carman expression can be rearranged into:

$$\frac{\Delta px}{L\rho u^2}\frac{\varepsilon^3}{1-\varepsilon}=180\frac{1-\varepsilon}{\mathrm{Re}} \tag{2.40}$$

This equation is valid for laminar flow (Re < 10), and the left hand side is the friction factor for flow through a porous medium. At higher Reynolds numbers an additional term is added to account for the increasing kinetic energy losses which, in the turbulent flow regime, are independent of Re. Numerous experimental data indicate that the friction factor-Reynolds number relationship which is valid for any flow regime is:

$$\frac{\Delta px_{sv}}{L\rho u^2}\frac{\varepsilon^3}{1-\varepsilon}=180\frac{1-\varepsilon}{\mathrm{Re}}+1.8 \tag{2.41}$$

This is the so-called Burke-Plummer equation. Pressure losses across beds of coarser (>50 to 100 μm) granular particles calculated from this equation should have an accuracy of about ± 50% (Macdonald *et al*, 1979), and the equation gives reasonable results over a wide porosity range 0.36 < ε < 0.92 (even though the Kozeny-Carman equation seems to hold over a narrower porosity range).

Example 2.3 Calculation of the pressure loss across a dual media filter.
Sea water (ρ_l = 1030 kg m^{-3}, μ = 0.001 Pa s) is fed to a deep bed coarse dual media filter at 0.04 m^3 s^{-1}. The filter is 1.7 m in diameter and consists of two layers of media, each 0.4 m deep; the lower layer has a porosity of 0.47 and is composed of garnet with a particle size of 500 μm. The upper layer has a porosity of 0.425 and is anthracite with a particle size of 1500 μm. Estimate the pressure loss across the filter.

Solution

The superficial velocity of the water in each layer is calculated from the volume flow rate:

$$u=\frac{0.04}{\pi 1.7^2/4}=0.0176 \quad \mathrm{m\,s}^{-1}$$

(a) Calculate the pressure drop across the upper layer:

The Reynolds number is: $\text{Re} = \dfrac{1030 \times 0.0176 \times 1500 \times 10^{-6}}{0.001} = 27.2$

Since there are significant inertial forces affecting the water flow ($\text{Re} > 1$), it is necessary to use equation (2.41) to calculate the pressure loss.

The modified friction factor is:

$$\frac{\Delta p x}{L \rho u^2} \frac{\varepsilon^3}{1-\varepsilon} = \frac{\Delta p \times 1500 \times 10^{-6}}{0.4 \times 1030 \times 0.0176^2} \frac{0.425^3}{0.575} = 0.00157 \Delta p$$

Using equation (2.41):

$$0.00157 \Delta p = 180 \frac{0.575}{27.2} + 1.8$$

$\therefore \quad \Delta p = 3570 \text{ Pa}$ across the upper layer

(b) Calculate the pressure loss across the lower layer: Following the above procedure, the Reynolds number is 9.1 and the modified friction factor is $0.00077 \Delta p$, and hence $\Delta p = 15950$ Pa.

(c) The total pressure loss across the filter bed is therefore 19.5 kPa. This pressure loss is across a "clean" bed; as the filter media collects contaminant, the resistance to flow and the pressure loss will rise.

2.4 Depth filtration (clarification)

A number of clarification filtration techniques capture the particles inside a porous medium rather than on its surface; such processes include deep bed, precoat, candle and cartridge filtrations. These utilise a deep filter medium such as a bed of fibrous or granular material, and filtration is effected by capture of the contaminant at the surfaces inside the deep bed; the surfaces are provided by collector particles or fibres that form the deep bed. Although the techniques are used for differing ranges of feed flow rate, feed concentration and scale of throughput, their successful operation depends on similar capture mechanisms.

2.4.1 Capture mechanisms

It is often assumed that if a feed particle contacts a collector surface, then the feed particle is removed from the flow. Although this may not be entirely correct at a fundamental level, it is a good engineering approximation for many processes and enables an understanding of the separation processes. One or more of the following capture mechanisms may be active during clarification filtration:

(i)   *Straining* occurs when a particle in the feed is larger than the pore or constriction through which it is attempting to pass; the size of the pore is related to the size and size distribution of the collector particles forming the porous medium in which the feed particles are to be captured, with smaller pore sizes being associated with smaller particle sizes in the medium.

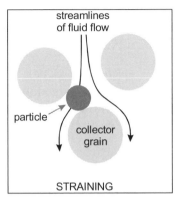

(ii) *Sedimentation or gravity effects* occur when the feed flow is downwards and the feed particles have a density that is greater than that of the liquid. Under such conditions the particles settle across flow streamlines as the latter distort to pass around the collector surfaces. The likelihood of deposition is characterised by

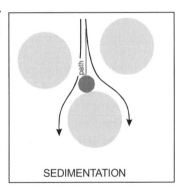

$$\frac{x^2\left(\rho_s - \rho\right)g}{18\mu u}$$

which is the ratio of the Stokes settling velocity (see equation (2.8)) to the velocity of the fluid approaching the deep bed (u). In a typical granular water filter, the Stokesian velocity is of the order 0.1 mm s^{-1}, the approach velocity is about 2 mm s^{-1}, and the Stokes number (see equation (2.43)) is about 0.05. This would indicate a low gravity effect, even for a quite dense particle.

(iii) *Interception* by the collector, which occurs when the feed particle moves with the fluid streamline that is one feed particle radius or less from the collector surface contacts the collector and is removed from the flow. This implies laminar flow conditions,

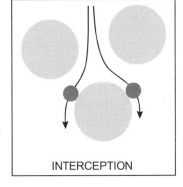

which are in fact most common in filtration systems, and that particles uniformly distributed through the liquid follow the fluid streamlines. Interception is characterised by the dimensionless ratio

$$I = \frac{d}{D} \tag{2.42}$$

where d and D are the particle and collector grain (or sometimes pore) diameters respectively. Greater values of I suggest more effective particle capture by interception; when $I \rightarrow 1$, straining becomes the dominant mechanism. For filtration of water and wastewater, typical values are $2 \times 10^{-4} < I < 1 \times 10^{-1}$.

(iv) *Inertial impaction* between a feed particle and a collector when the feed particle flows across the fluid streamlines due to its inertia. The fluid streamlines change direction many times as the fluid passes through the bed of collectors, but the feed particles generally have a density that is greater than the fluid and so their flow direction does not change so sharply as that of the fluid. The Stokes number can be used to characterise this mechanism, and is defined by

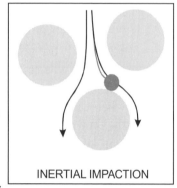

INERTIAL IMPACTION

$$St = \frac{x^2 \rho_s u}{18 \mu D} \tag{2.43}$$

St is independent of the density of the fluid, and inversely proportional to the collector grain (whether this be sand, filter aid, fibre, and so on) size D. Generally the pore size is proportional to the grain size, so St increases when the pore sizes are smaller. In water filtration, $u \approx 2$ mm s^{-1} and μ is 10^{-3} kg m^{-1} s^{-1}, so the value of St is between 2×10^{-9} and 1.5×10^{-3}, indicating a negligible collection efficiency. In contrast, in air filtration where the fluid viscosity in much lower St values are 10^3 to 10^4 times greater, making inertial impaction a more important parameter.

(v) *Diffusion* of the feed particles. All particles suspended in a fluid are subjected to random collisions by molecules of the surrounding fluid. Small particles may acquire sufficient momentum through the collisions to cause them to change their directions of flow; since the particles are bombarded by molecules in a random fashion, the resulting flow direction of the particle also changes randomly. This is commonly referred to as Brownian motion. When Brownian motion causes a feed particle to collide with a collector, the former is removed from the flow. These effects are particularly pronounced for particles whose size is smaller than about 1 μm, and can be characterised by the Peclet number defined by

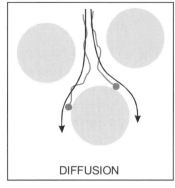

DIFFUSION

$$Pe = \frac{xu}{D} \tag{2.44}$$

where D is the particle diffusion coefficient. Values for Pe for 1 μm colloids separated by clarification filtration are in the range 10^{-3} to 10^{7}; lower values of the Peclet number indicate a greater chance of collection by diffusion.

(vi) *Hydrodynamic interaction* induced collision between a feed particle and a collector are less likely when flow in the filter bed is laminar. The Reynolds number, defined by equation (2.1) must be smaller than 4 to prevent excessive inertial effects in laminar flow; hence when Re > 4 the motion of suspended particles may be affected by eddies in the fluid, which in turn can lead to their contact with a collector.

HYDRODYNAMIC
INTERACTION

(vii) *Electrostatic interactions* occur due to the presence of electrical charges on the surfaces of the feed particles and the collector. Depending on the polarity of the charges, the inter-actions may be attractive or repulsive. In some filter types the collector may be given a high electrical charge during manufacture in order to promote electrostatic deposition of feed particles during usage. Electrostatic interactions play a lesser role in attracting particles to

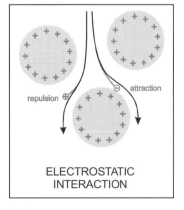

repulsion ⊕   ⊖ attraction

ELECTROSTATIC
INTERACTION

the collector surface, and are more important in causing attachment to the surface.

The collection efficiency, or put another way, the relative importance of the above mechanisms, is impossible to assess for almost all industrial applications of clarification filters. This is complicated by the coexistence of detachment forces, arising mainly from shear at the collector surfaces, that lead to a balance between the attachment and detachment mechanisms. However, cognisance of the mechanisms and their underlying forces is important as a qualitative tool for the interpretation of experimental data and practical information.

2.4.2 Modelling the capture process

The mathematical approaches to understanding granular filtration are ably developed by Tien (1989), but although knowledge of the surface chemistry and transport mechanisms in filtration systems is extensive it is still necessary to define and use an empirical removal efficiency. For a filter that is uniform at all depths, that is the collector size distribution and porosity is the same throughout, consideration of the filtration mechanisms suggests that every layer of collector particles should be equally as efficient as all others at removing particles from the feed. Hence, a filter coefficient λ is defined as (Iwasaki, 1937; Ives, 1973):

$$\lambda = -\frac{\delta c}{c}\frac{1}{\delta L} \tag{2.45}$$

where $-\delta c$ is the reduction of the concentration of particles passing through a layer of thickness δL. This equation can be rearranged into:

$$-\frac{dc}{dL} = \lambda c \tag{2.46}$$

Integrating equation (2.46) with the inlet ($L = 0$) concentration of particles represented by c_0 yields:

$$c = c_0 \exp(-\lambda L) \tag{2.47}$$

Equation (2.47) indicates that in a uniform filter more particles are deposited close to the filter inlet surface, and the number of particles in suspension declines logarithmically with the filter depth. Hence the top layer of the filter contains most of the deposit and the lower layers contain very little. The filter coefficient is smaller for larger collector grains, so in a filter whose collectors are size graded by backwashing – which results in the smaller collectors residing at the top of the filter bed – an even larger proportion of the feed particles are deposited near to the top if the filter is top fed. To prevent this occurring the filter may be bottom fed.

The second basic equation that is needed is a mass balance on the particles in suspension, which states that the particles removed from suspension are deposited in the filter pores. A filter element thickness ΔL with a cross-sectional area A, being fed with a suspension at a volumetric flow rate Q, experiences a loss of concentration $-\Delta c$ (volume by volume). During the time taken for fluid to flow through the element, Δt, the volume of deposited particles per unit filter volume (the specific deposit) increases by $\Delta \sigma_a$. Hence, we can write:

Volume of particles removed from suspension = $-\Delta c.Q.\Delta t$

Volume of particles increase in deposits = $\Delta \sigma_a.A.\Delta L$

Equating these terms yields:

$$-\frac{\partial c}{\partial L} = \frac{A}{Q}\frac{\partial \sigma_a}{\partial t} \tag{2.48}$$

It is more convenient to consider the volume of the deposit occupied in the pores of the filter, the effective specific deposit σ, where $\sigma = \beta \sigma_a$ and β is a bulking factor. The approach velocity v is defined by Q/A and equation (2.48) becomes:

$$-\frac{\partial c}{\partial L} = \frac{1}{\beta v}\frac{\partial \sigma}{\partial t} \tag{2.49}$$

During the clarification and deposition processes, the pores of the filter gradually clog. Consequently the removal efficiency changes and the exponential decline of equation (2.47) is not valid. Two general approaches have been made to deal with this problem: one allows for removal of deposit during the filtration process, known as the deposition and scour model (Mints, 1966); the second utilises a modified filter coefficient.

2.4.3 Deposition and scour

The pores of the filter bed are narrowed as deposits form in them, causing a rise in the interstitial velocity. The higher interstitial velocity cause an increase in the shear stress in the pores and on the deposited particles, and may lead to re-entrainment of some of the deposit. To account for these phenomena, Mints (1966) suggested using the following type of expression for filtration rates:

$$\frac{d\sigma}{dL} = \beta v \lambda c - a\sigma \qquad (2.50)$$

where a is defined as the scour coefficient. That is, the filtration rate is the net effect of the deposition rate and the re-entrainment rate, with the latter proportional to the specific deposit σ. Tien (1989) discusses alternative forms of this specific rate equation and recognises that the selection of a rate equation is arbitrary. The presence of more constants in the rate expression allows a better fit to experimental results.

2.5 Flow through fibrous and woven media

The filter medium is an essential component of any filtration equipment, yet the problem of relating the permeability term in equation (2.18) to the variables that define the flow characteristics of the medium is not well understood. A filter medium is an heterogeneous structure, with pores that have a non-uniform size and irregular geometry, and which may be unevenly distributed across the surface. This structure affects the way that the filter cake particles deposit and pack at the medium surface, and causes large variations in the distribution of liquid flow over the medium surface.

The relationship between liquid flow and the pressure difference across a fibrous or woven medium is given by Darcy's Law, equation (2.18). However, the problem remains to relate the permeability of a clean medium to its structural properties such as fibre diameter and weave construction parameters (e.g. warp and weft mesh counts).

2.5.1 Flow through nonwoven, random fibre media

Random fibre filter media (nonwoven media) have the appearance of felt or paper and are used generally in clarification processes; they are made from materials such as paper, cotton, wool and synthetic fibres. The media are available as belts, sheets or pads. Attempts made to relate pressure loss to flow rate have been semi-empirical approaches based on Darcy's Law; for flow of air through fibrous pads a relationship of the following form has been found (Davies, 1952):

$$\frac{\mu u L}{\Delta p} = k = \frac{x_f^2}{64(1-\varepsilon)^{1.5}\left[1+56(1-\varepsilon)^3\right]} \qquad (2.51)$$

where x_f is the fibre diameter in a pad of porosity ε. This equation is valid for the range $0.6 < \varepsilon < 1$, with $Re = \rho u x_f/\mu < 1$.

For air flow through felted materials Darcy's Law has been found to hold, but no general relationship for the permeability term has been found in terms of the medium structure. Values of k are specific to each material, as it is dependent on the geometric configuration of the cloth (length and shape of flow channels, porosity, etc). For certain felted materials (wool, rayon and cotton) the flow resistance k^* of the material is related to the cloth weight W_c (in grams per square centimetre) by $k^* \approx 5.01 \times 10^8 W_c$, where k^* has the units m^{-1} (Cunningham *et al*, 1954) and

$$u = \frac{\Delta p}{k^* \mu} \tag{2.52}$$

Comparisons between the permeabilities of different filter media may be made by applying equation (2.18) to measured flow rate versus pressure loss data, and for many practical purposes this would be adequate. Such an approach must, of course, be regarded as empirical and gives no insight into effects of the medium structure on its pressure loss characteristics.

2.5.2 Monofilament woven media

The basic warp yarn configurations of single layer, monofilament fabrics are shown in Figure 2.3. (In a monofilament fabric, each thread is made up from a single yarn.) A *plain weave* is the most basic, with a weft thread alternately passing over one warp thread and then under one warp thread. A *twill weave* produces a diagonal or twill line across the face of the fabric, caused by moving the yarn intersections one weft thread higher on successive warp yarns. A twill weave is designated 2/1, 2/2 or 3/1 depending on how many weft threads and warp threads

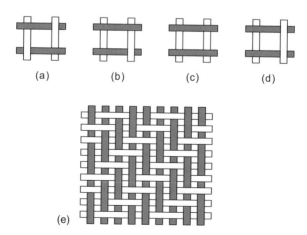

Figure 2.3 Basic configurations of monofilament weaves ((a) plain, (b) twill, (c) plain type I, (d) plain type II and (e) a 2/2 twill).

go over and under. A *satin weave* has a smooth surface caused by carrying the warp (or the weft) on the fabric surface over many weft (or warp) yarns. Intersections between warp and weft are kept to a minimum, just sufficient to hold the fabric firmly together.

Pedersen (1969) adopted an orifice analogy to interpret flow-pressure loss behaviour for monofilament media with various weave patterns. The five basic variables describing a cloth are:

(i)   the end count, ec (the number of warp yarns per unit length);

(ii) the warp yarn diameter d_2;

(iii) the pick count, pc (the number of weft (filling) yarns per unit length);

(iv) the weft yarn diameter d_1; and

(v)   the weave.

Pedersen was able to relate these to two variables descriptive of the cloth structure, the effective area of an orifice A_0 and the orifice perimeter W. A discharge coefficient was then defined for laminar flow as:

$$C_D = \sqrt{\frac{\rho u^2}{2\Delta p} \frac{(1-a^2)}{a^2}} \qquad (2.53)$$

where Δp is the pressure difference across the medium, u the superficial or approach velocity and a is the effective fractional open area of the pore, given by:

a = (effective area of the orifice)(warp yarns per centimetre)(weft yarns per centimetre)

$$= A_0 \times (ec) \times (pc) \qquad (2.54)$$

In order that a rigorous assessment of effects of the geometric configuration of any area of fabric can be made, it is necessary to make the following assumptions:

(i)   the filling yarns are straight;

(ii) the weaving is perfect;

(iii) the yarns are all cylindrical;

(iv) the warp yarns are perfectly straight between filling yarns; and

(v)   flow through any orifice is not influenced by flow through any other orifice.

Although these conditions are never fully met in practice, Pederson's model of a cloth enables the effective cross-sectional area of the orifice (A_0) (and therefore the effective fractional open area of the pore (a)) and the wetted perimeter of the orifice (W) to be determined analytically for both plain and twill type pores. For a plain pore ($a = a_p$):

$$a_p = \frac{\phi}{2} \ln\left[\frac{1+\sqrt{1+\phi^2}}{\phi}\right] + \frac{\sqrt{1+\phi^2}}{2\phi} - (ec)d_1 - (ec)(pc)d_1^2 \sqrt{\left(\frac{(ec)^{-1}-d_2}{d_1}\right)^2 + 1} \quad (2.55)$$

and

$$W_p = 2\sqrt{\left((ec)^{-1}-d_2\right)^2 + d_1^2} \; \xi\left(\left[1+\left(\frac{d_1}{(ec)^{-1}-d_2}\right)^2\right]^{-1/2}, \frac{\pi}{2}\right) + \frac{2}{pc}\sqrt{1+(d_1+d_2)^2(pc)^2} \quad (2.56)$$

For a 2/2 twill pore ($a = a_T$):

$$a_T = a_p + \frac{(ec)(pc)d_1^2}{2} \sqrt{\left(\frac{(ec)^{-1}-d_2}{d_1}\right)^2 + 1} + \frac{1}{2}(ec)(pc)d_1\left((ec)^{-1}-d_2\right) \quad (2.57)$$

and

$$W_T = \sqrt{\left((ec)^{-1}-d_2\right)^2 + d_1^2} \; \xi\left(\left[1+\left(\frac{d_1}{(ec)^{-1}-d_2}\right)^2\right]^{-1/2}, \frac{\pi}{2}\right) + \left((ec)^{-1}-d_2\right)$$

$$+ \frac{1}{(pc)}\left(1+\sqrt{1+\left(d_1+d_2\right)^2}\right)(pc)^2\right) \quad (2.58)$$

In the above equations, $\phi = [(ec)(d_1+d_2)]^{-1}$ and $\xi(k,\pi/2)$ is an elliptical integral of the second kind with variable k. Pedersen used air permeability data obtained from experiments on plain and 2/2 twill fabrics and obtained a good correlation between the discharge coefficient and the air Reynolds number, with an average error of about 12%.

More complex fabrics, 2/1 twills and 5/1 sateens, were studied by Rushton and Griffiths (1971, 1972) who also used water in the development of the correlations for the discharge coefficient as a function of Reynolds number. Water flow data in the range 1<Re<10 are represented by

$$C_D = 0.17\,\mathrm{Re}^{0.41} \quad (2.59)$$

for plain and twill fabrics, with a maximum error of 18%. The Reynolds number is defined by

$$\mathrm{Re} = \frac{v\,d_p\rho}{\mu} = \frac{u}{a}\frac{4A_0}{W}\frac{\rho}{\mu} = \frac{4u\rho}{W(ec)(pc)\mu} \quad (2.60)$$

where v is the effective velocity of the fluid in the pores of the fabric and d_p is the effective diameter of the orifice (i.e. pore). In the above definition of the Reynolds number it is noted that the pore velocity is related to the superficial velocity by $v = u/a$, and the pore diameter is related to the effective area of the orifice by $d_p = 4A_0/W$ where W is the wetted perimeter of the orifice. The relationship between ec, pc, W, A_0 and d_p is illustrated in Table 2.3.

Table 2.3 Monofilament cloth properties based on cloths supplied by P&S Filtration (Rushton and Griffiths 1971, 1987)

Cloth weave	Nominal aperture µm	ec cm⁻¹	pc cm⁻¹	d_1 µm	d_2 µm	k m²×10¹¹	W mm	A	A_0 mm²×10³	d_p µm
Plain	188	29.5	29.5	153	153	62.4	1.54	0.351	40.3	105
Plain	155	40.2	40.2	105	105	42.0	1.11	0.377	23.3	84
Plain	100	57.5	57.5	75	75	14.9	0.78	0.379	11.5	59
Plain	75	75.6	75.6	61	61	12.4	0.60	0.374	6.1	40
Plain	60	97.2	97.2	44	44	7.9	0.46	0.376	3.9	34
Plain	41	126	126	37	37	3.6	0.37	0.336	2.1	22
Plain	25	185	185	30	30	1.2	0.27	0.277	0.81	12
5/1 S		35.1	15.8	272	266	132.0	2.53	0.281	50.7	80
2/2 T		27.9	17.8	267	252	91.9	2.43	0.316	63.6	104
2/1 T		40.0	17.3	196	193	115.2	1.78	0.337	48.7	109
2/1 T		44.1	18.5	211	170	59.6	2.45	0.324	39.1	65

S = sateen; T = twill; d_p is calculated from $d_p = 4A_0/W$; ec,pc = end count and pick count respectively; d_1, d_2 = weft and warp diameters respectively.

In order to correlate the 2/1 twill with 5/1 sateen data, Rushton and Griffiths defined a 'flow cell' which is a repeated pattern in the cloth. For a 2/1 twill the cell is made up of six twill pores and three plain pores, and in a 5/1 sateen the cell comprises twenty four twill pores and twelve plain pores. Thus, for a 2/1 twill the weighting procedure for the effective fractional open area of the pore is given by:

$$a = \frac{2}{3}a_T + \frac{1}{3}a_P \qquad (2.61)$$

where a_T and a_P are the effective fractional open areas of the twill and plain pores. The average wetted perimeter of the orifice was a similarly weighted expression based on the plain and twill fabric pores

$$W = \frac{2}{3}W_T + \frac{1}{3}W_P \tag{2.62}$$

The monofilament pore characteristics are reproduced in Table 2.3; these can be applied to the analysis of any flow cell.

Example 2.4 Calculation of the pressure loss over a clean, monofilament, filter cloth.
A liquid (ρ_l = 1000 kg m^{-3}, μ = 0.00095 Pa s) is flowing with an approach velocity of 1.2 m s^{-1} through a filter cloth with a 2/1 twill weave construction. The warp and weft yarn counts are 40 and 17 yarns cm^{-1} respectively, and the corresponding yarn diameters are 193 and 196 µm. Estimate the pressure loss across the cloth.

Solution

Noting that ec = 40, pc = 17, d_2 = 193 µm and d_1 = 196 µm, the fractional open areas can be calculated from equation (2.61) using equations (2.55) and (2.57):

$$\phi = \frac{1}{40(0.0196+0.0193)} = 0.6427$$

$$a_P = \frac{0.6427}{2}\ln\left[\frac{1+\sqrt{1+0.413}}{0.6427}\right] + \frac{\sqrt{1+0.413}}{2\times0.6427} - 40\times0.0196$$

$$- 40\times17\times0.0196^2\sqrt{0.2908^2+1} = 0.263$$

$$a_T = 0.263 + \frac{40\times17\times0.0196^2}{2}\sqrt{0.2908^2+1}$$

$$+ \frac{40\times17\times0.0196\times(0.025-0.0193)}{2} = 0.437$$

$$a = \frac{2}{3}\times0.437 + \frac{1}{3}\times0.263 = 0.379$$

The wetted perimeter of the pore can be similarly calculated from equation (2.62); alternatively, to avoid evaluation of the elliptic integral W can be obtained from Table 2.3 as 1.78 mm.

The Reynolds number is given by equation (2.60) (converting to a consistent system of units):

$$\mathrm{Re} = \frac{4\times0.012\times1000}{0.00178\times(40\times10^2)\times(17\times10^2)\times0.00095} = 4.2$$

i.e. flow is laminar, so equation (2.53) can be used, from which the pore discharge coefficient is obtained as:

$$C_D = \sqrt{\frac{1000 \times 0.012^2}{2\Delta p} \frac{(1 - 0.379^2)}{0.379^2}} = \frac{0.655}{\sqrt{\Delta p}}$$

Using equation (2.59):

$$\Delta p = \left(\frac{0.655}{0.17 \times 4.2^{0.41}}\right)^2 = 4.6 \quad \text{Pa}$$

i.e. when the liquid approach velocity is 1.2 m s^{-1} the pressure loss across the clean filter cloth is 4.6 Pa.

2.5.3 Multifilament woven media

The yarn of a multifilament fabric is made up from many smaller fibres that are twisted together. The geometries of the fibres and yarns that make up a multifilament fabric are very complex, making interpretation of data for flow through the cloth difficult. Even in a fabric of apparently simple weave and construction, such as a plain weave, continuous filament cloth, flow takes place in the highly tortuous channels present in the yarns as well as around the yarns (Grace, 1956; Rushton and Griffiths, 1987). This causes the orifice analogy used for monofilament fabrics to fail when applied to multifilament cloths, particularly when the cloth is loosely woven.

McGregor (1965) found that the Kozeny-Carman model (equation (2.37)) provided a satisfactory description of fluid flow through woven textiles, when the permeability of the yarns themselves was taken into account and flow was in the laminar regime. McGregor suggested that, in order to describe the flow through an assembly of permeable yarns, it is reasonable to adopt Brinkman's (1948) approximate solution to the problem of calculating the fluid drag on an assembly of permeable spheres:

$$\beta = \frac{k}{k_1} = 1 + 1.80\frac{k_0}{k_1} + 2.68\sqrt{\frac{k_0}{k_1}} \qquad \text{for} \quad \frac{k_0}{d_y^2} < 0.0017 \qquad (2.63)$$

where k is the overall permeability of the cloth, k_0 is the permeability of the yarns in the cloth, k_1 is the permeability of the cloth if the yarns were impermeable cylinders and d_y is the yarn diameter.

2.6 Cake deliquoring – multiphase flow in porous media

One method of deliquoring or dewatering filter cakes is to displace the filtrate that has been retained in the cake voids by an immiscible fluid. The second fluid is usually a gas, with air being used most commonly. In the deliquoring process the cake can be considered as a matrix of solid particles in a liquid and gas mixture, and the mode of liquid removal reduces to one of two phase flow through the porous medium. One of the fluids, usually the liquid in the feed slurry (i.e. the filtrate), will tend to wet the solids in preference to the incoming gas. As a result the liquid, the wetting fluid, flows over the surface of the solids and prevents contact of the gas, the nonwetting fluid, with the solid. The gas flows through the remaining void space and contacts the retained liquid rather than the solids. The shape of the flow channels in the cake is considerably modified by the presence of the liquid and the extent of saturation of the cake by the liquid. The cake saturation, S, is defined by

$$S = \frac{\text{volume of liquid in the cake}}{\text{volume of voids in the cake}} \tag{2.64}$$

The liquid tends to smooth the surface of the channels through which the gas flows, and the flowing gas does not experience such sharp changes in direction as it would if the cake were completely dry.

In developing procedures for predicting the flow of two phases through porous media, it is necessary to consider the capillary forces in the bed in addition to the degree of saturation. Surface forces are present at the interface between the two fluids, which are in contact when flowing simultaneously through the medium; the surface tension force at the interface between the liquid and the solid usually acts to retain liquid in the finer pores of the cake.

When two immiscible fluids flow through a porous medium, each fluid can be considered to establish its own tortuous pathways through the cake voids. A unique set of channels would then correspond to each fluid at every degree of saturation. When the liquid saturation is high, the channels of gas are discontinuous; as the liquid saturation is reduced its associated channels break down and become discontinuous at the irreducible wetting fluid saturation (any flow of the wetting fluid ceases at the irreducible saturation). When the liquid saturation becomes very low the liquid film over the surface of the particles breaks down, and once this has happened deliquoring by displacement is no longer effective.

The notion of two phase flow enables Darcy's law to be readily applied to the problem, and a form of the law is required for each flowing phase. For the filtrate retained in the cake (subscript l) the equation is:

$$u_l = \frac{-k_l}{\mu_l}\frac{dp_l}{dz}$$

(2.65)

and for the gas (subscript g) it is

$$u_g = \frac{-k_g}{\mu_g}\frac{dp_g}{dz}$$

(2.66)

From equations (2.65) and (2.66) it should be noted that each phase has its own associated pressure, viscosity and permeability. The difference between the pressures of the two phases is the capillary pressure, which is a function of the radius of curvature of the interface between the phases, and this is in turn dependent on the saturation of each phase. The terms k_l and k_g are known as the effective permabilites of the liquid and gas phases respectively, and are related to the permeability of the saturated medium through:

$$k_{rl} = k_l / k \quad \text{and} \quad k_{rg} = k_g / k$$

(2.67)

where k_{rl} and k_{rg} are known as the relative permeabilities of the cake to the liquid and gas phases respectively.

2.6.1 Capillary pressure and threshold pressure

The capillary pressure acts to retain liquid inside the pores of the cake and is dependent on the saturation of the filter cake and the range of pore sizes that exist therein. The problem of developing a relationship between these parameters has arisen in other technologies, particularly in oil recovery processes, and the models used are common to these processes. The most useful relationship is obtained by measuring a capillary pressure curve for the porous medium or filter cake, and then plotting the capillary pressure against the saturation. After defining a reduced saturation, S_R, as:

$$S_R = \frac{S - S_\infty}{1 - S_\infty}$$

(2.68)

where S_∞ is the saturation at which flow of the liquid ceases (that is, at the wetting fluid irreducible saturation), the resulting data can generally be described by (Brooks and Corey, 1964; Wakeman 1976b, 1982):

$$S_R = \left(\frac{p_b}{p_c}\right)^\lambda \quad \text{for} \quad p_c \geq p_b \qquad (2.69)$$

where λ is the pore size distribution index which characterises the range of pore sizes in the filter cake or porous medium, and p_b is the minimum or threshold pressure needed to initiate displacement of liquid. When $S \leq S_\infty$ the liquid may be regarded as immobile, and further liquid removal from the cake (i.e. reduction of saturation) is usually only achieved by evaporative or drying techniques.

During formation of a filter cake, at the microscopic scale an infinite number of local packing arrangements of the particles may coexist in the overall structure of the cake. This suggests that λ may vary from point to point through the cake; a capillary pressure experiment averages these structural variations to yield a value for λ. Some values for the pore size distribution index and threshold pressure are shown in Table 2.4.

The threshold pressure increases as the mean size of particles in the cake decreases and as the size range increases; the value of λ does not follow any trend with particle size, probably because it is also dependent on the size range of the particles in the distribution. The variation in the value of λ is not large for smaller particles and the variation leads only to small variations in the relative permeability (Wakeman, 1982), which suggested $\lambda = 5$ to be a reasonable value for practical purposes.

2.6.2 Relative permeability and capillary pressure

Underlying the extension of Darcy's law for a single fluid to the simultaneous flow of two fluids, and related to the capillary pressure, is the relative permeability concept. Lloyd and Dodds (1972) showed how the flow of liquid from a filter cake could be calculated by the application of Darcy's law to elemental sections of the cake using a relative permeability model. A useful model of relative permeability was presented by Wyllie and Gardner (1958), and is obtained by extending the arguments presented in Section 2.2. Assume the filter cake is composed of a bundle of capillary tubes whose radii r have a distribution function $\alpha(r)$. Imagine that the bundle is cut into a large number of thin slices, and then rearrange randomly the short tubes in

Table 2.4 Pore size distribution index and threshold pressure data for randomly deposited porous media

Particle type	Mean size (μm)	λ	p_b (kPa)	Reference
Fine sand		3.7	14.8	Brooks and Corey (1964)
Silt loam		1.8	11.4	
Spheres	22	5.3	22.9	Wakeman (1975b)
(log-normal size	97	6.5	6.1	
distribution)	165	6.3	4.3	
	327	3.8	3.1	
	387	2.4	1.2	
Sand	137	5.0		Wakeman (1976b)
Vitrain	230	3.9		Wakeman (1982) – data
Galena	135	5.3		replotted from
	194	6.0		Harris et al (1963)
	230	8.0		
Coal	112	10.0		Wakeman (1982) – data
	230	3.8		replotted from Gray (1958)
	925	5.5		
Alumina trihydrate			4.5– 6.7	Puttock et al (1986)
Coal	>300	5.6		van Brakel et al (1984)
	225	9.5		
	195	2.5		
	60	3.2		
Sand	60	2.6	35	Carleton and Mackay (1988)
Diatomite		0.97	20	
Coal (froth concentrate)	250	0.7	10	Condie et al (1996)

each slice and reassemble the slices. The analysis is simplified by noting equation (2.68), and therefore considering the liquid to be part of the solid structure of the cake when $S \leq S_\infty$.

At a capillary pressure p_c the gas phase occupies all pores larger than size r', defined by:

$$r' = \frac{2\sigma}{p_c} \tag{2.70}$$

where σ is the interfacial tension at the gas-liquid interface. The liquid saturation is related to r' by:

$$S_R = \frac{\displaystyle\int_{r_1}^{r'} \pi\, r^2 \alpha(r)\, dr}{\displaystyle\int_{r_1}^{r_2} \pi\, r^2 \alpha(r)\, dr} \tag{2.71}$$

This equation indicates that $\alpha(r)$ can be determined from a plot of the capillary pressure against saturation.

Consider a single slice of the cake where the probability of there being a pore filled with liquid is nS_R. For two adjacent slices, the area common to a single pore of cross-sectional area πr^2 in one slice and all the pores filled with wetting fluid in the adjacent slice is $\pi r^2 nS_R$. The passage of liquid is thus assumed to be from an area πr^2 to a constricted area $\pi r^2 nS_R$. The constricted area can be visualised as a smaller pore of radius r'' such that

$$\lambda'\pi\,(r'')^2 = \pi\,r^2 nS_R \tag{2.72}$$

where λ' is a numerical coefficient dependent on $\alpha(r)$ and reflecting the manner in which the available total interconnected pore area is distributed. Equation (2.72) can be substituted into Poiseuille's equation (equation (2.24)) to give an expression describing the volumetric flow rate through the capillary:

$$Q = \frac{\pi\,(r'')^4}{8\mu}\frac{\Delta p}{L} = \frac{\pi\,r^4 n^2 S_R^2}{8\mu(\lambda')^2}\frac{\Delta p}{L} \tag{2.73}$$

The pore cross-sectional area available for flow is $\dfrac{1}{n}\displaystyle\int_{r_1}^{r_2}\pi r^2\alpha(r)\,dr$, and if λ' is independent of $\alpha(r)$ the discharge of liquid from the cake is given by:

$$u = \frac{nQ}{\displaystyle\int_{r_1}^{r_2}\pi r^2\alpha(r)\,dr} = \frac{n^3 S_R^2}{8\mu(\lambda')^2}\frac{\displaystyle\int_{r_1}^{r'}\pi r^4\alpha(r)\,dr}{\displaystyle\int_{r_1}^{r_2}\pi r^2\alpha(r)\,dr}\frac{\Delta p}{L} \tag{2.74}$$

Differentiating equation (2.71) with respect to r gives:

$$dS_R = \frac{r^2\alpha(r)\,dr}{\displaystyle\int_{r_1}^{r_2}r^2\alpha(r)\,dr} \tag{2.75}$$

Now, eliminate r from equation (2.74) using equations (2.75) and (2.70) to obtain the liquid flux leaving the cake:

$$u = \frac{n^3 S_R^2}{8\mu \,(\lambda')^2} \frac{\int\limits_{r_1}^{r'} r^2 \int\limits_{r_1}^{r_2} r^2 \alpha(r)\, dr\, dS_R}{\int\limits_{r_1}^{r_2} r^2 \alpha(r)\, dr} \frac{\Delta p}{L} = \frac{n^3 S_R^2 \sigma^2}{2\mu\,(\lambda')^2} \int\limits_{0}^{S_R} \frac{dS_R}{[p_c\,(S_R)]^2} \frac{\Delta p}{L} \qquad (2.76)$$

Comparing this expression with the extension of the Darcy equation (i.e. equation (2.65)), enables the effective permeability to the liquid to be written as:

$$k_l = \frac{n^3 S_R^2 \sigma^2}{2\,(\lambda')^2} \int\limits_{0}^{S_R} \frac{dS_R}{[p_c\,(S_R)]^2} \qquad (2.77)$$

and the permeability of the cake to be written as:

$$k = \frac{n^3 \sigma^2}{2\,(\lambda')^2} \int\limits_{0}^{1} \frac{dS_R}{[p_c\,(S_R)]^2} \qquad (2.78)$$

From these equations the relative permeability (k_{rl}) can be expressed in terms of the liquid saturation in the cake by dividing equation (2.77) by (2.78) and rearranging the result:

$$k_{rl} = S_R^2 \frac{\int\limits_{S_\infty}^{S} \frac{dS}{[p_c\,(S)]^2}}{\int\limits_{S_\infty}^{1} \frac{dS}{[p_c\,(S)]^2}} \qquad (2.79)$$

The relative permeability for the gas phase is obtained similarly:

$$k_{rg} = (1 - S_R)^2 \frac{\int\limits_{S_R}^{1} \frac{dS_R}{[p_c\,(S_R)]^2}}{\int\limits_{0}^{1} \frac{dS_R}{[p_c\,(S_R)]^2}} \qquad (2.80)$$

Equations (2.79) and (2.80) were also derived by Burdine (1953) using the hydraulic radius theory. These equations can be combined with equation (2.69) to give expressions for the relative permeabilities in terms of the pore size distribution index λ:

$$k_{rl} = S_R^{(2+3\lambda)/\lambda} = \left(\frac{p_b}{p_c}\right)^{2+3\lambda} \tag{2.81}$$

$$k_{rg} = \left(1 - S_R\right)^2 \left(1 - S_R^{(2+\lambda)/\lambda}\right) = \left(1 - \left(\frac{p_b}{p_c}\right)^{\lambda}\right)^2 \left(1 - \left(\frac{p_b}{p_c}\right)^{2+\lambda}\right) \tag{2.82}$$

2.6.3 Application to cake deliquoring

When Darcy's equations for the flow of each phase through the cake, equations (2.65) and (2.66) are combined with material balances (or continuity equations) for the liquid and gas phases respectively i.e.

$$\frac{\partial(\varepsilon S)}{\partial t} = -\frac{\partial u_l}{\partial z} \tag{2.83}$$

and

$$\frac{\partial(\varepsilon \rho_g (1 - S))}{\partial t} = -\frac{\partial(\rho_g u_g)}{\partial z} \tag{2.84}$$

A solution to the vacuum or pressure driven cake deliquoring problem can be obtained when appropriate boundary conditions are applied (Wakeman, 1982). In reaching the solution to equations (2.67) – (2.69), (2.79) and (2.80) are also used; the solutions are discussed and applied in Chapter 5. Baluais *et al* (1983) showed how an analytical solution to these equations, with some simplifying assumptions, could be used to calculate the deliquoring kinetics.

During deliquoring using centrifugal or gravitational forces, the only body force of significance is that acting on the liquid phase. A body force does exist in the gas phase and flow of the gas does occur as a result, but unlike deliquoring using vacuum or pressure there is no practical requirement to calculate the flow rate of the gas. The equation describing centrifugal drainage is (Wakeman and Vince, 1986a):

$$\frac{\partial S}{\partial t} = -\frac{k}{\varepsilon \mu_l} \left[k_{rl} \left(\rho_l \omega^2 - \frac{\partial^2 p_c}{\partial r^2} \right) + \left(\rho_l \omega^2 r - \frac{\partial p_c}{\partial r} \right) \left(\frac{\partial k_{rl}}{\partial r} + \frac{k_{rl}}{r} \right) \right] \tag{2.85}$$

where r is a radius coordinate measured from the centrifuge axis and ω is the rotational speed of the centrifuge. The equation describing drainage in a gravitational field is (Wakeman and Vince, 1986b):

$$\frac{\partial S}{\partial t} = \frac{k}{\varepsilon \mu_l} \left(\rho_l g - \frac{\partial p_c}{\partial z} \right) \frac{\partial k_{rl}}{\partial z} - \frac{k k_{rl}}{\varepsilon \mu_l} \frac{\partial^2 p_c}{\partial z^2} \tag{2.86}$$

In these expressions, S_R is given by equation (2.68), p_c by equation (2.69) and k_{rl} by equation (2.81). The above equations (2.85) and (2.86) have been solved numerically and their use is described further in Chapter 5. The model has been used for the analysis of pilot and industrial scale deliquoring processes and gives a good interpretation of the effects of changes in process operating conditions.

2.6.4 Other deliquoring models

Other deliquoring models that exist in the literature are identified in Table 2.5.

Table 2.5 Other models for cake deliquoring that appear in the literature

Model	Reference
Pipe flow analogy	Brownell and Katz (1947); Brownell and Gudz (1949)
Film drainage	Nenniger and Storrow (1958); Shirato et al (1983)

Brownell and Katz (1947) developed a calculation method based on an anology between pipe flow and flow in porous media. For laminar flow of a single fluid a Reynolds number and a friction factor are defined which include the sphericity and the size of the particles and the porosity of the filter cake:

$$\mathrm{Re} = \frac{\rho u x}{\mu \varepsilon^m} \quad \text{and} \quad f = \frac{2 x \Delta p \varepsilon^n}{L u^2 \rho}$$

where the exponents m and n are functions of the ratio of particle sphericity to bed porosity. From their experimental data Brownell and Katz produced charts to assign values to these exponents. The concept was extended to two phase flow by applying the relationship to each of the phases. The effective saturation, defined as

$$S_e = \frac{S - S_R}{1 - 2S_R + SS_R}$$

is included in their equations for the wetting phase, raised to a power y which is yet another function of porosity. Flow of the nonwetting phase is described by a similar relation which incorporates two further terms, "wetted porosity" and "wetted sphericity". Whilst wetted porosity was found to be a function of saturation, Brownell and Katz failed to establish a relation between wetted sphericity and saturation. For the gas phase, experiments are used to determine the wetted porosity and the wetted sphericity. The use of this approach in the calculation of gas flow in the deliquoring of filter cakes on rotary vacuum filters is described by Brownell and Gudz (1949). When the method is applied to air blowing on a pressure filter the calculated gas flow underestimates the experimental flow by a considerable margin (Rushton *et al*, 1996) and the predicted liquid flow rate compares poorly with the desaturation rate actually obtained. Extrapolation of this calculation method is not successful due to the difficulty of evaluating the required exponents outside the porosity limits investigated by Brownell and Katz.

During centrifugal or gravitational drainage, desaturating driving forces cause the liquid level in an initially saturated cake to recede from the cake surface. In the region between the cake surface and the level of the liquid bulk, liquid is retained in the voids and continues to drain in a filmlike way, until the supply of liquid (primarily from around the particle contact points) is insufficient to maintain the continuity of the film. At the end of the drainage process an equilibrium is reached, which is obtained when the weight of liquid below the level of the bulk is equal to the surface tension forces retaining liquid in the cake. This concept has been used to model cake deliquoring (Nenniger and Storrow, 1958; Shirato *et al*, 1983) in gravitational and centrifugal fields, and is illustrated in Figure 2.4. When the bulk liquid level is above the surface of the cake, $h_L > L$, h_L falls towards L during deliquoring and Darcy's law is applicable for the calculation of the cake permeability. When $L > h_L > h_c$, the liquid level lies inside the cake but it is above the capillary drain height which will be reached after a prolonged period of drainage. Liquid draining from the film in this region, into the saturated region below h_L, must be accounted for in the material

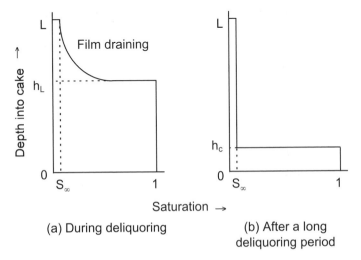

Figure 2.4 Idealised liquid distributions in gravity or centrifugally drained cakes.

balance for liquid in the cake. The variation of saturation with deliquoring time is given by:

$$S = 1 - \left(1 - \frac{h_L}{L}\right)\left[(1 - S_\infty) - \frac{2}{3}\frac{S_0\,(1-\varepsilon)}{\varepsilon}\sqrt{\frac{L\mu\left(1 - \dfrac{h_L}{L}\right)}{\rho g\,(t - t_0)}}\right] \qquad (2.87)$$

2.7 Cake washing – hydrodynamic dispersion

Flow of a miscible liquid through a filter cake, as is usually the case during displacement cake washing, can be envisaged by considering the migration of a tracer 'fluid particle' (Wakeman, 1975b). The downstream movement of the tracer particle is erratic, and its exact path is determined by the interplay of random processes of various kinds. Included amongst these are ordinary molecular diffusion, flow pattern effects caused by irregularities in the shape of the flow pores, mass transfer kinetics and sometimes electrokinetic phenomena. Dispersion of a front of wash liquid is a direct consequence of these processes and their influence on the random nature of each tracer particle's migration through the cake.

A random process is one in which individual movements occur unpredictably. The effects of this are best illustrated by ordinary molecular diffusion; if the motion of a large number of molecules of a given type were to be started at exactly the same time and location, each would undergo chaotic motion with some ending up on one side and some on the other side of a starting point. In this way an initially sharp concentration pulse would gradually become more and more dispersed.

In filter cake washing an analogous group of random processes leads to the dispersion of an initial step function input of wash liquor. Solute diffusion occurs continuously between the wash liquor and the retained filtrate. During flow through the cake, streamlines veer tortuously back and forth in an effort to find relatively unobstructed flowpaths. A fluid element following such a streampath will have its downstream velocity and its direction of flow fluctuate between wide limits because of the variable rate of fluid penetration through different size pores at different locations in the cake; the effect in a single idealized cell is shown in Figure 2.5a. A solute molecule being carried by the wash liquor in such a streampath will have its velocity changed erratically between high and low values. The velocity changes are random because

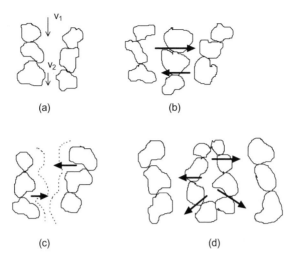

Figure 2.5 Solute dispersion is caused by (a) variation in pore size leading to a varying velocity of wash liquor and its associated solutes, (b) solute interchange between adjacent flow channels, (c) solute diffusion across a single flow channel, and (d) solute transfer out of stagnant pockets of fluid.

the structure of the channels controlling the flow pattern in the cake is random. Hence, some molecules, by following relatively open pathways, may acquire a greater than average downstream velocity, and others may be enmeshed in restricted channels and lag behind. The net result is dispersion of the solute in the wash liquor in the filter cake.

This phenomenon is further modified because solute molecules can acquire a variable downstream velocity by diffusing laterally into adjacent streampaths, as illustrated in Figure 2.5b. Dispersion of the wash liquor itself originates in the velocity inequalities of the flow pattern, but the extent of dispersion is governed largely by the interdiffusion which occurs between fast and slow streampaths.

The broad category of inequalities which lead to dispersion of the wash liquor originate in several different ways. In each interstitial flow channel a higher velocity exists near the centre and a lower velocity nearer the surface of the solid particles (Figure 2.5c). Diffusion of solute molecules through the wash leads to the exchange of molecules between these regions, thus giving rise to dispersion by molecular interchange across a single channel. Under some washing conditions, notably when washing porous particles or aggregates, solute may lie in the void space within the particulate structure. This solute is slow moving or stagnant, but surrounding voids may be occupied by mobile wash liquor (Figure 2.5d). Cakes can contain small regions of tightly packed particles which are bounded by a looser packing; the looser packing gives rise to larger interstitial channels in which higher wash liquor velocities occur. Also, velocity differences exist between the outer extremities of the cake where solid containing walls affect the particle packing density and fluid velocities.

In summary, hydrodynamic dispersion is the macroscopic outcome of the actual movements of solute and wash liquor through the pores of the cake and the various physical and chemical phenomena that take place within the pores. Two basic transport phenomena are involved: convection (sometimes called mechanical dispersion or convective diffusion) and molecular diffusion. The molecular diffusion contribution can take place in the absence of flow, and its effects on the overall dispersion are more significant at low flow velocities.

2.7.1 The equations of hydrodynamic dispersion

The transport equation describing dispersion processes in cake washing have been developed (Wakeman, 1986a), the solute material balance, which is in fact the diffusion equation with the molecular diffusivity replaced by an axial dispersion coefficient (omitting radial dispersion as this is negligible compared with axial dispersion), is written formally as:

$$\frac{\partial \phi}{\partial t} + v \frac{\partial \phi}{\partial z} = D_L \frac{\partial^2 \phi}{\partial z^2} \tag{2.88}$$

where ϕ is the concentration of solute in the liquid phase. This equation can be modified to account for the sorption or desorption of solute by the solid particles, as would be necessary when washing paper pulp, for example. When the concentration of solute is directly proportional to the solute concentration in the surrounding liquid (this only represents the simplest case of sorption and is not necessarily the best for all applications) equation (2.88) becomes (Sherman, 1964):

$$\left\{ 1 + K \left(\frac{1 - \varepsilon}{\varepsilon} \right) \right\} \frac{\partial \phi}{\partial t} + v \frac{\partial \phi}{\partial z} = D_L \frac{\partial^2 \phi}{\partial z^2} \tag{2.89}$$

where K is the proportionality constant relating the solute sorbed at the solid surface (ϕ_s) to the solute concentration in the liquid in the pores of the cake through the linear isotherm $\phi_s = K\phi$. Other isotherms with the general equation $\phi_s = K_0 + K\phi^{1/n}$ (where K_0 and n are constants) have been considered in relation to cake washing (Wakeman and Attwood, 1990). The initial and boundary conditions applicable for the solution of equation (2.89) are:

$$At \ t = 0, \quad z \geq 0 \quad \phi = \phi_0$$
$$\quad t > 0, \quad z = 0 \quad \phi = \phi_w \tag{2.90}$$

and the solution is:

$$\frac{\phi - \phi_w}{\phi_0 - \phi_w} = 1 - \frac{1}{2}\left\{ erfc\left(\frac{1 - \lambda W}{2\sqrt{\lambda W}}\sqrt{\frac{vL}{D_L}}\right) + \exp\left(\frac{vL}{D_L}\right)erfc\left(\frac{1 + \lambda W}{2\sqrt{\lambda W}}\sqrt{\frac{vL}{D_L}}\right)\right\} \tag{2.91}$$

where

$$\lambda = \left\{1 + K\left(\frac{1 - \varepsilon}{\varepsilon}\right)\right\}^{-1} \tag{2.92}$$

This equation is strictly only valid for washing saturated cakes, but in practice it also gives a good estimate when applied to deliquored cakes during the latter stages of washing, that is, for pore volumes of wash greater than about one. When there are no sorption effects $K = 0$ and $\lambda = 1$.

The term (vL/D_L) is properly interpreted in terms of the Reynolds number for wash liquor flow through the cake and the Schmidt number describing the ratio of the molecular diffusivity of momentum to the molecular diffusivity of mass, that is:

$$\frac{vL}{D_L} = \frac{\rho v x}{\mu}\frac{\mu}{\rho D}\frac{L}{x}\frac{D}{D_L} = Re \, Sc \frac{L}{x}\frac{D}{D_L} = Pe \frac{L}{x}\frac{D}{D_L} \tag{2.93}$$

where x is the particle diameter, L is the filter cake thickness, Sc is the Schmidt number and Pe is the Peclet number. Practical relationships between the ratio of the molecular diffusion coefficient to the axial dispersion coefficient and the Peclet number are considered in detail in Chapter 6. Equation (2.91) represents a family of curves when plotted as the concentration ratio ϕ/ϕ_0 (putting $\phi_w = 0$) against the wash ratio W, as shown in Figure 2.6.

The general shape of these washing curves is similar to those found in practice, with the exception that practical curves do not approach ϕ/ϕ_0 quite so rapidly. The reason for this is that it is usual for the wash liquor feed to form a pool on the surface of the cake, allowing back-diffusion of solute to occur into wash in the pool before the wash starts to penetrate the cake (Wakeman and Attwood, 1988), which invalidates

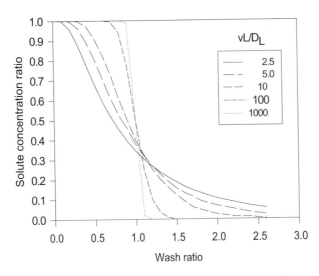

Figure 2.6 The instantaneous solute concentration in the wash effluent as calculated from the dispersion model with a step change of concentration at the cake inflow surface and with $K = 0$.

the boundary condition at $z = 0$ (see equation (2.90)). The problem is particularly severe when the cake is thin, as is usually the case in filtration. Sherman (1964) encountered a similar problem when he recognised the difficulty of obtaining a true step function at the wash inflow face in laboratory experiments. The back-diffusion is taken into account by using the more appropriate boundary condition (Wakeman and Attwood, 1988):

$$t > 0 \quad z = 0 \quad \phi = \phi_0 \, k_0 \, e^{-\gamma W} \tag{2.94}$$

where k_0 and γ are constants that characterise the back-diffusion and consequent mixing that occurs in the wash liquor layer lying over the cake surface. A numerical solution to the equations was given and it was also recommended, based on experimental data, that $k_0 = 0.15$ and $\gamma = 5.0$ were acceptable values for the constants. Using this boundary condition, Eriksson et al (1996) showed that the analytical equation describing the washing curve is:

$$\frac{\phi - \phi_w}{\phi_0 - \phi_w} = 1 - \frac{1}{2} \left\{ erfc \left(\frac{1 - \lambda W}{2\sqrt{\lambda W}} \sqrt{\frac{vL}{D_L}} \right) + \exp \left(\frac{vL}{D_L} \right) erfc \left(\frac{1 + \lambda W}{2\sqrt{\lambda W}} \sqrt{\frac{vL}{D_L}} \right) \right\}$$

$$+ k_0 \sqrt{\frac{vL}{\pi \lambda W D_L}} \int_1^\infty \exp \left(\frac{vL}{2D_l} - \gamma \left(1 - \frac{1}{\omega^2} \right) W - \frac{W vL\lambda}{4 D_L \omega^2} - \frac{\omega^2 vL\lambda}{4 W D_L} \right) d\omega \tag{2.95}$$

This model can be used in a predictive sense if the properties of the cake and the liquid are known.

2.7.2 Washing deliquored filter cakes

This complex problem requires models that simultaneously describe resaturation of the cake and mass transfer or dispersion of the solute within the cake voids. The saturation as a function of time and axial position in the cake may be calculated from the continuity equation (equation (2.83)) for the liquid written as:

$$\frac{\partial(\varepsilon S)}{\partial t} = -\frac{\partial(v \varepsilon S)}{\partial z} \tag{2.96}$$

and Darcy's equation describing the liquid movement relative to the particles, equation (2.65), written as:

$$v = -\frac{k \, k_{rl}}{\mu_l \, \varepsilon S} \frac{\partial p_l}{\partial z} \tag{2.97}$$

where k_{rl} is given by equation (2.81). These equations can be arranged into a convenient dimensionless form involving the reduced saturation (S_R) and the wash ratio (W):

$$\frac{\partial S_R}{\partial W} = \frac{S_{W=0} \beta_0}{v_{S=1}} \left(a S_R^{a-1} \frac{\partial S_R}{\partial z^*} \frac{\partial p_l}{\partial z^*} + S_R^a \frac{\partial p_l}{\partial z^{*2}} \right) \tag{2.98}$$

where $a = \dfrac{2 + 3\lambda}{\lambda}$, $\beta_0 = \dfrac{-k}{\varepsilon \mu_l L (1 - S_\infty)}$ and $z^* = \dfrac{z}{L}$ is the dimensionless distance in the cake. An appropriate form of the dispersion equation (equation (2.88)) is solved simultaneously, and this can also be written in terms of the same dimensionless variables as:

$$\frac{\partial \phi^*}{\partial W} = -\frac{S_{W=0}}{S_R + S_f} \frac{\partial}{\partial z^*} \left(\frac{1}{D_n} \frac{\partial S_R}{\partial z^*} \right) \phi^* \tag{2.99}$$

$$+ S_{W=0} \left(\frac{u}{u_{S=1}} - \frac{1}{D_n (S_R + S_f)} \frac{\partial S_R}{\partial z^*} \frac{\partial}{\partial z^*} \left(\frac{1}{D_n} \right) \right) \frac{\partial \phi^*}{\partial z^*} - \frac{1}{D_n} \frac{\partial^2 \phi^*}{\partial z^{*2}}$$

where $S_f = \dfrac{S_\infty}{1 - S_\infty}$ and $D_n = \dfrac{vL}{D_L}$; $\phi^* = \dfrac{\phi}{\phi_0}$ is the dimensionless instantaneous concentration of solute in the wash effluent from the cake.

The above equations are solved with the boundary conditions stated on equations (2.90) and (2.94), together with

$$S_R = \text{constant} \quad \text{for} \quad W = 0, \quad z^* \geq 0$$
$$S_R = 1 \quad \text{for} \quad W > 0, \quad z^* = 1$$

(2.100)

to obtain a solution (Wakeman and Attwood, 1990). The general form of washing curve obtained is shown in Figure 2.7; following an initial displacement of retained liquor from the voids (AB on Figure 2.7), the duration of which is dependent on the initial saturation of the cake, a sharp decline in the solute concentration is observed in the solute concentration as the retained liquor is diluted by wash liquor (BC). This is soon followed by a slowing of the rate of solute removal from the cake as wash and retained liquors are mixed in the pores, and thereafter the shape of the curve (DE) of the solute concentration variation with the number of pore volumes of wash used is similar to that obtained when washing a saturated cake. The wash ratio associated with washing a deliquored cake is much larger, because of the smaller amount of liquor in the cake at the start of washing.

High initial saturations (≥ 0.5) result in a slight increase in wash liquor usage; deliquoring must result in the cake saturation being reduced to below 0.5 in order to reduce the amount of wash liquor required. As the initial saturation is decreased below 0.5, the volume of wash liquor

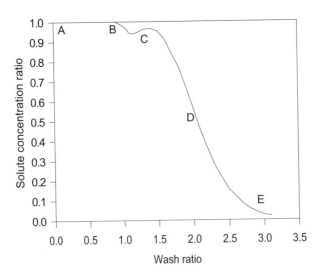

Figure 2.7 The instantaneous solute concentration in the wash effluent calculated from the combined dispersion and resaturation model, with the cake initially 50% saturated, $S_\infty = 0.3$ and $vL/D_L = 50$.

required is reduced, the resaturating wash liquor front becomes sharper, displacement is reduced and mass transfer effects rapidly become significant.

2.7.3 Other models for cake washing

The alternative mechanisms proposed to describe the washing process are varied, and some may be relevant to particular types of washing. These are listed in Table 2.6.

Table 2.6 Other models for cake washing that appear in the literature

Model	References
Film model	Kuo (1960); Kuo and Barrett (1970)
Side channel model	Han (1967); Han and Bixler (1967); Fasoli and Melli (1971); Wakeman (1972, 1973, 1974); Wakeman and Rushton (1974)
Diffusion from microporous particles	Michaels et al (1967); Wakeman (1976a); Eriksson et al (1996); Rasmuson and Neretnieks (1980); Rasmuson (1985)

If it assumed that the initial displacement of filtrate from the cake pores by the wash liquor creates a channel for wash flow, with a stagnant film of solute bearing liquor remaining at the surfaces of the pore, then mass transfer from a uniform film could be envisaged. This approach was adopted by Kuo (1960), and the model often gives a good interpretation of washing data when the model equations are fitted to experimental data. The fit is obtained by adjusting values for mass transfer coefficients which are unknown in advance of carrying out experiments, which limits the model as a predictive tool.

The structure of a filter cake can be analysed in terms of flow pores through which wash liquor passes, and which have dead volumes distributed along their length. There is no bulk flow into or out of these so-called blind side channels that contain the residual filtrate. The solute in the trapped filtrate is transferred by diffusion into the wash liquor, where it is transported through the cake by plug flow. The model might be envisaged to represent washing of drained cakes, and comparison with experimental data has shown a match over part of the washing curve.

Diffusion from porous or aggregated particles has been modelled, and the mathematical problem is similar to that describing adsorption processes. In these cases, solute is considered to be washed from micropores in the cake by diffusion. This sometimes gives rise to two rates of mass transfer, one from the intraparticle diffusional process and the other from the removal of solute from the particle surfaces and from between neighbouring particles. The resulting equations generally require that a detailed characterisation of the particles is carried out in order that sufficient information is available to use the model quantitatively, and even then a fitting procedure needs to be adopted to evaluate mass transfer coefficients.

The above models have unknown parameters which have to be obtained from experimental measurements before they can be applied in practice, making their application difficult. It is possible that the film model and the micropore model could both be developed to the extent that the mass transfer coefficients could be suitably correlated, as the axial dispersion coefficient has been for the dispersion model, but this has not been done. Also, the blind side channel model appears to describe only limited cases of washing. For these reasons, the dispersion model is preferred as a practical tool.

2.8 Expression – deliquoring by compression

The moisture content of solid/liquid mixtures (either suspensions or cakes) can sometimes be reduced by subjecting the mixture to a mechanical compressive force, and thereby removing liquid by squeezing; although this technique reduces the mass of liquid in the mixture its saturation remains at unity (100%) throughout the operation. When the starting mixture is a suspension the initial mechanism of liquid removal is filtration, followed by consolidation after a critical solids content in the mixture has been exceeded; the equations describing filtration are applicable to this initial stage of liquid removal. If the initial solids content exceeds the critical value (which is unique for each mixture), then expression is accomplished entirely by a consolidation process.

The equations that describe consolidation are derived from Darcy's law, most conveniently written in terms of a moving coordinate ω that represents the volume of solids (per unit area) between a plane and the filter medium. The apparent velocity u of the liquid viewed from this moving material is:

$$u = \frac{1}{\mu \alpha \rho_s} \frac{\partial p_l}{\partial \omega} = -\frac{1}{\mu \alpha \rho_s} \frac{\partial p_s}{\partial \omega} \qquad \text{(cf. equation (2.20)) (2.101)}$$

where α is the local specific resistance to flow of liquid in the cake. The mass balance of liquid states that the change of the voids ratio e is related to the liquid velocity and the consolidation time t_c by:

$$\frac{\partial e}{\partial t_c} = \frac{\partial u}{\partial \omega}$$
(2.102)

Combining equations (2.101) and (2.102) leads to:

$$\frac{\partial p_s}{\partial t_c} = C_e \left[\frac{\partial^2 p_s}{\partial \omega^2} - \frac{1}{\alpha} \frac{d\alpha}{dp_s} \left(\frac{\partial p_s}{\partial \omega} \right)^2 \right]$$
(2.103)

where C_e is the modified average consolidation coefficient defined by

$$C_e = \left[\mu \alpha \rho_s \left(-\frac{de}{dp_s} \right) \right]^{-1}$$
(2.104)

If C_e is assumed to be constant, combining equations (2.101) with (2.102) leads to the diffusion equation used by Terzaghi (1948) to describe consolidation in soil mechanics:

$$\frac{\partial p_s}{\partial t_c} = C_e \frac{\partial^2 p_s}{\partial \omega^2}$$
(2.105)

The solution of this equation for constant pressure expression of a semi-solid material (that is, a solid/liquid mixture whose solids content is greater than the critical value) is written in terms of the average consolidation ratio U_c:

$$U_c = \frac{L_1 - L}{L_1 - L_\infty} = 1 - \sum_{n=1}^{\infty} \frac{8}{(2n-1)^2 \pi^2} \exp\left(-\frac{(2n-1)^2 \pi^2}{4} T_c \right)$$
(2.106)

where L_1, L and L_∞ are the thicknesses of the cake at $t_c = 0$, t_c and ∞ respectively, and where T_c is a dimensionless consolidation time defined by:

$$T_c = \frac{i^2 C_e t_c}{\omega_0^2}$$
(2.107)

Whilst equation (2.106) generally gives a good description of consolidation during its earlier stages, poor agreement with experiment is a common problem during the later stages. To overcome this, the concept of secondary consolidation due to creep effects in the mixture

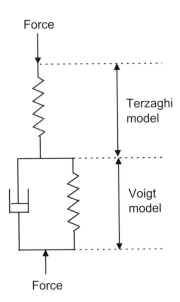

Force

Terzaghi model

Voigt model

Force

Figure 2.8 The Terzaghi-Voigt rheological model for consolidation.

was introduced. Because the variation of the voids ratio e within the cake is caused by changes in both the local solids compressive pressure p_s and the creep effect of the materials, e is a function of both p_s and the consolidation time t_c. It is widely assumed that the rheological behaviour of secondary consolidation is described by the Voigt element in Figure 2.8, and the average consolidation ratio is then given by:

$$U_c = \frac{L_1 - L}{L_1 - L_\infty} = (1-B)\left[1 - \exp\left(-\frac{\pi^2}{4}T_c\right)\right] + B\left[1 - \exp(\eta t_c)\right] \qquad (2.108)$$

for constant pressure expression of filter cakes, and by:

$$U_c = \frac{L_1 - L}{L_1 - L_\infty} = (1-B)\left[1 - \sum_{n=1}^{\infty}\frac{8}{(2n-1)^2 \pi^2}\exp\left(-\frac{(2n-1)^2 \pi^2}{4}T_c\right)\right]$$
$$+ B\left[1 - \exp(\eta t_c)\right] \qquad (2.109)$$

for constant pressure expression of semi-solid materials (Shirato *et al*, 1974).

The quantity B in equations (2.108) and (2.109) is defined by the ratio $V_{sc(max)} / V_{c(max)}$ where $V_{sc(max)}$ is the liquid volume squeezed by the secondary consolidation and $V_{c(max)}$ is the total liquid volume squeezed from the mixture during the entire expression. η is an empirical constant that characterises the creep of the materials; both B and η can be determined from the latter stages of a plot of $\ln(1 - U_c)$ versus t_c.

2.9 Conclusions

This chapter sets out the principles behind the various processes (flow in the filter medium, cake formation and consolidation, cake deliquoring and cake washing) that occur during cake filtration operations. Even though the fluid dynamics underlying filtration processes are complex, models to interpret the processes all stem from Darcy's law. When Darcy's law is combined with appropriate continuity equations a fluid dynamic based model for each process is formed. It should be recognised that this approach to modelling simplifies considerably the microscopic phenomena that actually occur during flow in filter cakes and filter media. Some of the phenomena are described in Section 2.7, but other physicochemical factors such as surface charge effects and particle or filter medium hydro-phobicity/ -philicity do not appear in any of the models. They are not included in the models because of the inherent difficulties behind describing their effects in a formulaic way, nonetheless their importance in practical filtration problems should not be overlooked. More sophisticated models exist in the research literature for some of the filtration processes, but these are not included in this chapter which concentrates on those modelling approaches that have been developed in such a way that they are able to provide process engineering solutions to practical problems.

The development of Darcy's law to describe filter cake formation is an essential feature in understanding cake filtration, and therefore constitutes the entire subject of Chapter 4.

In this chapter the fundamentals of flow in woven and non-woven filter media, two phase flow during cake deliquoring, hydrodynamic dispersion in cake washing and filter cake consolidation are elaborated to form the basic equations from which engineering solutions can be developed. Practical forms of these solutions are then obtained in Chapters 5, 6 and 7 to describe cake deliquoring, cake washing and cake consolidation. In those chapters examples are given of use of the solutions in process engineering calculations.

3 Gravity clarification and thickening

The general term 'settling' describes all types of particle fall through a (gaseous or liquid) fluid under a gravitational force. 'Sedimentation' is used to describe settling phenomena in suspensions, where particles or aggregates are suspended by hydrodynamic or particle-particle interaction forces, compression being absent. When the suspension or solid-liquid mixture is in compression, that is, the particle aggregates are compressed by layers of aggregates lying above them, the term 'subsidence' is sometimes used. Sedimentation and subsidence can be, and often are, two stages of the same settling process, with sedimentation being followed by compression. Initially, the particles or aggregates settle with return (upward) flow of liquid between the downward moving particles; eventually the aggregates come into close proximity of one another and those at lower levels feel compressive forces due to the presence of those higher in the mixture.

In order to increase the settling rate of smaller particles, the force applied to the particles must be increased. One way to do this is to use a centrifugal force. Whereas in a gravity force field particle motion is upwards or downwards, depending on whether the particle is less or more dense than its suspending liquid, in a centrifugal field the motion is radial through the liquid either inwards or outwards, again depending on whether the particles are less or more dense than the liquid.

3.1 Basics of gravity settling

Particle size, particle density and fluid viscosity are the primary factors to be considered in a sedimentation process, but suspension concentration and particle shape can also have a significant influence. Suspensions with particle diameters of the order of microns settle too slowly for most practical operations; to increase their settling rate the individual particles are aggregated or flocculated into larger collections of particles known as *flocs*.

3.1.1 Batch sedimentation

The mechanism of sedimentation is most simply described by reference to what happens in a batch settling test as solids settle from a slurry in a glass cylinder (Figure 3.1). Figure 3.1(a) shows a newly prepared,

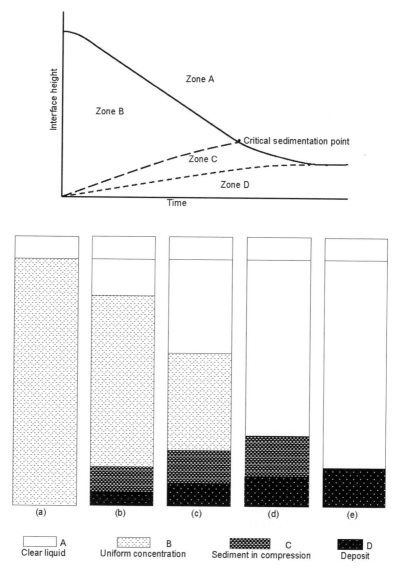

Figure 3.1 Sedimentation of a flocculated suspension.

flocculated, slurry with a uniform concentration of solid particles throughout the cylinder (Zone B). At the start of settling, all particles begin to fall towards the bottom of the cylinder and are assumed to rapidly reach their terminal settling velocity under hindered settling conditions (see Chapter 2). If all particles or aggregates in the original suspension were of the same size and density, then the *mud line*, the line separating A and B would be sharp. The presence of fine or non-flocculated solids in the suspension blurs the interface between A and B.

Several zones of concentration are soon formed (Figure 3.1(b)). Zone B is one in which all aggregates or flocs retain their position relative to their neighbours, and the zone therefore maintains the concentration of the original suspension during the settling process until all particles have moved into Zone C. A region of clear supernatant liquid forms at the top of the suspension (Zone A), whilst a deposit of flocs and larger particles forms at the bottom (Zone D). Immediately above the lowest zone is a transition layer that is intermediate in density between the deposit and the initial suspension, Zone C, which is a volume of suspension in compression; many texts follow closely the assumptions made by Coe and Clevenger (1916) that divided the compression zone into two. Coe and Clevenger concerned themselves with the settling of non-colloidal particles, but a flocculated suspension does not usually exhibit quite so clearly the characteristics that allow two sub-zones to be observed in the compression zone.

As sedimentation continues, zones A and D both grow larger whilst B becomes smaller. Eventually, a point is reached when B and C disappear and all solids are in D – this is the so-called *critical sedimentation point* (Figure 3.1(e)), where a single distinct interface forms between the clear liquid and the deposit. From this point, the sedimentation process consists of a slow compression of the deposit, with liquid being forced upwards through the pores between the solids into the clear zone. Flocs or aggregates are compressed by the weight of others that have settled above them. Hence, in the compression region settling is no longer a result of hydrodynamic forces. The deposit (often referred to as *sludge* or *sedimented pulp*) usually possesses small mounds and craters (*volcanoes*) on its surface – these are the visible ends of the channels through which the upwards flowing liquid has passed, with the mounds caused by the small particles that have flowed with the rising fluid.

3.1.2 Types of sedimentation

Particles can settle in different manners dependent largely on their concentration and their tendency to flocculate. The various types of

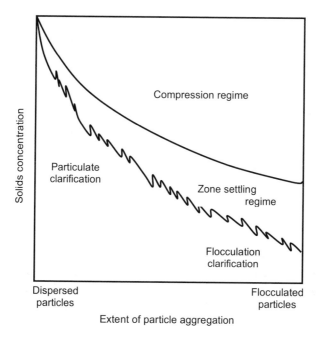

Figure 3.2 Regimes of sedimentation (Fitch, 1986).

sedimentation, represented in Figure 3.2, make different demands on the size and shape of equipment used for settling. The left hand side of the diagram represents particles with little tendency to cohere, the right hand side those between which the cohesive forces are strong. The state of dispersion of the particles, and hence the type of settling, can be affected by the addition of flocculants or coagulants to the feed suspension.

Clarification settling occurs at low concentrations of solids (<1 to 2% v/v). On average, the particles are far apart and free to settle individually with few collisions occurring between particles. Faster settling particles can overtake slower ones. If the particles are cohesive, the settling rate is determined by the size of the aggregate (commonly referred to as the floc) rather than by the primary size of the particle, with increased settling rates being associated with larger aggregates. Clarification behaviour is easily identified in a batch settling test; The upper layer of fluid clears only slowly, and a visible interface located at the surface of the settled solids starts at the bottom of the column of suspension and moves upwards.

In concentrated suspensions the particles are in closer proximity to one another, and ultimately there is a concentration where most particles are in contact with their near neighbours. If they cohere, a networked suspension is formed. Within the network, particles are constrained and the solids settle with a sharp interface between the pulp and the supernatant. The slurry is in a consolidated or zone settling regime, and exhibits line settling behaviour – this is readily seen in a batch settling test where a visible interface starts at the top of the column of suspension and moves downwards.

At very high concentrations the pulp structure can be so firm that it has quite a high compressive strength, with each layer of pulp solids being able to provide mechanical support for the layers above. Such support that is propagated from the bottom of the vessel slows subsidence of the layers and each layer is subjected to a solids stress (or squeeze). Under these circumstances, the pulp is said to be in compression.

3.2 Clarification

The so-called long tube test is commonly used to determine the clarification zone requirements for a process scale clarifier. The test is intended to represent a cylindrical element extending from the top to the bottom of a clarifier unit, and the purpose of the test is to determine combinations of pool depth and detention time that will give a specified removal of solids. The apparatus is a glass or plastic tube from 3 to 4 m long and 50 to 75 mm diameter, with sampling taps located at 200 to 300 cm intervals along its length; the top tap is about 100 mm below the upper end of the tube. The sampling tap is a glass or plastic tube with a closing tap. The length of the apparatus determines the maximum overflow rate that can be tested with any given detention time.

A quantity of slurry sufficient to fill the apparatus is prepared, mixing in any flocculants, coagulants or other pre-treatment chemicals; this is poured into the long tube to establish an upper level of suspension at the topmost outlet. The timer is then started and the column is allowed to settle until a visually acceptable clarity is achieved. After flushing the sampling tube, each successive outlet from top to bottom is sampled quickly. The test is repeated for several detention times, t_d.

For each detention time test, the data are recorded as in Table 3.1. The first column shows the sampling point distance H below the original top surface; the second column gives the residual solids concentration in the section lying above the sampling point. For example, the fourth sample was taken at a sampling point 1 m below the original top surface, and contains the supernatant that occupied the column between the depths of 0.75 and 1 m at the moment the samples were taken ($t_d = 0.5$ h). The third column is calculated and gives the average concentration of supernatant above the corresponding sampling point. Thus, the average concentration above the 1 m depth after half an hour settling is $(42 + 49 + 49 + 53)/4 = 48.3$ ppm, which equates to the overflow concentration if all the supernatant from a column 1 m high were recovered in half an hour. This therefore corresponds to an *ideal* settling basin having a detention time of 0.5 h and an overflow rate of 1 m/0.5 h = 2 m h^{-1}.

Table 3.1 Example of long tube test data ($t_d = 0.5$ h; $c_0 = 408$ ppm) (Fitch, 1986)

Cumulative depth, H (m)	Sample concentration (ppm)	Cumulative average concentration (ppm)	Equivalent overflow rate, u_0 (m h^{-1})
0.25	42	42.0	0.5
0.50	49	45.5	1.0
0.75	49	46.7	1.5
1.00	53	48.3	2.0
1.25	52	49.0	2.5
1.50	54	49.8	3.0
1.75	52	50.1	3.5
2.00	52	50.4	4.0
2.25	54	50.8	4.5
2.50	60	51.7	5.0
2.75	68	53.2	5.5
3.00	80	55.4	6.0

Clarification zone requirements are determined by the overflow volume flux, (m^3 m^{-2} h^{-1}), and the detention time, t_d. They are related to the clarifier pool area and depth by:

$$u_0 = \frac{Q_0}{A} \tag{3.1}$$

$$t_d = \frac{V}{Q_0} = \frac{AH}{Q_0} = \frac{H}{u_0} \tag{3.2}$$

In these equations, Q_0 is the volume of overflow per unit time, A is the plan area of the settling chamber of the pool (in tray, tube, or laminar settlers or whenever the pool is bounded by inclined surfaces, A is the area of the chamber projected onto a horizontal plane), V is the volume of the clarification zone, and H is the depth of the clarification zone (measured in the direction of the gravity force). The meanings of the

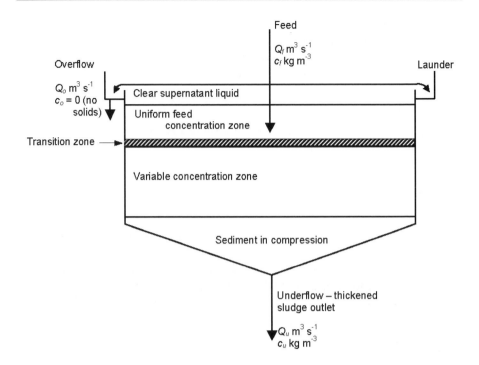

Figure 3.3 Settling zones in a continuous thickener. All zones are rarely present simultaneously, but the concept is used to model thickener performance.

terms overflow and underflow used in clarification are the same as those identified for thickeners in Figure 3.3, and the pool area refers to the cross-section of the settling chamber.

Example 3.1 Calculation of the design basis for a clarifier following a long tube test.
The following data are either specified or are available from a long tube test:

Overflow rate, Q_0	$= 130 \text{ m}^3 \text{ h}^{-1}$
Feed solids concentration	$= 300 \text{ mg dm}^{-3}$
Maximum allowable overflow solids concentration	$= 10 \text{ mg dm}^{-3}$
Overflow velocity, u_0	$= 6 \text{ m h}^{-1}$
Detention time, t_d	$= 1.2 \text{ h}$

Solution

Pool volume: $V = Q_0 t_d = 130 \times 1.2 = 156 \text{ m}^3$

Pool area: $\quad A = \dfrac{Q_0}{u_0} = \dfrac{130}{6} = 21.7 \text{ m}^2$

Pool depth: $\quad H = \dfrac{V}{A} = \dfrac{156}{21.7} = 7.2 \text{ m}$

Pool diameter: $\quad D = \sqrt{\dfrac{4A}{\pi}} = \sqrt{\dfrac{4 \times 21.7}{\pi}} = 5.3 \text{ m}$

3.3 Thickening

Industrial sedimentations are invariably carried out as continuous processes in equipment called thickeners. A thickener is essentially a settling tank, usually circular in shape, and fitted with a rake to move the settled deposit to a central underflow discharge. The clear liquid is taken off at the top of the vessel by an overflow launder. Figure 3.3 shows a simplified diagram of a gravity thickener. Examples of industrial thickeners are shown in Figures 3.4 and 3.5, which illustrate

Figure 3.4 A view of two 50 m diameter thickeners. The central driveheads on each thickener can be seen, which drive the submerged rotating arms that convey the settled solids to the point of discharge and provide channels within the compressing pulp to allow upwards release of suspension liquid to promote high underflow concentrations. The walkway gives access to the driveheads and enables observation of the thickener performance (Dorr-Oliver Eimco).

Figure 3.5 Two 15 m diameter deep cone thickeners. Deep cone thickeners are essentially conical in shape with a cylindrical section and are particularly suited for separation of large flocs which settle rapidly and form a sludge that compacts under the solids pressure to produce an underflow that has a high solids concentration, and leaves a clean overflow (Dorr-Oliver Eimco).

the large diameters of gravity sedimentation equipment that are due to the slow settling rates of smaller particles, as well as the high throughput requirements in many industrial applications.

In a batch sedimentation process, as in Figure 3.1, the heights of the various zones vary with time. It is generally envisaged that the same zones exist in a thickener, but at steady state operation the height of each zone is constant. Whereas the solids handling capacity of a thickener is controlled largely by the required solids underflow concentration, the liquid handling capacity depends on the finer particles in the feed and the required degree of overflow clarity. However, thickener throughput is in most cases limited by the underflow rather than the overflow.

3.3.1 Relating batch test data to thickeners

The capacity of thickeners is based on their ability to perform both thickening and clarifying at the required rates. The area of the unit controls the time allowed for settling of solids through the liquid for a given feed rate, and determines the clarification capacity. The depth of the thickener controls the time permitted for thickening the sludge, and determines the thickening capacity.

Correct functioning of a continuous thickener requires that the rate at which solids settle through every zone must be at least fast enough to accommodate the solids that are flowing into the zone. In the upper part of the thickener the slurry is relatively dilute, and settling more rapidly. In the lower part, the slurry density and solids concentration are higher; even though the settling velocity of individual particles may be small, the solids capacity rate per unit area is higher than at some intermediate point in the thickener. The velocity of the solids has been decreased by the high concentration of solids. Hence, considering levels in the thickener starting from the top, the mass velocity decreases and then increases as the solids settle.

In the design of a thickener, therefore, for a specified quantity of slurry the minimum cross-sectional area that will allow passage of the solids is found at the limiting concentration, intermediate between the feed and underflow concentrations. This cross-sectional area must be sufficiently large to enable solids to pass at a rate equal to or greater than the feed rate. If the area is not large enough, the material balance at this level is satisfied only by the accumulation of solids, which results in the limiting zone travelling upwards in the thickener. Determination of the minimum cross-section is a key part in the analysis of settling data, and requires (i) identification of the concentration at which the mass rate of settling of solids is a minimum, (ii) determination of the settling velocity at this concentration, and (iii) estimation of the amount of liquid flowing upward at the limiting level. Once the minimum cross-section has been calculated, the retention time in the compression zone can be evaluated.

The behaviour of concentrated suspensions during sedimentation has been analysed by Kynch (1952), and applied to the problem of continuous thickener design by Talmage and Fitch (1955). Most practical analyses of thickener behaviour are based on developments of Kynch's work, so from this point of view it is worth understanding his analysis. The analysis is largely based on continuity considerations and is underpinned by the assumptions that:

(i)   particle concentration is uniform across any layer in the sedimenting suspension;

(ii)   there are no wall effects (which can become more important in small scale tests);

(iii)   there is no differential settling of particles as a result of differences in shape, size or composition;

(iv)   the settling velocity of the particles depends only on the local concentration of the particles;

(v)   the initial concentration is uniform (after a short time, it increases towards the bottom of the suspension); and

(vi) the settling velocity tends to zero as the concentration approaches a limiting value corresponding to that of the sediment layer deposited at the bottom of the vessel.

Kynch (1952) showed that the rate of upward propagation $(\overline{v}_L)$ of a limiting zone is constant if throughput exceeds its capacity and is a function of the solids concentration. If v_L is the settling velocity of the particles in the limiting layer, and c is the solids concentration (mass of solids per unit volume of slurry), then a material balance in the limiting zone is written as:

$$\overline{v}_L = c\frac{dv_L}{dc} - v_L \tag{3.3}$$

Since the velocity is assumed to be a function only of concentration, $v_L = f(c)$, it follows that

$$\frac{dv_L}{dc} = f'(c) \tag{3.4}$$

and equation (3.3) becomes

$$\overline{v}_L = cf'(c) - f(c) \tag{3.5}$$

Since c is constant in the capacity limiting layer, so are $f(c)$ and $f'(c)$, and so $\overline{v}_L$ is also constant. Since $\overline{v}_L$ is constant in the rate limiting zone, it may be used to determine the solids concentration at the upper boundary of the layer from a batch settling test (see Figure 3.1).

In a batch settling test, let c_0 and z_0 be the initial concentration and height of the suspended solids. The weight of solids in the suspension is $c_0 z_0 A$, where A is the cross-sectional area of the cylinder in which the test is being performed. The limiting layer, assuming that it exists, must start at the bottom of the cylinder and move upwards towards the clear liquid interface (see Figure 3.1). If the concentration of the limiting layer is c_L and the time for it to reach the interface is t_L, the mass of solids passing through the layer is $c_L(v_L + \overline{v}_L)At_L$. Since this layer starts at the bottom of the cylinder and moves upwards to meet the clear liquid interface, this mass must equal the total solids present in the cylinder. Hence

$$c_L \left(v_L + \bar{v}_L \right) A t_L = c_0 z_0 A \qquad (3.6)$$

If z_L is the height of the interface at t_L, then

$$\bar{v}_L = \frac{z_L}{t_L} \qquad (3.7)$$

Substituting the value of $\bar{v}_L$ from equation (3.7) into equation (3.5), and noting that $\bar{v}_L$ is constant, gives

$$c_L = \frac{c_0 z_0}{z_L + v_L t_L} \qquad (3.8)$$

The batch settling data from a laboratory test may be plotted as the height of the interface as a function of the sedimentation time. From this plot, the value of $\bar{v}_L$ is the slope of the curve at $t = t_L$, and the tangent of the curve at t_L intersects the ordinate at z_i, as shown in Figure 3.6. The slope of this tangent is

$$v_L = \frac{z_i - z_L}{t_L} \quad \text{or} \quad z_i = z_L + t_L v_L \qquad (3.9)$$

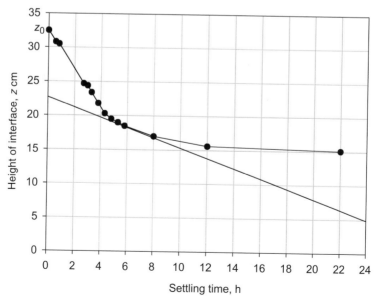

Figure 3.6 Height of the interface as a function of the settling time for the data in Example 3.2.

Combining equations (3.8) and (3.9) gives

$$c_L z_i = c_0 z_0 \qquad (3.10)$$

from which it follows that z_i is the height which the suspension would occupy if all the solids were present at a concentration c_L, and this is the minimum concentration at which boundary layers around neighbouring particles interfere.

The settling velocity as a function of solids concentration may also be obtained from a single sedimentation curve by determining the slope of the tangents to the curve at various values of t. The values of the intercepts (z_i) are then used in equation (3.10) to obtain the corresponding concentration.

Example 3.2 Calculation of the settling rate versus solids concentration relationship from batch settling data.

A single batch settling test was carried out on a mine tailings slurry. The interface between the clear liquid and the suspended solids (the mud line) was measured as a function of the settling time, and the results are tabulated in Table 3.2. The initial slurry contained 30% w/w solids with a density of 2760 kg m^{-3} in water (density 1000 kg m^{-3}). A 1 litre graduated cylinder with a diameter of 60 mm was used for the test. Prepare a graph showing the relationship between settling rate and solids concentration.

Solution

The mass percent of solids in the slurry ($100 \times s$) is related to the mass concentration (c' (kg of solids)/(m^3 of solids + m^3 of liquid)) by

$$c' = \frac{1}{\dfrac{1}{\rho_s} + \dfrac{1}{\rho}\dfrac{1-s}{s}} \qquad (3.11)$$

Hence, the solids concentration in the feed slurry is

$$c_0 = \frac{1}{\dfrac{1}{2760} + \dfrac{1}{1000}\dfrac{1-0.3}{0.3}} = 371 \ \text{kg m}^{-3}$$

Using the test data, the height of the interface (z) is plotted as a function of time (t) (Figure 3.6). From the solids concentration of the initial slurry

$$c_0 z_0 = 371 \times 32.5 = 12057.5 \ \text{kg cm m}^{-3}$$

Table 3.2 Data measured during a jar settling test for Example 3.2

Time (h)	Volume of slurry below the mud line (cm³)	Height of mud line (interface)*, z (cm)
0	1000	32.5
0.55	950	30.8
0.8	940	30.5
2.7	760	24.7
3	750	24.4
3.3	720	23.4
3.8	670	21.8
4.3	625	20.3
4.8	600	19.5
5.3	585	19.0
5.8	570	18.5
8	525	17.0
12	480	15.6
22	465	15.1

*Calculated from the volume of slurry below the mud line (column 2) and the cross sectional area of the cylinder (30.8 cm²).

From equation (3.10),

$$c = \frac{12057.5}{z_i}$$

Now draw tangents to the settling curve (for illustrative purposes only one tangent is shown on Figure 3.6; the y-axis intercept value of the tangent is z_i and settling velocity is the gradient of the tangent, after 6 hours $z_i = 22.7$ cm and the settling velocity is 0.742 cm h⁻¹) and find the slope and intercept values of each tangent. Other points found in the same way are tabulated in Table 3.3 and plotted in Figure 3.7.

3.3.2 Thickener area

The cross-sectional area that would be required of a thickener is determined by the concentration layer in the batch test requiring the maximum area to pass unit mass of solids. Hence, for the thickener shown in Figure 3.3, the solids material balance is:

$$Q_u = \frac{Q_f c_f}{c_u} \tag{3.12}$$

Table 3.3 Solution to Example 3.2

t (h)	z_i (cm)	v (cm h^{-1})	c (kg m^{-3})
0	32.5	2.826	371
2	32.5	2.826	371
4	32.5	2.826	371
5	25.0	1.131	482.3
6	22.7	0.742	531.2
7	21.2	0.589	568.7
8	20.5	0.470	588.2
12	18.2	0.341	662.5

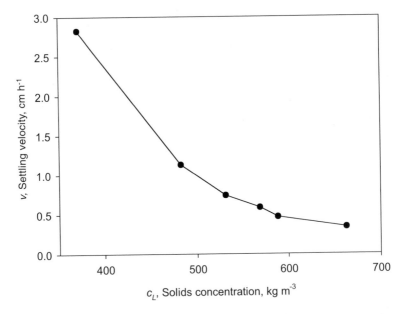

Figure 3.7 Relationship between the settling velocity and solids concentration in Example 3.2.

An overall liquid balance gives

$$Q_f\left(\rho_f - c_f\right) = Q_o\rho + Q_u\left(\rho_u - c_u\right) \tag{3.13}$$

Elimination of Q_u between these equations and rearranging the result gives

$$Q_o = Q_f c_f \left[\frac{\rho_f}{c_f} - \frac{\rho_u}{c_u}\right]\frac{1}{\rho} \tag{3.14}$$

Dividing both sides of this equation by the thickener cross-sectional area gives the upward linear velocity of the clarified liquid as

$$\frac{Q_0}{A} = \frac{Q_f c_f}{A}\left[\frac{1}{c_f} - \frac{1}{c_u}\right]\frac{\rho_{av}}{\rho} \tag{3.15}$$

where, for the purposes of modelling the process, the suspension feed and underflow densities are assumed to be the same and equal to ρ_{av}. Strictly, the suspension density is related to the densities of the solid and liquid components by

$$\rho_{sus} = \rho_s(1-\varepsilon) + \rho\varepsilon \tag{3.16}$$

where ε is the volume fraction of liquid in the suspension (referred to as the porosity in a filter cake or thickened sludge).

To prevent solids overflowing with the clarified liquid, the upward velocity of the liquid must be equal to or less than the settling velocity of the solids, therefore Q_0/A is replaced by v. Also, equation (3.15) may be written in terms of the capacity limiting layer (Talmage and Fitch, 1955), hence

$$Q_f c_f = Q_L c_L \tag{3.17}$$

and

$$\frac{Q_L c_L}{A} = \frac{v}{\left[\dfrac{1}{c_L} - \dfrac{1}{c_u}\right]\dfrac{\rho_{av}}{\rho}} \tag{3.18}$$

For various values of v and c, equation (3.18) can be used to estimate $Q_L c_L/A$; the smallest value of $Q_L c_L/A$ determines the minimum thickener area required.

Example 3.3 Calculation of thickener area.

The slurry tested in Example 3.2 is fed to a thickener at a rate of 50 tonnes of dry solids per hour ($Q_f c_f$) to produce a thickened sludge with a concentration of 600 kg m^{-3} (c_u). Estimate the thickener area required.

Solution

Since the slurry conditions entering the thickener are the same as those in the previous illustration, the calculations developed in Example 3.2

are used here. Using the relationship between v and c_L for this slurry, Figure 3.7, and equation (3.18), Table 3.4 can be prepared.

Table 3.4 Determination of the minimum value of $\dfrac{Q_L c_L}{A}$ in Example 3.3

v (cm h^{-1})	c_L (kg m^{-3})	$\dfrac{1}{c_L}$ (m^3 kg^{-1})	$\dfrac{1}{c_L}-\dfrac{1}{c_u}$ (m^3 kg^{-1})	$\dfrac{Q_L c_L}{A}$ (cm h^{-1})/(m^3 kg^{-1})
0.45	600	0.00167	0	∞
0.5	582	0.00172	0.000053	5018
0.6	568	0.00176	0.000093	3432
0.7	546	0.00183	0.000163	2284
0.8	526	0.00190	0.000233	1826
0.9	514	0.00195	0.000283	1692
1.0	498	0.00201	0.000343	1551
1.1	487	0.00205	0.000383	1528
1.2	477	0.00210	0.000433	1474
1.3	472	0.00212	0.000453	1525
1.5	459	0.00218	0.000513	1555
2.0	426	0.00235	0.000683	1558

In this example, the average density of the feed and underflows is taken as the arithmetic mean of the solid and liquid densities (this is equivalent to assuming a voidage of 0.5 in the slurry). Hence, $\rho_{av} = 0.5(2760 + 1000) = 1880$ kg m^{-3}.

From Figure 3.8, the minimum of $\dfrac{Q_L c_L}{A}$ is about 1474 (cm h^{-1})/(m^3 kg^{-1}) at 1.2 cm h^{-1}, and $c_L = 477$ kg m^{-3} (from Figure 3.7 or Table 3.4). Since no solids leave in the overflow a solids mass balance (equation (3.17)) gives

$$Q_L c_L = 50 \text{ tonnes h}^{-1} = 50,000 \text{ kg h}^{-1}$$

and

$$A = \frac{50000}{1474} \times 100 = 3392 \text{ m}^2$$

That is, the required thickener area would be 3392 m^2.

3.3.3 Thickener depth

The depth of the thickening zone increases as the underflow rate is

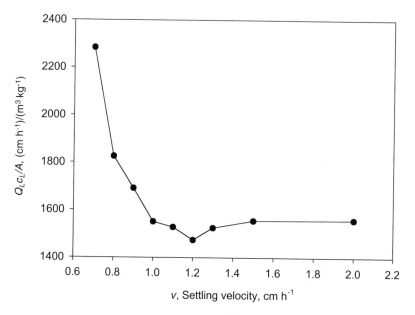

Figure 3.8 Determination of the minimum value of $\dfrac{Q_L c_L}{A}$ in Example 3.3.

decreased. In most cases the time for compression of the sediment is large compared with the time taken for the critical settling conditions to be reached (that is, the time taken for the critical concentration to be reached – that concentration at which the slurry in the thickening zone begins to compress). The critical concentration is often not clearly defined, but Roberts (1949) suggested a way to estimate the critical time (compression point). This involves plotting $(z - z_\infty)$ on a log scale versus time on a linear scale, when the compression part of the settling curve then appears linear. On this plot, the intersection of the linear compression zone and the falling rate period gives an estimate of the compression point. Other practical methods of estimation of the compression point are reviewed by Fitch (1986).

The required volume for the compression zone is the sum of the volumes occupied by the solids and by the liquid:

$$V_{cz} = \frac{Q_f c_f}{\rho_s}(t - t_c) + \frac{Q_f c_f}{\rho} \int_{t_c}^{t} \frac{M_l}{M_s} dt \qquad (3.19)$$

where V_{cz} is the compression zone volume, $Q_f c_f$ is mass feed rate of solids to the thickener, $(t - t_c)$ is the retention time in the compression zone, M_l is the mass of liquid in the compression zone, and M_s is the mass of solids in the compression zone.

After the volume of the compression zone has been calculated, the depth of the zone can be evaluated by dividing the volume by the thickener area calculated by equation (3.18).

Example 3.4 Calculation of thickener depth.

For the conditions set out in Example 3.3, estimate the depth of the thickener required to carry out the thickening operation.

Solution

From Figure 3.6, $z_\infty \approx 14.6$ cm ($\equiv 450$ cm^3). Table 3.5 can be prepared using Table 3.2, and noting that

$$\frac{M_l}{M_s} \approx \frac{\rho_{av}}{c} - 1 = \frac{\text{mass of liquid in suspension with a density } \rho_{av}}{\text{mass of solids in the suspension}}$$

From Figure 3.9, the critical time (t_c) is estimated by determining the time that corresponds to $z'_c = (z_0 + z'_0)/2z_0$ – this time is found to be 2.75 h, noting that $z/z_0 = 1$, $z_0 = 32.5$ cm and $z'/z_0 = 0.126$. From Table 3.5, a time of about 8.6 h is required to produce a concentration of solids equal to 600 kg m^{-3}, thus the retention time in the compression zone is $(8.6 - 2.75)$ h, that is, 5.85 h.

Table 3.5 Calculated data for the estimation of the thickener depth in Example 3.4

t, (h)	$(z - z_\infty)$ (cm)	$\dfrac{z - z_\infty}{z_0 - z_\infty}$	$\dfrac{M_l}{M_s} \approx \dfrac{1880}{c} - 1$	$\displaystyle\int_{2.75}^{8.6} \dfrac{M_l}{M_s} dt$
0	17.9	1.0	4.067	
0.55	16.2	0.905	4.067	
0.8	15.9	0.888	4.067	
2.7	10.1	0.564	4.067	
*2.75			4.067	0
3	9.8	0.547	4.067	1.0167
3.3	8.8	0.492	4.067	1.2201
3.8	7.2	0.402	4.067	2.0335
4.3	5.7	0.318	3.649	1.9290
4.8	4.9	0.274	3.087	1.6840
5.3	4.4	0.246	2.783	1.4675
5.8	3.9	0.218	2.606	1.3472
8	2.4	0.134	2.196	5.2822
*8.6			2.142	1.3014
12	1.0	0.056	1.838	
22	0.5	0.028		$\Sigma = 17.2816$
∞	0	0		

(* These rows have been inserted after calculation of the critical and retention times.)

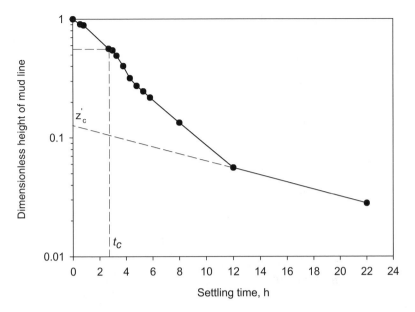

Figure 3.9 A Roberts (1949) plot of the settling characteristics for Example 3.4.

Equation (3.19) is used to estimate the compression zone volume, by graphically integrating the data in Table 3.5.

$$\frac{Q_f c_f}{\rho_s} = \frac{50000}{2760} = 18.116 \ \mathrm{m^3_{solids} \ h^{-1}}$$

$$\frac{Q_f c_f}{\rho} = \frac{50000}{1000} = 50 \ (\mathrm{kg_{solids} \ h^{-1}})/(\mathrm{kg_{liquid} \ m^{-3}_{liquid}})$$

$$(t - t_c) = 5.85 \ \mathrm{h}$$

Therefore

$$V_{cz} = 18.116 \times 5.85 + 50 \int_{2.75}^{8.6} \frac{M_l}{M_s} dt$$

$$= 105.98 + 50 \times 17.2816$$

$$= 970.06 \ \mathrm{m^3}$$

The thickener area is 3378 m² (from Example 3.3), and so the required compression zone height is 0.29 m.

In a thickener design, allowances to be added to the compression zone height are the bottom pitch, a storage capacity and submergence of the feed.

3.3.4 Zone sedimentation models

The Kynch model described in the previous section is an example of so-called zone sedimentation models. An earlier model of this type was developed by Coe and Clevenger (1916), but experimentally it is inconvenient due to multiple tests having to be carried out. In their method, suspensions (pulps) of various concentration between the thickener feed and underflow must be prepared by progressive decantation and dilution, which is likely to alter the characteristics of the suspension. This is particularly so if the suspension is flocculated, when the floc structure may change as a result of shearing or mixing. The Coe and Clevenger method applies to the constant rate part of the batch settling curve and assumes, like the Kynch theory, that the settling rate is only a function of the local concentration of solids. Hence, both models apply strictly only to ideal suspensions. These models are best suited to unflocculated suspensions, although they have been found to give reliable predictions for settling of activated sludges (Dick and Ewing, 1967).

Talmage and Fitch (1955) used the principle of constancy of the upward velocity of a zone of constant concentration to obtain the solids concentration of the mud line. The corresponding velocity can be measured from the slope of the settling curve at the appropriate point. To obtain the area requirement of a thickener, the concentration demanding maximum area must be determined; their approach differs from the Kynch analysis in that the need to obtain the concentration-velocity plot is eliminated. Hassett (1965) further developed the ideas of Talmage and Fitch through the use of a modified batch settling plot by dividing the axes by $c_0 z_0$. Supporting and conflicting evidence has been produced for both the Kynch (1952) and Talmage and Fitch (1955) models by Gaudin and Fuerstenau (1962) who found that: (i) the iso-concentration lines, when plotted on a settling curve, were not straight, thus apparently rejecting the Kynch and Talmage and Fitch assumptions; and (ii) all experimental data for the relationship between concentration and settling rate lay on a single curve, thus confirming the Kynch and Talmage and Fitch assumption that $v_L = f(c)$.

3.3.5 Compression subsidence models

In flocculated systems, the flocs are essentially a networked structure linking together the solid particles, and may have a very high porosity and be of considerable size. Even at low concentrations, therefore, flocs may not settle as separate entities. As settling proceeds a compressive stress is developed in the networked structure, so the flocs are not supported solely by their hydrodynamic drag. Concentrated suspensions, even in an unflocculated state, are less compressible than flocculated structures but exhibit similar tendencies. The thickening of such suspensions depends greatly on the characteristics of compression subsidence rather than particle settling.

In Example 3.4 the Roberts (1949) plot was used, which has the underlying assumption that the interstitial liquid eliminated in the compression regime is proportional to the amount present, that is

$$\frac{d\left(M^* - M_\infty^*\right)}{dt} = -kM^*$$

(3.20)

where M^* is the mass of liquid per unit mass of solids and the subscript ∞ indicates "at infinite time". For the compression part of a batch settling test, after integration this equation becomes

$$\ln\left(z - z_\infty\right) = -k\left(t - t_0\right)$$

(3.21)

and a plot of $\ln(z - z_\infty)$ against t should be a straight line.

A suspension in compression should exhibit a compressive yield value, σ, which is a function of the concentration. Also, the permeability, k, would exist and after an initial period during which the floc structure develops, this too would be a function of concentration. These arguments underpin the compression models proposed by Gaudin and Fuerstenau (1962), Michaels and Bolger (1962), Kos (1974), Adorján (1975), and Landman and White (1992).

Kos (1974) showed that the permeability in continuous thickening is not a function of concentration alone, as proposed by Michaels and Bolger (1962), but is best represented by $k = k(c,v)$ where v is the settling rate. Scott (1970) found that the solids concentration was low over a range of thickener depth in which concentration varied considerably; over this range Michaels and Bolger predicted zone settling, which does not permit a range of concentrations in steady state continuous thickening.

3.3.6 Zone settling vs. compression subsidence models

The consensus of opinion seems to be that the Talmage and Fitch (1955) under-designs thickener capacity and therefore over-designs in terms of area, while the Coe and Clevenger (1916) method overestimates capacity and therefore under-designs in terms of area (Scott, 1968; Carman and Steyn, 1965). Fitch (1975) used the Kos model to make some sense of this situation, recognising that the Kos model is probably a better representation of reality than the Coe and Clevenger model. According to Kos, solids stress gradients reduce settling rates in two ways: (i) they support part of the weight of the suspended solids, leaving less force to create liquid movement, and (ii) they reduce the dynamic pressure gradient, and therefore reduce channelling. In the Coe and Clevenger zone settling tests there is little solids stress (σ) gradient initially, even when $\sigma \neq 0$. If $v \neq f(c)$ the Kynch theory is violated and the Talmage and Fitch construction cannot correctly predict the concentration at the mud line. However, the critical or rate limiting concentration occurs in a region of high concentration gradient, $\partial c/\partial z$, irrespective of its value. It should therefore have a lower settling rate than would be expected in a continuous thickener where $\partial c/\partial z$ is normally lower, providing $\sigma \neq 0$. The conservative approach of the Talmage and Fitch method is obviously safer from a design point of view.

There are mixed views about the importance of depth and suspension height in a thickener. Robins (1964) and Yoshioka (1957), using a flux concept to determine thickener requirements, considered that the depth of a thickener had no effect on its operation and that the process imposed only an area limitation on design. This assumption was based on steady state conditions, and it is more generally accepted that a finite depth is necessary to allow for fluctuations in the feed conditions (Akers, 1974). Also, pretreating the feed with polymeric flocculants leads to a sludge with deformable characteristics for which sediment depth is of great importance, which suggests that compression models may be more appropriate for design.

Thickener throughput is controlled by area, but underflow concentration is controlled by retention time (Coe and Clevenger, 1916; Moncrieff, 1964; Talmage and Fitch, 1955). However, in continuous thickening the underflow concentration is controlled by the sludge pump-off rate, which affects the retention time. Lower underflow rates are associated with a greater thickener depth and a greater weight of solids per unit area. Figure 3.3 shows that several regions can exist in a thickener. The depth of the feed concentration zone will only pose a design problem if feed dilution is so great that a networked solids structure is not formed. If there is compression, the depth of the sediment zone will need to be determined (often this is up to 1 m in a 3 m deep thickener). Also, from the practical point of view, the feed entry to a thickener should be as quiescent as possible to prevent classification and subsequent formation of a floc bed.

3.4 Conclusions

There are two major settling modes in a thickening process – zone sedimentation and compression subsidence. In the gravity thickening of biological sludges there is a strong emphasis on the compression mode of thickening, although the design of thickeners for metallurgical and chemical suspensions relies heavily on behaviour in the settling zone. It is common to employ the theory of Talmage and Fitch to predict thickener area requirements from laboratory tests. However, when the feed suspensions are flocculated it is probably more appropriate to model the process as one of compression subsidence.

4 Filtration: cake formation

Industrial process filters are available in a wide variety of forms ranging from versatile units capable of handling different filtration applications to those restricted to use with specific fluids and process conditions, from very small scale to very large scale, and for either continuous or batch operation. Continuous filters are essentially capable of carrying out several of the separation functions of cake formation, cake compression, cake deliquoring, cake washing and discharge, serially without interrupting flow of the process feed, whereas batch filters may be able to perform the same range of functions but need to be stopped in order to discharge the cake. The filter surfaces on which separation takes place – the filter media – may be orientated horizontally or vertically, and may be either planar or cylindrical. The driving force may be pressure – either a positive pressure or a vacuum – or centrifugal.

Examples of industrial filters are shown in Figures 4.1 to 4.5. In these photographs can be seen examples of continuous filters (the vacuum belt filter, the rotary vacuum filter, the disc filter, and the vertical diaphragm filter press) and of a batch filter (the pressure vessel filter), as well as filters with planar and curved filtering surfaces.

The above and other factors have led to a bewildering choice in filter design, but even so the underlying principles of all cake filters are similar. In cake filtration a particulate deposit, the cake, accumulates on the surface of the filter medium. After an initial period of deposition the cake itself starts to act as the filter medium whilst further particles are deposited. This process continues until the pressure drop across the cake exceeds the maximum permitted by economic or technical considerations, or until the filtrate flow rate falls to an unacceptable level. The most important factor in cake filtration is the permeability of the filter cake, often interpreted through a measure of its specific resistance (introduced through equations (2.20) to (2.23)). Those

Figure 4.1 92 m² roller discharge rotary vacuum filter (RVF); the filter is a cylinder whose internals are divided into sections by a rotary valve and the pipework design to enable different sectors of the drum to be used for slurry pick-up and cake formation, cake deliquoring and cake washing. The term "roller discharge" refers to the fact that the filter cloth passes adjacent to a small diameter roller to effect discharge of the filter cake (Dorr-Oliver Eimco).

Figure 4.2 280 m² disc filter; a vacuum inside each disc causes cake formation as the disc rotates through the slurry trough, and the cake is deliquored by air suction in the uppermost sectors of the disc. The cake is generally discharged by blade or wire scrapers on each side of the disc (Dorr-Oliver Eimco).

Figure 4.3 Vacuum belt filter; an endless filter cloth forms a belt that is driven around two end rollers and over a sequence of suction boxes. The feed is introduced at one end of the filter, a cake is formed that is then sequentially deliquored and washed as many times as is required, enabling countercurrent washing operation when a high cake purity is sought (Madison Filter).

Figure 4.4 Vertical diaphragm, semi-continuous, filter press; a continuous filter cloth is pulled by rollers across filter plates mounted vertically one above the other, in each of which a filter cake is formed under pressure (Larox).

Figure 4.5 Vertical leaf pressure vessel filter; the surfaces of the leaves are the filter medium, and the whole leaf assembly retracts on an overhead rail. Each leaf has a filtrate outlet that is plugged into a common filtrate manifold, while side bars act as supports (LFC).

factors which determine the permeability, the porosity of the filter cake and the size of the particles in the cake, together with the particle size distribution and the state of aggregation of the particles, are fundamental parameters which dictate the ease with which any filtration process will be accomplished.

Particles larger than about 50 μm do not usually aggregate to a significant extent in suspension, and when filtered form a cake whose properties are quite insensitive to any increase in pressure applied to the surface of the cake (provided, of course, that the pressure is not so great as to cause fracture of the particles themselves). Such a cake is said to be *incompressible*. On the other hand, finer particles have a tendency to aggregate (aggregation may arise as a result of the chemical composition of the suspension or it may be induced by the addition of coagulants or flocculants prior to filtration) and the porosity, and hence permeability, of the cake is often dependent on the applied pressure. Such cakes are said to be *compressible,* and may be formed from small particles of organic, inorganic or biological matter. It should not be inferred from the above that larger particles will always form incompressible cakes or that smaller ones will always form compressible cakes, as either form of cake may occur. The occurrence depends on the physicochemical properties of the particles and liquids forming the suspension.

This chapter develops Darcy's law in a way that is applicable to carrying out process engineering calculations on the rate of filtration; much of the mathematical development and interpretation stems from the works of Ruth (1946) and Tiller (1953 to 1975). Their work formulated the filtration problem so that two design or scale-up parameters evolved, the specific resistance of the cake (α) and the cake porosity (ε). Many subsequent studies of filtration have been aimed at understanding what suspension properties and filter operating conditions affect these two parameters.

4.1 Relationships between process variables

Relationships between process variables have been developed from equations that are strictly only valid for incompressible filter cakes, and compressible systems are then described by modifying the relationships. In the following the basic incompressible cake filtration equations are formulated.

The process of particle deposition upon a filter medium of resistance R is depicted in Figure 4.6. A slurry containing particles with a mean diameter x in a liquid with a viscosity μ is shown flowing towards the medium at a superficial velocity u. The pressure in the slurry is p, and as fluid passes through the filter medium frictional drag at the surfaces of the fibres in the medium causes a drop in fluid pressure to the level p_0. In many practical cases this is taken as the atmospheric pressure existing outside the filter, but exceptions to this occur when extended filtrate

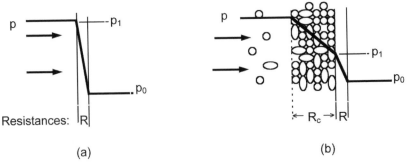

Figure 4.6 Pressure distributions with (a) clean liquid flowing through the medium of resistance R, and (b) during filtration when a cake of resistance R_c is being formed; the pressure gradient over the filter medium is greater when there is no cake deposition.

pipes create a further pressure loss as in, for example, rotary vacuum filters.

Figure 4.6(a) shows the condition which exists when particles are about to arrive at the filter medium through which clear fluid is passing under the pressure difference $(p - p_0)$. In all but a few exceptional cases laminar flow conditions exist and the filtrate flux u is related to the pressure difference by an appropriate form of Darcy's Law, equation (2.18):

$$u = \frac{p - p_0}{\mu R} \tag{4.1}$$

During deposition of a small quantity of solid material, Figure 4.6(b), the filtrate flux falls due to the increased resistance to fluid flow. The pressure at the cake-medium interface drops to p_1 and the instantaneous filtrate flux is given by:

$$u = \frac{p_1 - p_0}{\mu R} = \frac{\Delta p_m}{\mu R} \tag{4.2}$$

where Δp_m is the pressure drop over the medium.

The flux u is also given by an analogous form of Darcy's Law for flow through the cake:

$$u = \frac{p - p_1}{\mu R_c} = \frac{\Delta p_c}{\mu R_c} \tag{4.3}$$

where Δp_c is the pressure drop over a cake of thickness L and resistance R_c. Hence the total pressure drop $\Delta p = \Delta p_c + \Delta p_m = (p - p_0)$ is:

$$\Delta p = u\mu(R_c + R) \tag{4.4}$$

It is often assumed that the medium resistance remains constant during filtration, in spite of the likelihood of medium compression and particle penetration into its pores causing the resistance to increase. Comparison of equations (4.1) to (4.4) with the more usual form of Darcy's Law (equation (2.18)) shows that R is defined by the ratio of the thickness to the permeability of the filter medium, and R_c is similarly defined for the cake, and both have dimensions L^{-1}.

In experimental filtration studies it is more convenient to measure the cumulative volume of filtrate V discharging from the filter, and convert

V to a volume flow rate of filtrate dV/dt. Filtrate flux and flow rate are related by:

$$u = \frac{1}{A}\frac{dV}{dt} \qquad (4.5)$$

where A is the area of the filter, and the instantaneous filtrate flux is obtained from equation (4.4) as:

$$\frac{1}{A}\frac{dV}{dt} = \frac{\Delta p}{\mu(R_c + R)} \qquad (4.6)$$

The resistance of the cake is directly proportional to the mass of dry solids deposited per unit area of filter w, and the proportionality constant α provides a definition of the specific cake resistance:

$$R_c = \alpha\,w \qquad (4.7)$$

Since the respective dimensions of R_c and w are L^{-1} and ML^{-2}, it follows from equation (4.7) that α has the dimensions LM^{-1}. Also, the mass deposited M_s is related to the volume of filtrate by:

$$M_s = wA = c(V + V') \qquad (4.8)$$

where c is the mass of solids (per unit volume of feed liquid) in the slurry to be filtered and V' is the volume of liquid retained in the cake after filtration. In many instances $V'\!<\!<\!V$, when a good approximation for the cake mass may be calculated from the observed filtrate volume. However, V' becomes significant for the filtration of more highly concentrated suspensions or when loosely packed cakes are obtained at low pressures, and must then be taken into account. If the porosity (ε) and total cake volume (AL) are known, V' can be calculated directly from $V' = \varepsilon AL$. More often than not such data are not known when a filtration problem is defined initially, in which case the concentration is more conveniently corrected as below.

The true relationship between the mass of dry cake M_s and the filtrate volume V is derived from a material balance. V volumes of filtrate are produced during the formation of a cake which contains M_s mass units of dry solids and has a moisture or liquor content m defined by the mass ratio:

$$m = \frac{\text{mass of wet cake}}{\text{mass of dry cake}} \tag{4.9}$$

Noting that the mass fraction of solids in the feed slurry is

$$s = \frac{\text{mass of solids}}{\text{mass of slurry}} \tag{4.10}$$

a liquid material balance can be written in terms of m_{av} and s:

$$\frac{\text{mass of liquid in feed}}{\text{mass of solids}} = \frac{\text{mass of moisture in cake}}{\text{mass of solids}} + \frac{\text{mass of filtrate}}{\text{mass of solids}}$$

i.e.

$$\frac{1-s}{s} = (m-1) + \frac{\rho_l V}{wA} \tag{4.11}$$

where ρ_l is the filtrate density. This mass balance assumes that all feed suspension is filtered to form a cake, that is, no feed remains in the filter as suspension. After rearrangement, equation (4.11) leads to:

$$M_s = wA = \frac{\rho_l s}{1 - ms} V \tag{4.12}$$

The term $\rho_l s/(1 - ms)$ may be considered as a concentration corrected for the presence of liquid in the cake. Equation (4.12) is usually used in preference to equation (4.8).

Substituting for R_c in equation (4.6) using (4.7), and using equation (4.12), gives the basic filtration equation to be solved for specific operating conditions:

$$\frac{1}{A}\frac{dV}{dt} = \frac{A\Delta p}{\mu(\alpha cV + AR)} \tag{4.13}$$

where $c = \rho_l s/(1 - ms)$.

Two typical experimental curves of the filtrate volume V collected during filtration are shown in Figure 4.7. For a cake formed at a constant pressure, and where the pressure difference over the filter is constant, the general form of the curve resembles a parabola. Different V vs. t data result at different pressure differences; it has already been inferred that the V vs. t relationship is influenced by particle size and concentration, cake porosity and permeability, and pH and the state of

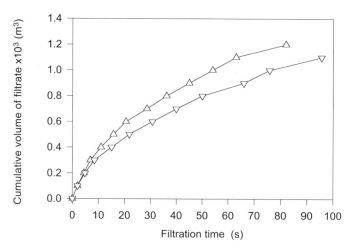

Figure 4.7 Variation of the filtrate volume collected with time, at constant 69 kPa (∇) and 138 kPa (Δ), during filtration of calcium silicate suspensions.

aggregation of the particles. This curve can be further affected by the choice and condition of the filter medium and operational characteristics of particular types of filters, giving a very wide range of variables which each play their part in controlling the rate of filtration.

The filtration rate (i.e. the slope of the V vs. t curve) diminishes as the cake grows; in a constant pressure process a maximum rate is obtained at the start of filtration when no cake has formed. The manner in which q (= dV/dt) changes with time is important in certain types of filters. For example, carrying out cake formation in a batch filter for longer than a certain filtration time might be uneconomic because q has become too low.

Equation (4.13) is the basis of analysis for filtration test data and filter process design, but requires modification to meet specific filter or test operational characteristics before it can be used.

4.2 Compressible cake filtration

Many filter cakes are compressible to some extent, and increases in filtering pressure lead to less porous and more resistant cakes. For such systems data are needed which relate the specific resistance α, and the porosity, ε, to the pressure difference Δp before laboratory experiments can be used in calculations on plant performance, or else data are needed at the particular pressure at which the plant filtration is to be carried out. Aspects of the effects of pressure on cake formation have

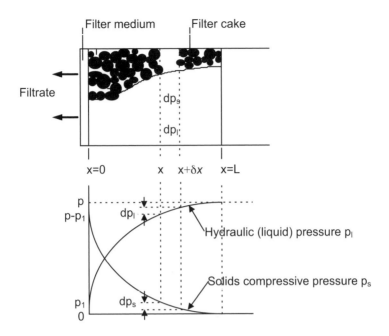

Figure 4.8 The variation of hydraulic pressure and solids compressive pressure through a filter cake; the hydraulic pressure varies from p_1 at the surface of the filter medium to the pressure in the slurry at the cake surface, and the sum of the hydraulic and solids compressive pressures is equal to the pressure being delivered by the pumping system.

been the subject of many studies, e.g. Carman (1938), Grace (1953), Ruth (1946), Shirato *et al* (1987), Tarleton *et al* (1997a,b), Tiller (1975), Wakeman (1978) and Walas (1941).

4.2.1 Pressure conditions in a compressible filter cake

The conditions in a compressible filter cake are depicted in Figure 4.8. Flow is from right to left, originating in the slurry and exiting from the medium. The porosity has a maximum value ε_0 at the cake surface, and decreases to a minimum ε_1 at the medium. The hydraulic pressure varies non-linearly from the applied pumping pressure p at $x = L$ (the distance of the cake-slurry interface from the filter medium) to p_1 at the medium where $x = 0$. The fluid flows through the interstices of the cake and exerts a drag on each particle. Surface forces due to frictional flow result in stresses on each particle, and the force is then transmitted to other particles through an interconnecting network. Particles may be in point contact, but for force transmission through the particulate phase contact is not essential and particle connectivity may be provided by components such as polymers in the solution phase surrounding the particles.

For the low solids velocities normally encountered in filtration the pressure at any plane in the cake is equal to the sum of the pressure in the liquid phase and the compressive drag pressure in the particulate phase, enabling a force balance to be written as:

$$Ap = Ap_l + F_s \qquad (4.14)$$

where F_s is the accumulated frictional drag (a combination of skin and form drag) developed at the surface of the particles, and in the simplest sense is communicated through the points of contact of the particles. During the formation of a compressible cake its porosity varies with both filtration time (t) and distance measured from the filter medium (x), making the liquid pressure and the solids compressive pressure both functions of time and distance. The total pressure may also vary with time. Defining the *compressive drag pressure* as $p_s = F_s/A$ leads to:

$$p(t) = p_l(x,t) + p_s(x,t) \qquad (4.15)$$

p_s acts through the solids network and is analogous to the *effective pressure* used in soil mechanics. If the particles are in point rather than area contact, the hydraulic pressure p_1 may be assumed to be effective over the entire cross-sectional area A of the cake. The pressure loss in the liquid over any section of cake is therefore exactly equal to the change of the compressive drag pressure in the solids in that section, i.e. differentiating equation (4.15):

$$\left(\frac{\partial p_s}{\partial x} \right)_t + \left(\frac{\partial p_l}{\partial z} \right)_t = 0 \qquad (4.16)$$

or,

$$dp_s + dp_l = 0 \qquad (4.17)$$

Equations (4.15) to (4.17) simply state that drag pressure increases as the hydraulic pressure decreases. Cake structure is determined by the way particles are initially deposited and then move into new positions as load in the form of fluid drag is applied. Changes in porosity with applied pressure are of importance for the process design of filters, but the analysis developed here does not recognise that any such changes can occur.

4.2.2 Working relationships for compressible filtration

A working equation for compressible cake filtration is obtained by rewriting equations (4.3), (4.7) and (4.17) as:

$$\frac{dp_l}{dw} = -\frac{dp_s}{dw} = \mu\alpha u \tag{4.18}$$

Since α is a function of p_s, equation (4.18) is rearranged to:

$$-\mu\, u \int_0^w dw = \int_{p-p_1}^0 \frac{dp_s}{\alpha} \tag{4.19}$$

where the limits ($w = 0$, $p_s = p - p_l$) and ($w = w$, $p_s = 0$) are at the medium and the cake-slurry interface respectively. The pressure p_l is given by equation (4.2). Integrating equation (4.19) with the approximation that u is constant throughout the cake (and equal to the filtrate flux) leads to:

$$\mu\, uw = \int_0^{p-p_1} \frac{dp_s}{\alpha} \tag{4.20}$$

In accordance with equation (4.7) the average specific resistance α_{av} is defined by the harmonic mean of the local resistance values through the depth of the cake. If the local resistance is defined by $\alpha = \alpha_0 p_s^n$, then α_{av} is:

$$\frac{p-p_1}{\alpha_{av}} = \int_0^{p-p_1} \frac{dp_s}{\alpha} = \int_0^{p-p_1} \frac{dp_s}{\alpha_0\, p_s^n} = \frac{1}{\alpha_0(1-n)}\Delta p_c^{1-n} \tag{4.21}$$

$$\alpha_{av} = \alpha_0(1-n)\Delta p_c^n$$

where $p - p_1 = \Delta p - \Delta p_m = \Delta p_c$. Substituting for the integral in equation (4.20) and eliminating p_1 using equation (4.2) produces

$$\mu\, uw = \frac{(p-p_0) - u\mu R}{\alpha_{av}} \tag{4.22}$$

Solving for the filtrate flux yields:

$$\frac{1}{A}\frac{dV}{dt} = u = \frac{\Delta p}{\mu(\alpha_{av}w + R)} \tag{4.23}$$

which is analogous to equation (4.6) for an incompressible filtration. In terms of filtrate volume equation (4.23) becomes:

$$\frac{1}{A}\frac{dV}{dt} = \frac{A\Delta p}{\mu(\alpha_{av}cV + AR)}$$

(4.24)

where $c = \rho_l s/(1 - ms)$ and $\alpha_{av} = \alpha_0(1 - n)\Delta p_c^n$. The above derivation is not rigorous because it has been assumed that the liquid flux and the area over which the forces act are constant throughout the thickness of the cake, but in most instances it is a good working relationship to use in process design calculations. When thick or high resistance cakes are being formed the assumption $\Delta p_c \approx \Delta p$ is often employed; in many practical instances it is essential to make this assumption, which is widely valid for industrial calculations.

4.2.3 Effects of pressure on filter cake properties

Many empirical equations can be developed to describe the variations of α and ε with p_s. When applying the equations to practical problems it is usual to make the simplification that $p_s \approx \Delta p$. Over limited pressure ranges the local $\alpha(p_s)$ or $\alpha(\Delta p)$ and $\varepsilon(p_s)$ or $\varepsilon(\Delta p)$ relationships can be linearised using logarithmic coordinates and represented by "power law" type relationships:

$$\alpha = \alpha_0 p_s^n$$

(4.25)

and

$$\varepsilon = \varepsilon_0 p_s^{-\lambda}$$

(4.26)

where α_0 and ε_0 are the specific resistance and porosity at unit applied pressure respectively. These are obtained, together with n and λ, by extraploation and calculation from experimental data. Although equations (4.25) and (4.26) can be criticised on fundamental grounds, they are useful in process calculations.

Sometimes it is more convenient to work in terms of the cake solids volume fraction rather than the cake porosity, whence the equation generally used is:

$$(1-\varepsilon) = (1-\varepsilon)_0 p_s^\beta$$

(4.27)

where $(1 - \varepsilon)_0$ is the solids fraction at unit applied pressure. The index β gives a measure of the potential for deliquoring the cake by compression, which is important in the assessment of equipment such as variable chamber presses and tube presses for particular applications. Deliquoring mechanically by squeezing the cake or sucking or blowing gas is much cheaper than thermal drying.

The data in Table 4.1 may be taken as a guide to α and ε values for particular materials as particle size, solution pH, and other relevant properties were not always completely specified by the investigators. The value of n, the compressibility index, is indicative of the extent to which a filter cake may compress when it is subjected to a compressive force; when $n \sim 1$ the cake would be regarded as very compressible, $n \sim 0.5$ suggests a moderately compressible cake, and $n < 0.2$ suggests that the cake is almost incompressible. For any specific design problem it is always advisable to measure data for the suspension to be filtered. There are inherent dangers associated with using the data outside of the pressure ranges quoted, highlighted by the fact that some ε_0 are greater than unity since they are empirical values obtained from curve fitting procedures. Although α_0 and ε_0 have a mathematical meaning, no physical significance can be attached to their values and there are considerable experimental difficulties behind obtaining real values for the parameters.

4.2.4 Other factors affecting filter cake properties

The specific resistance of filter cakes is affected by many variables, and not merely the applied pressure as implied by Table 4.1. Instead of equation (4.26) for the porosity, an alternative is to use an expression with the following form for the voids ratio:

$$e = e_0 - b \log p_s \qquad (4.28)$$

The porosity is related to the voids ratio by $\varepsilon = e/(1 + e)$. Although equations (4.25) to (4.28) are acceptable for practical use, it should be observed that they have an incorrect limiting form as $p_s \rightarrow 0$; at this limit the equations suggest that α, ε, $(1 - \varepsilon)$ and e all equal zero. To overcome this problem, the following alternative expressions in which the limiting values are finite can be used:

$$\alpha = \alpha_0 (1 + p_s)^n \qquad (4.29)$$

$$e = e_0 - b \log(1 + p_s) \qquad (4.30)$$

Table 4.1 Specific resistances and equilibrium porosities of some filter cakes (from Shirato et al, 1987)

Particle type	α_0 (m kg^{-1}kPa^{-n})	n	ε_0	λ	p_s (kPa)	Investigator
Alumina	2.37x10^8	0.30	0.96	0.010	7–689	Grace
			0.73	0.0047	82–489	Lindquist
Aluminium hydroxide	3.32x10^{11}	0.34			173–689	Carman
Asbestos			1.06	0.017	52–689	Tiller
Calcium carbonate			0.895	0.017	1–48	Ruth
			0.755	0.036	70–345	Walas
– in distilled water	8.93x10^9	0.20	1.04	0.033	7–689	Grace
– in Na$_4$P$_2$O$_7$ solution	4.69x10^{10}	0.13	0.795	0.013	34–689	Grace
Carbon			0.845	0.021	82–758	Lindquist
Carbonyl iron			0.425	0	14–6895	Grace
Celite			1.05	0.017	82–414	Lindquist
Cement			0.82	0.058	150–340	Shirato
	2.22x10^{10}	0.298			100–880	Shirato
Clay						
– colloidal	7.43x10^{11}	0.16			173–689	Carman
– Hara gairome	1.44x10^{11}	0.612			100–880	Shirato
– Mitsukuri gairome	7.82x10^9	0.669			100–880	Shirato
Copper oxide			0.505	0.021	82–589	Lindquist
Ferric hydroxide	2.59x10^{11}	0.39			173–689	Grace
Ferric oxide			1.12	0.037	70–345	Walas
Hyflo Super–Cel			0.995	0.014	10–689	Tiller
Kaolin			0.88	0.045	10–689	Tiller
– in Al$_2$(SO$_4$)$_3$ solution	4.76x10^{10}	0.27	0.9	0.049	7–689	Grace
– in Na$_4$P$_2$O$_7$ solution			0.79	0.031	7–689	Grace
– Hong Kong pink	6.48x10^9	0.485	1.0	0.047	1–880	Shirato
– Korean			1.03	0.06	50–340	Shirato
– Shinmei			0.98	0.046	6–690	Shirato
Magnesium carbonate			1.1	0.011	70–345	Walas
Magnesium hydroxide	1.35x10^{10}	0.47			173–689	Carman
Talc, in Al$_2$(SO$_4$)$_3$ solution	7.05x10^8	0.51	1.39	0.054	7–689	Grace
Titanium dioxide						
– in distilled water	1.27x10^{10}	0.32	1.12	0.038	7–689	Grace
– in HCl solution	9.29x10^{11}	0.058			7–689	Grace
Zinc sulphide	1.48x10^9	0.92	1.36	0.047	7–689	Grace

where e_0 and α_0 are the values of e and α at $p_s = 0$ (Wakeman, 1993). In practice, it does not matter which equations are used provided they are used consistently.

The order of the effects of particle size and porosity variations on specific resistance can be demonstrated using the definition of α (equation (2.23)) combined with the Kozeny-Carman permeability (equation (2.37)), to give

$$\alpha = \frac{180}{\rho_s x^2} \frac{1-\varepsilon}{\varepsilon^3} \tag{4.31}$$

Table 4.2 shows that small changes in either particle size or porosity lead to significant changes in α; for example, an increase in particle size by a factor of 2 leads to a 4 fold reduction of α, or a change in porosity from 0.4 to 0.5 leads to a 2.3 fold reduction of α. It is reasonable to assume that in a filter cake either the 'mean' particle size or the porosity (or both) may vary from one region to another, leading to heterogeneities in cake structure that manifest themselves in local variations of specific resistance. From a processing point of view, greater rates of filtration may be expected with larger particles in the feed. The cakes formed have a lower specific resistance (that is, they are more permeable) and they possess a larger pore volume (which is synonymous with wetter cakes).

Table 4.2 The effect of particle size and porosity on the cake specific resistance ($\rho_s = 2500$ kg m^{-3}) calculated using equation (4.31)

Particle size, x (mm)	Porosity, ε	Specific resistance, α (m kg^{-1})
1	0.4	6.7×10^{11}
2	0.4	1.7×10^{11}
10	0.4	6.7×10^{9}
100	0.4	6.7×10^{7}
2	0.4	16.9×10^{10}
2	0.5	7.2×10^{10}
2	0.6	3.3×10^{10}
2	0.7	1.6×10^{10}

The nature of filtration of fine particle suspensions is dependent on pH and particle surface charge (i.e. the magnitude of particle-particle interactions, represented by the ζ-potential in Table 4.3), particle size and particle shape as well as the applied pressure. At high or low values of ζ-potential there is a net repulsion between the particles which are well dispersed in the suspension, but at a zero ζ-potential (the so called isoelectric point) the particles tend to aggregate. Some experimental data are given in Table 4.3; particle size has a major effect on compaction which, in order for cakes to be formed with the same moisture content, requires greater forces and becomes more protracted as the size is reduced. As particle size is decreased interparticle repulsive forces become more appreciable compared with gravitational and imposed mechanical forces, and the permeability of a filter cake is lowered. The final moisture content of a cake is affected to only a small extent by particle size provided the imposed force is sufficiently large to overcome the repulsive forces. There is no model currently available for the prediction of specific resistance and voids ratio values of fine particle suspensions, so the importance of obtaining data from reliable experiments for such suspensions is emphasised. It is usually poor practice to infer data from a different feed suspension, since factors such as the ionic strength that affect the ζ-potential and hence the aggregate size in suspension are not usually identical from one suspension to another.

Table 4.3 Variation of the coefficients in the constitutive equations (4.29) and (4.30) with particle size and pH for particles suspended in distilled water with pH adjusted with HCl or NaOH (Wakeman, 1993; Wakeman et al, 1991)

Particle type	pH	ζ (mV)	x (μm)	α_0 (m kg^{-1} MPa^{-n})	n	e_0	b	p_s (MPa)
Anatase	4.0	−5	0.5	7.16×10^{11}	0.222	1.681	0.134	0.3–11
	7.0	−39	0.5	5.35×10^{12}	0.027	5.916	0.318	0.3–11
	9.1	−47	0.5	3.50×10^{12}	0.111	8.787	1.222	0.3–11
Aragonite	11.0		1.7	1.38×10^{11}	0.227	4.286	0.687	0.2–1.0
	11.0		3.3	4.79×10^{10}	0.214	4.722	0.812	0.2–1.0
	11.0		8.2	5.98×10^{9}	0.260	6.558	1.239	0.2–1.0
	11.0		10.0	5.57×10^{8}	0.551	8.704	1.825	0.2–1.0
Calcite	10.0		2.3	8.35×10^{10}	0.246	4.951	0.619	0.13–2.1
China clay	2.9	−15	5.0	8.32×10^{10}	0.649	2.695	0.499	0.3–20
	5.2	−29	1.5	7.60×10^{10}	0.648	2.968	0.546	0.3–20
	5.2	−29	5.0	4.02×10^{11}	0.458	2.521	0.446	0.3–20
	5.2	−29	8.1	5.57×10^{10}	0.581	2.491	0.469	0.3–20
	10.2	−57	5.0	7.53×10^{11}	0.613	2.567	0.469	0.3–20
Hydromagnesite	9.9		16.4	1.04×10^{10}	0.564	12.00	2.636	0.3–11
Wollastonite	9.2		12.7	1.00×10^{10}	0.100	2.208	0.230	0.13–2.1

The effect of the ionic strength of mono- and tri- valent electrolytes on the specific resistance and voids ratio of filter cakes is demonstrated in Table 4.4. The specific resistance decreases (and hence the rate of filtration increases) with electrolyte addition due to particle aggregation. The greater effectiveness of the trivalent ion (Al^{3+}) as a coagulant is shown by the greater reduction of specific resistance, but addition of excess electrolyte causes the resistance to increase. The greater strength of the aggregates and the more open cakes formed by addition of the trivalent ions is reflected by the higher voids ratios (and hence cake moisture contents).

Table 4.4 Effect of solution conductivity on specific cake resistance and equilibrium voids ratio for anatase ($x \approx 0.5$ μm) filter cakes filtered at pH 7.05 with $\Delta p = 1.01$ MPa (Wakeman, 1993)

Solution conductivity (mS cm^{-1})	Equilibrium voids ratio, e_∞	Cake specific resistance, α_{av} (m kg^{-1})
Particles in distilled water		
0.2	1.906	5.25×10^{12}
Monovalent electrolyte: HCl		
6.0	2.209	3.72×10^{12}
11.0	1.683	2.04×10^{12}
35.0	1.567	1.82×10^{12}
50.0	1.567	1.80×10^{12}
Trivalent electrolyte: $Al_2(SO_4)_3$		
1.0	2.013	2.71×10^{12}
2.0	2.267	1.52×10^{12}
3.0	2.619	1.08×10^{12}
5.9	2.325	1.36×10^{12}
11.6	2.619	1.66×10^{12}
20.0	2.601	2.24×10^{12}

The effect of particle shape is illustrated through Tables 4.5 and 4.6. Particle shape affects the volume and surface area of a particle, and hence the specific surface. For some regular geometries of particles, the effect of the major dimensions of the specific surface are shown on Table 4.5. For particles with the same characteristic dimension a (the diameter in the case of a spherical particle, or the length of a side for a cubic particle, for example), and assuming the same cake porosity and solids density, the variation of specific surface of the particle and specific resistance of the filter cake is illustrated in Table 4.6. It is readily seen that fibrous particles have a lower specific surface than most other shapes, whilst flakey particles have a higher specific surface. This then ranks the specific resistance, with cakes of fibrous particles having the lowest resistance. Such considerations, however, give a misleading picture as fibrous particles tend to pack to a higher porosity in filter cakes: assuming the porosity is 0.8 rather than 0.5 changes the cake specific resistance from 1.6×10^9 to 1.6×10^8 m kg^{-1}. This would suggest that suspensions of fibrous particle should be easier to filter than other particle shapes, but in practice this is often not the case. Fibrous particles of very high aspect ratio, that is, a high length to diameter ratio, tend to ball up during filtration to a greater or lesser extent depending on the type of filter being used, whilst in other filter types the fibres may have a greater tendency to mat on the filter cloth. Both cases lead to handling problems with the filter cake, and often reduced rates of filtration.

Table 4.5 Effect of particle shape on specific surface

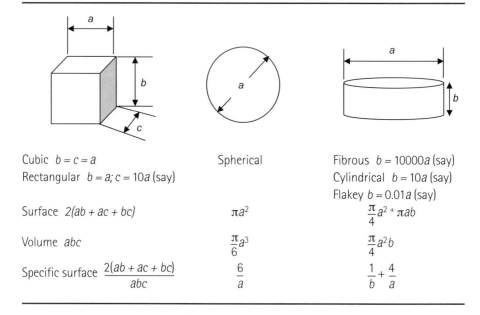

Cubic $b = c = a$	Spherical	Fibrous $b = 10000a$ (say)
Rectangular $b = a;\, c = 10a$ (say)		Cylindrical $b = 10a$ (say)
		Flakey $b = 0.01a$ (say)
Surface $2(ab + ac + bc)$	πa^2	$\dfrac{\pi}{4}a^2 + \pi ab$
Volume abc	$\dfrac{\pi}{6}a^3$	$\dfrac{\pi}{4}a^2 b$
Specific surface $\dfrac{2(ab + ac + bc)}{abc}$	$\dfrac{6}{a}$	$\dfrac{1}{b} + \dfrac{4}{a}$

A further consideration when determining the specific resistance of a filter cake is the age of the feed suspension; many suspensions, particularly those composed of fine particles or biosolids, tend to change in storage – an effect known as ageing. This effect is shown in Figure 4.9 for filtration of an activated sludge, where the specific resistance has increased by almost an order of magnitude during 5 days storage; this is quite typical of activated sludges. For the data shown, the increase of specific resistance is largely attributable to the increase in specific surface (also shown on Figure 4.9) as the floc structure changed during the storage period. As a rule of thumb, filter cakes formed from organic materials are more compressible and hence more difficult to filter than those formed from inorganic materials.

Table 4.6 Effect of particle shape on specific resistance of the filter cake

Particle shape	Specific Surface S_0 (m^{-1})	Specific resistance (m kg^{-1}) $$\alpha = \frac{5S_0^2}{\rho_s}\frac{1-\varepsilon}{\varepsilon^3}$$ [a = 10 μm; ε = 0.5; ρ_s = 2000 kg m^{-3}]
Fibrous	$\dfrac{4.0001}{a}$	1.6×10^9 [when ε = 0.8, α = 0.16 × 10^9]
Cylinder	$\dfrac{4.1}{a}$	1.7×10^9
Rectangular	$\dfrac{4.2}{a}$	1.8×10^9
Sphere	$\dfrac{6}{a}$	3.6×10^9
Cube	$\dfrac{6}{a}$	3.6×10^9
Flake	$\dfrac{104}{a}$	1100×10^9

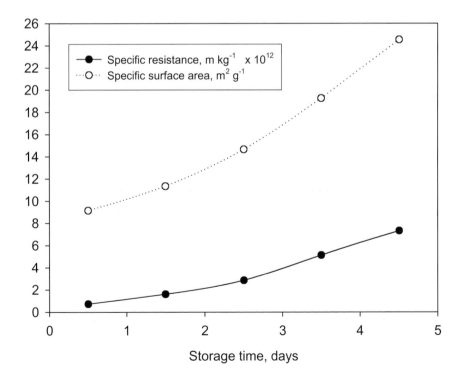

Figure 4.9 The specific surface area and cake specific resistance for activated sludge from Aså Rensningsanlæg as a function of anaerobic storage time (Sørensen and Wakeman, 1996).

The average specific resistance is a useful tool in the sizing and process design of filters; it also gives a first approximation of filtration behaviour of a suspension, which is summarised in Table 4.7.

Table 4.7 The relative ease of filtration as characterised by the magnitude of the specific resistance

Ease of separation	α_{av} (m kg^{-1})
Very easy	$\leq 10^9$
Easy	10^{10}
Moderate	10^{11}
Difficult	10^{12}
Very difficult	$\geq 10^{13}$

4.2.5 Expressions for the solids mass deposited

Substitution of equation (4.12) into (4.24) gives the relationship between the solids mass deposited and the process variables:

$$\frac{dw}{dt} = \frac{c\Delta p}{\mu(\alpha_{av} w + R)} \tag{4.32}$$

This expression may be used in preference to equation (4.24) in filtration calculations.

4.2.6 Expressions for suspension concentration

The term used for the effective solids concentration in the suspension, c, in equations (4.24) and (4.32) has the units (kg of solid)/(m^3 of filtrate liquid). However, specifications of practical problems or results from experimental data may be expressed using alternative descriptions for solids concentration and it is often necessary to convert these into the appropriate form. This concentration has already been expressed in terms of the wetness of the cake (m in equation (4.9), which for a compressible cake becomes m_{av} since there is a porosity variation through the thickness of the filter cake) and the mass fraction of solids in the suspension (s in equation (4.10)) as:

$$c = \frac{\text{mass of solids}}{\text{volume of liquid}} = \frac{\rho_l}{\dfrac{1-s}{s} - (m_{av} - 1)} = \frac{\rho_l s}{1 - m_{av} s} \tag{4.33}$$

Solids concentration c is related to solids volume fraction in the suspension, V_s, by:

$$c = \frac{1}{\frac{1}{\rho_s}\left(\frac{1}{V_s}-1\right)-\frac{m_{av}-1}{\rho_l}} \qquad \text{since} \qquad s = \frac{\rho_s}{\rho_s + \rho_l\left(\frac{1}{V_s}-1\right)} \qquad (4.34)$$

Solids concentration c is related to the more conventional definition of solids concentration c' (kg of solids)/(m³ of solids + m³ of liquid) by:

$$c = \frac{1}{\frac{1}{\rho_s}\left(\frac{\rho_s}{c'}-1\right)-\frac{m_{av}-1}{\rho_l}} \qquad \text{since} \qquad c' = \rho_s V_s \qquad (4.35)$$

s and c' are related by:

$$s = \frac{c'}{c' + \rho_l\left(1-\frac{c'}{\rho_s}\right)} \qquad (4.36)$$

4.2.7 Relationships between cake thickness and filtrate volume

The mass of dry solids deposited on a unit area of filter is related to the cake thickness L by:

$$w = \rho_s(1-\varepsilon_{av})L \qquad (4.37)$$

where ρ_s is the true density of the solid. Using equation (4.12) the filtrate volume can now be related to cake thickness by:

$$V = \frac{1-m_{av}\,s}{s}\frac{\rho_s}{\rho_l}(1-\varepsilon_{av})AL \qquad (4.38)$$

The mass of wet cake per unit mass of dry cake m_{av} is related to the average porosity of the cake ε_{av} by:

$$m_{av} = 1 + \frac{\rho_l\,\varepsilon_{av}}{\rho_S(1-\varepsilon_{av})} \qquad (4.39)$$

Either m_{av} or ε_{av} can be eliminated from equation (4.38). Eliminating m_{av} yields

$$V = \frac{\rho_s}{\rho_l}\left(\frac{1-s}{s}-\varepsilon_{av}\left[\frac{1-s}{s}\frac{\rho_s}{\rho_l}+1\right]\right)AL \qquad (4.40)$$

or eliminating ε_{av} gives

$$V = \frac{\rho_s(1 - m_{av}s)}{s[\rho_s(m_{av} - 1) + \rho_l]} AL \qquad (4.41)$$

As most process filters are designed on the basis of cake thickness, equations (4.40) and (4.41) are important for converting V to L in formulae relating V to Δp and t.

4.3 Basis for filtration calculations

A filter may be operated under a variety of conditions, dependent on the methods used to develop the necessary driving pressure gradient for the process. When slurry is fed to the filter from a feed vessel where the space above the suspension is subjected to a source of compressed gas (e.g. air), or when the volume downstream of the filter plate is connected to a vacuum source, filtration is accomplished at a constant pressure differential. In this case the rate of filtration decreases due to an increase of the cake thickness and, consequently, flow resistance (it should be noted that the specific resistance remains fairly constant as the cake thickness increases).

When slurry is fed to a filter using a positive displacement pump filtration is performed under (more or less) constant flow rate conditions, whilst the pressure drop over the filter increases. Since the specific resistance of the filter cake may be pressure dependent, it is clear that different results may be obtained operating at constant pressure when compared with constant rate.

Use of centrifugal pumps to deliver slurry to a filter is widespread in practice. In these cases the feed rate decreases as the cake resistance builds-up, and filtration is carried out under conditions of variable pressure difference and variable rate; the relationship between pressure and rate is decided by the pump characteristic curve.

The various modes of operation are indicated on Figure 4.10. The constant pressure case is represented by a vertical line on Figure 4.10(a) and a horizontal line on Figure 4.10(b), whilst constant rate filtration is shown by a horizontal line on Figure 4.10(a) and a line of constant gradient on Figure 4.10(b). The centrifugal pump case is shown by curves on both plots, and on Figure 4.10(a) it is actually the shape of the pump characteristic curve. In some cases the pump curve can be approximated by a constant rate process followed by a constant pressure. The arrow on each line indicates the direction of increasing filtration time.

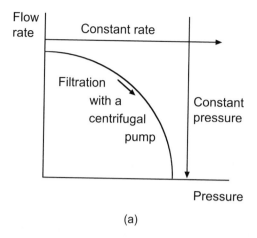

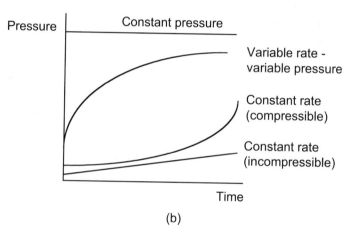

Figure 4.10 Pressure-flow relationships in filtration; the arrows point in the direction of increasing filtration time.

4.3.1 Constant pressure operation

When Δp is maintained constant during filtration equation (4.24) may be integrated directly to give a relationship between filtrate volume and time:

$$\int_{t_i}^{t} dt = \int_{V_i}^{V} (K_1 V + K_2) dV \qquad (4.42)$$

i.e.

$$\frac{t - t_i}{V - V_i} = \frac{K_1}{2}(V + V_i) + K_2 \qquad (4.43)$$

where $K_1 = \alpha_{av} c \mu / A^2 \Delta p$ and $K_2 = \mu R / A \Delta p$. Plotting $(t - t_i)/(V - V_i)$

against $V + V_i$ may lead to a straight line from which cake and medium properties can be evaluated; it is sometimes convenient to make the start of filtration coincide with the start of the integration, whence $t_i = V_i = 0$.

Figure 4.12 in Example 4.1 shows a plot of (t, V) data in the form suggested by equation (4.43). Non-linearities close to the start of filtration can often be observed on such plots and are probably due to the fact that K_1 and K_2 are not truly constant in the initial stages of filtration. When the cake is very thin the main part of the total pressure drop is over the filter medium. As the cake becomes thicker a larger proportion of the total pressure drop is lost over the cake, and cake resistance then dominates the medium resistance. This figure is a typical constant pressure difference filtration plot; the gradient of the plot is $K_1/2$, from which α_{av} can be evaluated; the intercept on the t/V axis at $V = 0$ is K_2, which gives a value for R. Constant pressure tests are the most frequently used method of obtaining laboratory data for scale-up calculations. The test may be carried out either with a positive pressure or under vacuum; provided the same pressure difference is applied, the output from the tests should be the same. On the other hand, the orientation of the filter surface (upwards or downwards facing) can have a marked effect on the data obtained, due mainly to particle settling in the suspension during filtration.

Non-linearities in the plot of the (t, V) data (not exhibited by the data on Figure 4.12) can often also be seen towards the end of filtration, where the curve through the data points tends to rise sharply. This is due to one or more of several factors. (i) If the filtration is being carried out in a fixed volume filtering chamber, then when the chamber is full of cake the filtrate volume collected in any time interval is reduced; (ii) if a cake has been formed and the feed slurry is completely used the applied pressure difference then serves to deliquor the cake, and again the filtrate volumetric flow rate is considerably reduced. If settling of the solids has occurred during the test, then t/V tends to become constant. General aspects of where non-linearities occur on plots of t/V versus V are summarised on Figure 4.11.

Example 4.1 Calculation of the cake specific resistance from constant pressure test data.

Estimate the cake specific resistance and the medium resistance from the data shown in Tables 4.8 and 4.9 obtained from a constant pressure filtration test on a calcium silicate suspension using a laboratory scale filter press (Hosseini, 1977).

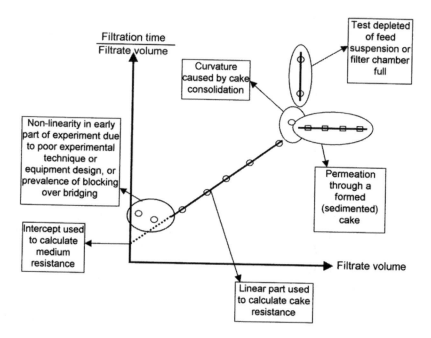

Figure 4.11 Examples of where non-linearities are often observed in the plot of t/V versus V and the reasons for their occurrence. Care must always be taken when carrying out the experiment to obtain the data, as experimental methodology can also be the cause on non-linearities. Sometimes these can make the linear part of the data so small as to be quite useless.

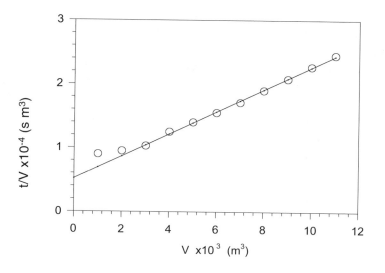

Figure 4.12 A t/V vs. V plot of the constant pressure data in Example 4.1.

Table 4.8 Data measured for Example 4.1

Mean particle size	6.5 μm
Filtration area	0.0429 m²
Filtration pressure	69 kPa
Filtrate viscosity	0.001 Pa s
Filtrate density	1000 kg m⁻³
Solids density	1950 kg m⁻³
s	0.0495
m_{av}	3.405

Table 4.9 Filtrate volume and filtration time data measured for the test in Example 4.1

Filtration time, t (s)	Filtrate volume, V (m³)	t/V (s m⁻³)
0	0	—
9	0.001	0.90×10⁴
19	0.002	0.95×10⁴
31	0.003	1.03×10⁴
50	0.004	1.25×10⁴
70	0.005	1.40×10⁴
93	0.006	1.55×10⁴
120	0.007	1.71×10⁴
152	0.008	1.90×10⁴
187	0.009	2.08×10⁴
227	0.010	2.27×10⁴
270	0.011	2.45×10⁴

Solution

From the time-volume data calculate t/V, and then plot t/V against V (see Figure 4.12). Fit a straight line through the linear part of the plot. In this case the first two data points have been disregarded, and linear regression gives $t/V = 1.742×10^6 V + 0.5186×10^4$.

From the plot: Slope $= 1.742×10^6$ s m⁻⁶

$$c = \frac{\rho_l s}{1 - m_{av} s} = \frac{1000 \times 0.0495}{1 - 3.405 \times 0.0495} = 59.5 \quad \text{kg m}^{-3}$$

$$\alpha = \frac{2A^2 \Delta p \times slope}{c\mu} = \frac{2 \times 0.0429^2 \times 69000 \times 1.742 \times 10^6}{59.5 \times 0.001} = 7.4 \times 10^9 \quad \text{m kg}^{-1}$$

i.e. the cake specific resistance is $7.4×10^9$ m kg⁻¹ at 69 kPa.

From the intercept on the t/V axis: Intercept $= 5186$ s m^{-3}

$$R = \frac{A \Delta p \times \text{intercept}}{\mu} = \frac{0.0429 \times 69000 \times 5186}{0.001} = 1.54 \times 10^{10} \quad \text{m}^{-1}$$

i.e. the medium resistance is 1.54×10^{10} m^{-1} at 69 kPa.

Since the medium resistance parameter is usually determined by extrapolation of the graph to the ordinate, erroneous (sometimes negative!) values of R can be obtained. The method of obtaining the filter medium resistance from the intercept value can lead to unrealistic values if the medium clogging is significant and continues throughout the course of the filtration. Even though the medium resistance is a significant part of the total resistance to filtration, the average specific resistance of the filter cake obtained from the gradient of the plot does not appear to be in error (Hosten and San, 1999). In cases where the medium resistance term is zero or negative a simpler relationship is sometimes used to analyse the plot of the (t, V) data:

$$\frac{t - t_i}{V - V_i} = \frac{K_1}{2}(V + V_i) \tag{4.44}$$

Penetration and embedding of particles into the interstices of the medium frequently occur, causing the medium resistance to increase most noticeably at the start of filtration. In some cases the increase in resistance is progressive, giving a medium which needs cleaning after several uses. On the (t, V) data plot non-linearities towards the end of filtration (see Figure 4.11) occur when the test equipment has a confined filter chamber volume, e.g. a pilot scale plate-and-frame press; a rapid increase of the slope of the curve indicates that the filter chamber has filled (effectively the available filter area has dropped to a very low value) and further increases in V are very small, which also occurs when the slurry batch has all been filtered.

The correct value of the specific resistance is obtained from the linear region on Figure 4.12. Alternatively, cake and medium resistances can be obtained by plotting the (t, V) data according to equation (4.24), i.e. by plotting dt/dV against V. Similar non-linearities can occur when plotting the data in this way, and data are significantly more sensitive to small fluctuations in test conditions. The end of filtration and the start of cake compression is, however, more easily identified on this plot by a sudden and sharp increase of the slope of the graph – as with many graphical methods of analysing experimental data, data scatter can present a problem and should be minimised during the experiment. The integrated form (i.e. equation (4.43)) produces a smoother change of the slope in the same circumstances.

In some scale-up problems it is not necessary to determine α_{av} from K_1, but the filterability coefficient:

$$F_K = A^2 K_1 \tag{4.45}$$

is a sufficient parameter for practical purposes.

4.3.1.1 Neglecting medium resistance in constant pressure filtration
It is frequently considered appropriate or expedient to neglect the medium resistance R, but the error incurred by making this assumption should be estimated (Tiller, 1975). If R is neglected, the calculated time of filtration to produce a volume V of filtrate is given by:

$$t = \frac{\mu \alpha_{av}}{2c\,\Delta p} w^2 \tag{4.46}$$

The value of t calculated by equation (4.46) will be smaller than a value calculated when R is included. Subtracting equation (4.46) from the integrated form of (4.32) gives the error which will be introduced by neglecting R:

$$\text{error in } t = \frac{\mu R}{c\,\Delta p} w \tag{4.47}$$

The fractional error is given by the error divided by t:

$$\text{fractional error} = \left(1 + \frac{w}{2}\frac{\alpha_{av}}{R}\right)^{-1} \tag{4.48}$$

If the percentage error is to be < 5% then $w > 40R/\alpha_{av}$.

Example 4.2 Calculation of the average specific resistance vs. filtration pressure relationship.
The specific resistance data in Table 4.10 were obtained from a series of constant pressure filtration experiments. The specific resistance data in the Table are values obtained in experiments conducted at constant filtration pressures, therefore the data as measured are averaged (α_{av}) over pressure up to the filtration pressures. Assuming the data can be represented by equation (4.21) ($\alpha_{av} = \alpha_0(1 - n)\Delta p_c^n$) and that $\Delta p \approx \Delta p_c$, calculate the relationship between the average specific resistance and the filtration pressure.

Solution

Develop the α_{av} vs. Δp relationship for this suspension by plotting the data on logarithmic coordinates as in Figure 4.13. Using the above

Table 4.10 Specific resistance data obtained for Example 4.2

Filtration pressure (kPa)	Specific resistance (m kg^{-1})
70	1.4×10^{11}
104	1.8×10^{11}
140	2.1×10^{11}
210	2.7×10^{11}
400	4.0×10^{11}
800	5.6×10^{11}

assumptions, equation (4.21) leads to $\alpha_{av} = \alpha_0(1 - n)\Delta p^n$. A regression gives the relationship as $\alpha_{av} = 1.43 \times 10^{10}\Delta p^{0.602}$, and hence

$$\alpha_0 = \frac{1.43 \times 10^{10}}{1 - 0.602} = 3.59 \times 10^{10} \quad \text{m kg}^{-1} \text{ kPa}^{-0.602}.$$

Alternatively, and potentially less accurately, since the data form a straight line two sets of data may be used to estimate the $\alpha = \Delta p$ relationship. Using the first and last points, $1.4 \times 10^{11} = \alpha_0(1 - n)(70)^n$ and $5.6 \times 10^{11} = \alpha_0(1 - n)(800)^n$. Solving these equations simultaneously gives $\alpha_0 = 2.90 \times 10^{10}$ and $n = 0.569$.

A similar sequence of calculations can be carried out to obtain the constants that describe the porosity variation with pressure in equations such as (4.26).

4.3.1.2 Stepped pressure tests
It has been suggested from time to time (see, for example, Usher et al (2001)) that the variation of specific resistance with pressure can be obtained from a single experiment in which the applied pressure is increased incrementally during a filtration. The method involves starting

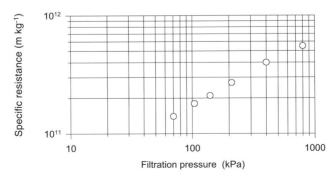

Figure 4.13 Plot of the specific resistance vs. filtration pressure data in Example 4.2.

the filtration at a low applied pressure and measuring a sufficient amount of the filtrate volume versus time data to enable calculation of the specific resistance at that pressure. The applied pressure is then incremented and a new set of data obtained that correspond to the new pressure; this sequence is repeated so that data are obtained over the full range of pressures.

The value of such tests appears to be dependent on the characteristics of the suspension being filtered, and in general the procedure cannot be recommended for evaluating filter cake properties or for predicting filtration rates and calculating filter sizes (Tarleton and Hadley, 2003). During a stepped test the cake formed in the earlier stages of the test is done so at low pressures, and it is subsequently compacted by further deposition of solids and higher hydraulic pressures (which may be transmitted as a force through the particulate network). Not only is the nature of the compressive force transmission to those layers of cake closest to the filter medium (which are well known to have a controlling influence on the overall rate of filtration) quite different from what occurs in an industrial filtration, but the stabilisation of the cake by the fluid flow (Heertjes, 1957) is also entirely different.

4.3.2 Constant rate operation (Tiller, 1955)

When slurry is fed to a filter by a positive displacement pump the rate of filtration is constant, i.e. $q = dV/dt =$ constant. In constant pressure filtration the rate of filtration decreases as the cake builds up, and it is customary to obtain volume-time data from tests. In constant rate filtration, the volume is linear in time, and the pressure-time relationship is measured. The initial pressure rise is linear and this is followed by a curve with an increasing gradient. With slightly compressible materials the approximate linearity may extend over a considerable pressure range, while with highly compressible materials curvature in the Δp vs. t relationship may be observed from the start of filtration. Constant rate filtration is characterised by

$$\frac{dV}{dt} = \frac{V}{t} = \text{constant} = q$$

when (V, t) data are measured from the start of filtration, and the data are analysed by use of equation (4.24) in the form:

$$\Delta p = \frac{\mu c \alpha_{av}}{A^2} qV + \frac{\mu R}{A} q \tag{4.49}$$

which may be alternatively written as:

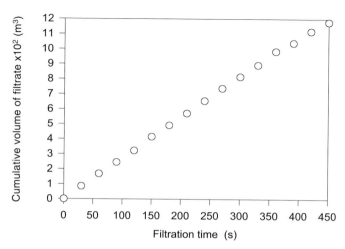

Figure 4.14 During constant rate filtration the filtrate volume collected varies linearly with the filtration time (data obtained from a plate and frame press (Rushton et al, 1996)). Compare the form of this graph with those plotted on Figure 4.7 for filtration at constant Δp.

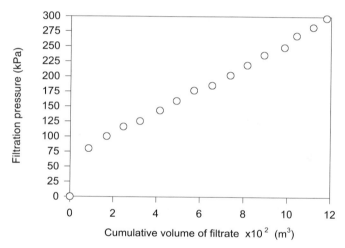

Figure 4.15 Variation of filtration pressure with filtrate volume for the data shown in Figure 4.14, plotted according to equation (4.49); the specific resistance is obtained from the gradient of the line drawn through the data points.

$$\Delta p = \frac{\mu c\alpha_{av}}{A^2}q^2 t + \frac{\mu R}{A}q \tag{4.50}$$

This suggests that a plot of Δp against V or t will be linear, as shown on Figures 4.14 to 4.16. Linear relationships will only be obtained if $\left(\frac{\mu c\alpha_{av}}{A^2}\right)$, $\left(\frac{\mu R}{A}\right)$ and q are all invariant; if α_{av} is a function of Δp then data will give a curve as suggested on Figure 4.10.

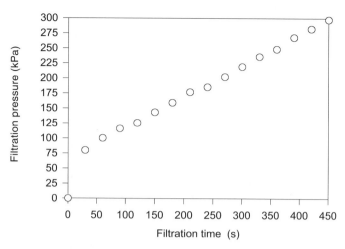

Figure 4.16 Variation of filtration pressure with filtration time for the data shown in Figure 4.14, plotted according to equation (4.50); the specific resistance is obtained from the gradient of the line drawn through the data points.

The equations further suggest that if $\left(\Delta p - \dfrac{\mu R}{A} q\right)$ is plotted against $q^2 t$, then data from tests at different flow rates may be represented by a single line. This is useful for predicting effects of flow rate, and for calculations relating to optimisation of the filter cycle.

4.3.3 Variable pressure–variable rate operation (Tiller, 1958)

If both pressure and rate vary it is necessary to impose the pump characteristics on the equations. No simple formulae can be obtained to relate Δp to t, and a relatively simple numerical integration method must be used. Equation (4.27) can be rearranged to:

$$V = \frac{1}{\alpha_{av} c}\left(\frac{A^2 \Delta p}{\mu \dfrac{dV}{dt}} - AR\right) \tag{4.51}$$

The filtrate rate dV/dt is a function of Δp, and α_{av} is a function of $p - p_1 (= \Delta p - \Delta p_m)$. It is therefore possible to construct a relationship between V and Δp; substituting for α_{av} yields:

$$V = \frac{1}{\alpha_0 (1-n) c \Delta p_c^n}\left(\frac{A^2 \Delta p}{\mu \dfrac{dV}{dt}} - AR\right) \tag{4.52}$$

For practical calculations the simplification $\Delta p_c \approx \Delta p$ is often used. When V has been obtained as a function of Δp and dV/dt, the filtration time can be obtained by integration of:

$$t = \int_0^V \frac{dV}{q} \tag{4.53}$$

In this section three modes of operation are referred to, namely *constant pressure*, *constant rate* and *variable pressure-variable rate*. Specific resistance (α_{av}) values calculated from these approaches have different meanings. From a *constant pressure* test an α_{av} value corresponding to a particular Δp is obtained, and by running a sequence of tests at different pressure differences the α_{av} vs. Δp relationship can be developed. From a *constant rate* test the resulting α_{av} value is an average over the range of pressure from the start to the end of the test. Constant rate test data are therefore less easy to interpret and use for process calculations, and are really only useful if the plant pump and filter are to operate at the same conditions as the test pump and filter. Experiments run under the *variable pressure-variable rate* mode also have limited use as it is not possible to extrapolate the α_{av} vs. Δp relationship from them. However, it is most important to be able to predict the performance of filter plant operating in this way, and for this the (Δp, α_{av}) data are essential. When measuring the α_{av} vs. Δp relationship, in the same experiments the variation of the ratio of the mass of wet cake to mass of dry cake m (which may be used to calculate the porosity, ε) are usually evaluated and the above discussion is equally as valid for these parameters.

4.4 Filtration tests to determine cake formation data

Filtration tests can be carried out with quite simple apparatus to determine cake formation data. The objectives of the tests are often twofold. Firstly, cake formation rate is required for preliminary equipment selection purposes, and secondly specific resistance or filterability coefficient data are needed for filter sizing and filtration rate calculations.

Figure 4.17 shows a vacuum filter apparatus that can be used to obtain the data. The drainage characteristics of the normal ceramic, perforate bottom, Buchner funnel are poor and make it unsuitable for these tests; it is preferable to use a filter funnel with a porous sintered bottom, or ideally a top-fed leaf assembly equipped with a filter cloth. To simplify

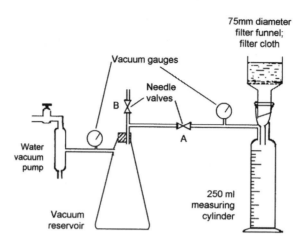

Figure 4.17 Test apparatus showing a top-fed leaf filter for obtaining vacuum filtration data.

subsequent analysis of the data it is usual to operate the apparatus at a constant pressure difference (or vacuum) (that is, at a constant pressure when one side of the filter is at atmospheric pressure) – see Section 4.3.1; if a constant flow experiment is carried out, data analysis is fairly straightforward but the pressure variation with time must also be regulated and measured during the test – see Section 4.3.2. An equivalent pressure driven apparatus is easily devised for obtaining data using higher pressure differences across the filter. More sophisticated vacuum or pressure leaf apparati can be used if a well equipped separations laboratory is already available. However, the use of complicated equipment at the early stage of slurry assessment is probably not justified, but it is wise to use the same type of filter medium in the tests as is used at full scale so that any medium effects are incorporated into the experimental results.

The general procedure to obtain the filtration data is as follows:

1. Determine the solids concentration in the slurry sample before carrying out the filter test(s).

2. With the needle valve A in the vacuum line closed, adjust the bleed needle valve B to give the required level of (constant) vacuum. This is likely to be in the region of 50 to 70 kPa (380 to 530 mm Hg). In the wastewater treatment industry the standard used is 49 kPa (386 mm Hg).

3. Pour the well stirred slurry sample into the filter funnel containing the filter cloth and open the needle valve A in the vacuum line so that a pre-set vacuum level is achieved as rapidly as possible in the

graduated cylinder. It is better not to pour the sample directly onto the cloth, but to feed it onto a perforated plate located 2 to 3 cm above the cloth. The plate acts as a distributor to (i) prevent solids approaching the pores in the cloth at an unrealistically high velocity and thereby causing unexpected plugging of the pores, and (ii) spread the feed over the full area of the funnel, enabling formation of a cake of more uniform thickness.

4. Monitor the filtration test by recording the filtration rate by measuring the volume of filtrate collected at various time intervals. The intervals between recording the measured volumes need not be constant but may be increased progressively to compensate for the gradual drop in filtrate flow rate.

5. If the cake form rate is too slow (e.g. of the order of cm h^{-1}) it may be desirable to add flocculants or pretreatment chemicals to the slurry and repeat the filtration tests as appropriate. For slow filtering slurries it may be necessary to use smaller measuring cylinders.

6. When ten or more sets of volume-time readings have been obtained fully open both valves to break the vacuum. There should be some surplus unfiltered slurry visible on top of the cake at this stage. If there is not surplus slurry lying on the cake it is likely that the cake will have started to deliquor and subsequent cake moisture measurements will be erroneously low, whilst leaving the surplus on the cake will lead to incorrectly high moisture measurements.

7. Pipette the excess slurry from the surface of the filter cake.

8. If possible, measure the thickness of the filter cake (m_{av}).

9. Remove as much of the cake as possible from the filter, weigh it, dry it and reweigh it. From these measurements calculate the ratio of the mass of wet cake to the mass of wet cake.

Example 4.3 Calculation of the cake formation rate from constant pressure test data.

The data in Table 4.11 were obtained from a leaf filter with an area of 45 cm^2 using a pressure difference of 70 kPa. The ratio of the mass of wet cake to mass of dry cake was measured and found to be 1.34, the densities of the solid and liquid phases are 2640 and 1000 kg m^{-3}, and the solids concentration in the feed slurry is 132 kg m^{-3}. Estimate the cake formation rate from these data.

Table 4.11 Time-volume data for Example 4.3

Filtration time, t (s)	Filtrate volume, V (ml)	Cake thickness[†] (cm)
0	0	0
170	141	0.324
275	200	0.460
340	230	0.529
390	252	0.580
457	285	0.656
527	320	0.736
589	341	0.784
660	370	0.851

[†]Calculated data.

Solution

The mass fraction of solids in the slurry is obtained using equation (4.36):

$$s = \frac{132}{132 + 1000\left(1 - \dfrac{132}{2640}\right)} = 0.122$$

Cake thickness is related to m, s and V in equation (4.41):

$$V = \frac{2640(1 - 1.34 \times 0.122)}{0.122[2640(1.34 - 1) + 1000]}\, 45L = 429.3L$$

i.e. $L = 0.0023V$ cm

The cake thicknesses shown in the last column in the above table are calculated from this expression. The rate of cake formation decreases with time in a constant pressure filtration; in this case it decreases from 0.0019 cm s^{-1} over the period 0 to 170 s down to 0.0013 cm s^{-1} over the period 0 to 660 s;

i.e. the cake formation rate is on the order of 0.001 cm s^{-1}.

The method of using the data to determine cake specific resistances is demonstrated in Example 4.1.

4.5 Filtration onto cylindrical elements

The equations in the previous sections have all assumed that the filter area is constant, which is strictly only true where the filter design forces the cake deposition to be uniform. If a cake is deposited on either the inner or the outer surface of a cylindrical element, the filtration area increases with the radius of the filtering surface. To allow for this the filtration equations require modification if area growth is appreciable. Brenner (1961) studied the growth of a cake in a three dimensional filtration on a circular leaf, and Shirato *et al* (1967) developed a term which they defined as the *effective filtration area factor*. For two and three dimensional filtration involving a circular or square plate of radius or half-length r, the area factors j_{II} and j_{III} were found experimentally for four different suspensions to be:

$$j_{II} = 1 + 0.801 \left(\frac{L}{r}\right)^{1.22} \tag{4.54}$$

$$j_{III} = 1 + 1.47 \left(\frac{L}{r}\right)^{1.20} \tag{4.55}$$

These area factors are incorporated as a multiplying factor on the area term in the filtration equations; for example, equation (4.24) would become:

$$\frac{1}{A}\frac{dV}{dt} = \frac{jA\Delta p}{\mu (\alpha_{av} cV + AR)} \tag{4.56}$$

where j is either j_{II} or j_{III}.

Hermia and Brocheton (1993) developed useful relationships for filtration in candle filters which have a cylindrical filtration surface; although precoat filtration was being studied, the resulting equations are equally useful during cake filtration with or without a precoat. Letting r_0, e_p and e_c be respectively the radius of the candle, the precoat thickness and the thickness of the cake produced, and letting a_p be the precoat dosage expressed as the mass of filter aid per unit filter area, a mass balance on the precoat gives:

$$a_p A_0 = N\pi h \rho_p \left[(r_0 + e_p)^2 - r_0^2\right] \tag{4.57}$$

where A_0 is the candle filter area ($=2N\pi r_0 h$), ρ_p is the bulk density of the precoat layer on the filter, h is the candle length and N the number of candles in the filter. After defining $\delta = 2/r_0\rho_p$, it follows from equation (4.57) that:

$$r_0 + e_p = r_0\sqrt{1 + \delta a_p} \qquad (4.58)$$

A mass balance on the cake produced during filtration is written as:

$$cV = N\pi h\rho_c\left[(r_0 + e_p + e_c)^2 - (r_0 + e_p)^2\right] \qquad (4.59)$$

where ρ_c is the bulk density of the cake. After defining $\delta_c = 2/r_0\rho_c$ and noting that the cake radius is $r = r_0 + e_p + e_c$, equations (4.58) and (4.59) can be combined to give:

$$r = r_0\sqrt{1 + \delta a_p + \delta_c\frac{cV}{A_0}} \qquad (4.60)$$

and consequently the available filtration area is:

$$A = A_0\sqrt{1 + \delta a_p + \delta_c\frac{cV}{A_0}} \qquad (4.61)$$

4.5.1 Constant pressure filtration

Starting from equation (4.24) and substituting equation (4.61) yields:

$$(\mu\alpha_{av}cV + \mu A_0 R)\frac{dV}{dt} = A^2\Delta p = A_0^2\left(1 + \delta a_p + \delta_c\frac{cV}{A_0}\right)\Delta p \qquad (4.62)$$

Integrating between the limits (t_i, t) and (V_i, V) at constant Δp and rearranging the results yields:

$$\frac{t - t_i}{V - V_i} = \left(\frac{K_1}{2}(V + V_i) + K_2\right)(1 + \delta a_p) +$$
$$\left(\frac{K_1}{3}(V^2 + VV_i + V_i^2) + \frac{K_2}{2}(V + V_i)\right)\frac{\delta_c c}{A_0} \qquad (4.63)$$

where $K_1 = \alpha_{av}c\mu/A_0^2\Delta p$ and $K_2 = \mu R/A_0\Delta p$. If no precoat is used, $a_p = 0$. Unlike filtration onto a planar surface where a plot of $(t - t_i)/(V - V_i)$ against $(V - V_i)$ shows a linear relationship, such a plot for filtration onto a cylindrical surface is non-linear. A plot of $(t - t_i)/$

$(V - V_i)$ against V is expected to be parabolic and a curve fitting procedure can be used to evaluate the constants K_1, K_2 and ρ_p or ρ_c when a_p and either ρ_p or ρ_c are known.

4.5.2 Constant rate filtration

The relationship between pressure and filtrate flow rate is given by equation (4.50). Writing this equation for an infinitesimal pressure increase and substituting equation (4.61) into the result gives:

$$dp = \mu \alpha_{av} c \left(\frac{q}{A}\right)^2 dt = \mu \alpha_{av} c \left(\frac{q}{A_0}\right)^2 \frac{dt}{1 + \delta a_p + \delta_c \dfrac{cV}{A_0}} \tag{4.64}$$

which, after integration between (p_0, p) over $(0, t)$ at constant flow rate, becomes:

$$p = p_0 + \frac{\mu \alpha_{av}}{\delta_c} \left(\frac{q}{A_0}\right) \ln\left(1 + \frac{\delta_c c}{1 + \delta a_p} \frac{q}{A_0} t\right) \tag{4.65}$$

If no precoat is used $a_p = 0$. Equation (4.65) indicates that, in the case of constant rate filtration onto a cylindrical surface, the pressure difference across the cake increases according to the logarithm of time.

Example 4.4 Comparison of the pressure rise during constant rate filtration, when filtration is carried out on plane and cylindrical surfaces.

The data for this problem are for beer filtration using a precoat and body feed and are given in Table 4.12 (Hermia and Brocheton, 1993). For the two candle diameters specified above and for a planar filtering surface, calculate the pressure rise with the volume of beer filtered.

Table 4.12 Data for Example 4.4

Initial pressure	p_0	10^4 Pa
Beer viscosity	μ	3×10^{-3} Pa s
Body feed dosage	c_b	1 kg m^{-3}
Precoat dosage	a_p	1.5 kg m^{-3}
Specific flow rate	q/A_0	1 m^3 m^{-2} h^{-1}
Specific resistance	α_{av}	10^{11} m kg^{-1}
Filter aid bulk density	ρ_p	350 kg m^{-3}
Candle diameter	$2r_0$	20 or 30 mm

Solution

In beer filtration the concentration of impurities in the beer is negligible in comparison with the concentration of filter aid used as a body aid, hence $c = c_b$ (the body aid concentration) and $\rho_c = \rho_p$.

For the candle filter the pressure variation is given by equation (4.65). Substituting values into this equation for the 20 mm candles gives:

$$p = 10^4 + \frac{3 \times 10^{-3} \times 10^{11}}{\delta_c} \left(\frac{1}{3600} \right) \ln \left(1 + \frac{\delta_c \times 1}{1 + \delta \times 1.5} \frac{t}{3600} \right)$$

where $\delta_c = \delta = 2/(r_0 \rho_c) = \dfrac{2}{10^{-2} \times 350} = 0.5714$. Also, noting that $\dfrac{V}{A_0} = \dfrac{q}{3600 A_0} t$, the expression between filtration pressure and filtrate volume becomes:

$$p = 10^4 + 1.46 \times 10^5 \ln \left(1 + 0.3077 \frac{V}{A_0} \right) \tag{A}$$

The equivalent expression for the 30 mm candles is:

$$p = 10^4 + 2.19 \times 10^5 \ln \left(1 + 0.2424 \frac{V}{A_0} \right) \tag{B}$$

The relationship for the planar surface, when the medium resistance term is negligible in comparison with the cake resistance term, is obtained from equation (4.50) as:

$$p = 10^4 + \mu c \alpha_{av} \left(\frac{q}{A_0} \right)^2 t = 10^4 + \mu c \alpha_{av} \left(\frac{q}{A} \right) \frac{V}{A_0}$$
$$= 10^4 + 3 \times 10^{-3} \times 1 \times 10^{11} \left(\frac{1}{3600} \right) \frac{V}{A_0} = 10^4 + 8.33 \times 10^4 \frac{V}{A_0} \tag{C}$$

The pressure rise with filtrate volume (i.e. volume of beer filtered), equations (A), (B) and (C), is plotted on Figure 4.18 for each filter type. The data show that the pressure rise is slower for the candle filters, and is slowest for the 20 mm diameter candles.

Solutions to filtration problems on candle filters when using variable pressure-variable rate require numerical techniques and are best carried out with the aid of specialist filtration solving software such as Filter Design Software (2005).

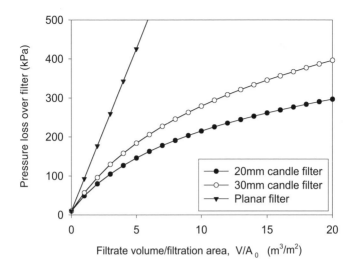

Figure 4.18 The rise of the pressure loss across candle and planar filters during constant flow rate filtration of beer, using precoat and body aid to assist the filtration.

4.6 Filtration in centrifuges

An alternative technique for applying the filtration driving force is to use a centrifugal field; the basic component of a filtering centrifuge is a rotating perforated drum into which suspension is fed. The feed is accelerated up to the speed of the drum, thereby acquiring a centrifugal force. Provided the solids have a density that is greater than the density of the liquid, they are accelerated to a higher radial velocity in the drum. The solids are retained inside of the drum whilst the liquid passes through the perforations; depending on the centrifuge design, the solids are discharged whilst the drum is either moving or stationary. The names of centrifuges are indicative of the developmental stages of the various machine types; early types were relatively simple machines that suffered from imbalance problems that result from unequal loading and had names like buffer centrifuges (foundations mounted on rubber buffers) and pendulum centrifuges (rigid mounting in a housing which is hung like a pendulum). The next generation names were generally associated with the ways that solids were transported or conveyed through the drum, like the peeler centrifuge (which utilises a knife to discharge the cake whilst the drum is rotating at high speed), the worm screen centrifuge (where solids are transported using a spiral mounted in the drum), or the pusher centrifuge (where solids are pushed through the drum). These were followed by other centrifuges such as screen decanters that combine sedimentation and filtration in the single unit, or so-called "siphon centrifuges" that add a siphon effect to increase the filtration pressure and therefore increase capacity. There are many other centrifugal filters besides these; two types of centrifugal filter are shown in Figures 4.19 and 4.20 – an inverting bag centrifuge and a horizontal peeler centrifuge.

Fundamentals of the theory of fluid flow through porous media have not been used generally for design calculations, but reliance has been placed on practical experience and empirical approaches. Nonetheless, Haruni and Storrow (1952a, b; 1953a, b) described the experimental and analytical techniques developed to assess the validity of both the flow mechanism and the equations previously proposed for centrifugation by Burak and Storrow (1950). They checked the flow rate equation by testing the effect of major variables and by observing the applicability of functions derived from the flow equations. They also presented permeability data obtained from cakes formed and tested in a small cell acting as part of a centrifuge basket under both centrifugal and hydrostatic heads. Filtration permeabilities were found for samples cut from the centrifuge. The results show a maximum difference of 20%, a figure later confirmed by Valleroy and Maloney (1960). Oyama and Sumikawa (1954) also confirmed that specific resistances obtained in centrifuges were larger than those obtained by constant pressure filtration, under the same formation pressure. Rushton and Spear (1975) demonstrated that specific resistances obtained from permeation experiments were inadequate for the prediction of cake form or load times.

The following sections consider a simple centrifuge geometry, such as would be associated with a peeler centrifuge. The axis of rotation of the basket may be either horizontal or vertical.

4.6.1 Centrifugation rate and filtrate flow rate

Consider a differential cylindrical element of filter cake between radii r and $r + dr$, as shown in Figure 4.21. The area available for filtrate flow is:

$$A = 2\pi r h \tag{4.66}$$

and the mass of solids contained in the element is:

$$dM_s = \rho_s(1-\varepsilon_{av})2\pi r h dr \tag{4.67}$$

The pressure drop in the fluid caused by frictional effects as the liquid flows through the element is given by Darcy's law (equation (4.18)):

Figure 4.19 Inverting bag centrifuge; a high product yield is produced with thick or thin cakes and cake discharge is effected by the cloth being turned inside out whilst the bowl is rotating, reducing the likelihood of filter cloth blinding. This type of centrifuge can meet clean room requirements, be totally enclosed and gas tight to maintain an inert atmosphere (Heinkel).

Figure 4.20 The horizontal peeler centrifuge has developed into a validatable filter for operation in the ultra-clean conditions of the pharmaceutical industry. The front end casing is shown in its open position that facilitates safe operator inspection of the interior and to ensure total cleanliness with no dead areas (Thomas Broadbent & Sons).

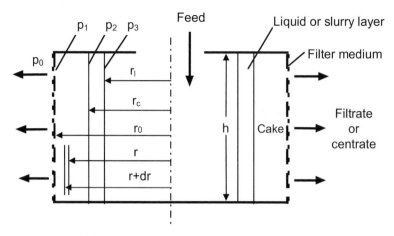

Figure 4.21 Diagrammatic cross-section through a centrifuge basket showing the filter medium, cake and liquid layers, and defining the notation used for the analysis of centrifugal filtration.

$$dp_l = \mu\alpha_{av}\frac{q}{A}\frac{dM_s}{A} = \frac{q\mu\rho_s\alpha_{av}(1-\varepsilon_{av})}{2\pi h}\frac{dr}{r} \tag{4.68}$$

The pressure due to a centrifugal body force acting on the liquid flowing through the cake is:

$$dp_r = \rho_l\omega^2 r dr \tag{4.69}$$

The actual pressure differential over the element of cake is the difference between the centrifugal body force and the frictional force:

$$-dp = dp_l - dp_r \tag{4.70}$$

Therefore, over the centrifuge cake from radius r_c to r_0 the pressure difference is:

$$-\int_{p_2}^{p_1} dp = \frac{q\mu\rho_s\alpha_{av}(1-\varepsilon_{av})}{2\pi h}\int_{r_c}^{r_0}\frac{dr}{r} - \rho_l\omega^2\int_{r_c}^{r_0} r dr \tag{4.71}$$

i.e. $\quad -(p_1 - p_2) = \frac{q\mu\rho_s\alpha_{av}(1-\varepsilon_{av})}{2\pi h}\ln\left(\frac{r_0}{r_c}\right) - \frac{\rho_l\omega^2}{2}\left(r_0^2 - r_c^2\right)$

The frictional pressure loss in the filter medium as filtrate passes through is obtained from Darcy's law as:

$$p_1 - p_0 = \frac{q\mu R}{2\pi r_0 h} \tag{4.72}$$

The hydraulic pressure arising from the existence of the supernatant liquid (or slurry) layer is:

$$p_2 - p_3 = \rho_l \omega^2 \int_{r_l}^{r_c} r dr = \frac{\rho_l \omega^2}{2} \left(r_c^2 - r_l^2 \right)$$

(4.73)

Combining equations (3.72) and (3.73) and assuming that the pressure inside the centrifuge basket is the same as that on the outside (that is, $p_3 = p_0$) leads to:

$$-\left(p_1 - p_2 \right) = \left(p_2 - p_3 \right) + \left(p_0 - p_1 \right) = \frac{\rho_l \omega^2}{2} \left(r_c^2 - r_l^2 \right) - \frac{q \mu R}{2 \pi r_0 h}$$

(4.74)

Setting equation (4.71) equal to equation (4.74) gives, after rearrangement, an expression for the filtrate volumetric flow rate flowing from the centrifuge:

$$q = \frac{\dfrac{\rho_l \omega^2}{2} \left(r_0^2 - r_l^2 \right)}{\dfrac{\mu \rho_s \alpha_{av} (1 - \varepsilon_{av})}{2 \pi h} \ln \left(\dfrac{r_0}{r_c} \right) + \dfrac{\mu R}{2 \pi r_0 h}}$$

(4.75)

It is often more convenient to write this expression in terms of the mass of solids filtered, M_s. The mass of solids in the cake is:

$$M_s = \rho_s (1 - \varepsilon_{av}) \pi h \left(r_0^2 - r_c^2 \right)$$

(4.76)

Substituting for $\rho_s (1 - \varepsilon_{av})$ in equation (4.75) gives:

$$q = \frac{\dfrac{\rho_l \omega^2}{2} \left(r_0^2 - r_l^2 \right)}{\alpha_{av} \mu M_s \dfrac{\ln (r_0/r_c)}{2 \pi h \left[\pi h \left(r_0^2 - r_c^2 \right) \right]} + \dfrac{\mu R}{2 \pi r_0 h}}$$

(4.77)

The denominator of this expression can be rewritten in terms of an average flow area in the cake (A_{av}), a logarithmic mean area of flow (A_{lm}) and the bowl or medium area (A_0) by noting (Rushton et al, 1996):

$$\frac{2 \pi h \left[\pi h \left(r_0^2 - r_c^2 \right) \right]}{\ln \left(r_0/r_c \right)} = \frac{2 \pi h (r_0 + r_c)}{2} \frac{2 \pi h (r_0 - r_c)}{\ln \left(r_0/r_c \right)} = A_{av} A_{lm}$$

(4.78)

Hence, the filtrate volumetric flow rate is written as:

$$q = \frac{\dfrac{\rho_l \omega^2}{2}\left(r_0^2 - r_i^2\right)}{\dfrac{\alpha_{av}\mu M_s}{A_{av} A_{lm}} + \dfrac{\mu R}{A_0}} \equiv \frac{\Delta p}{\dfrac{\alpha_{av}\mu M_s}{A_{av} A_{lm}} + \dfrac{\mu R}{A_0}} \tag{4.79}$$

This equation suggests that plots of t/V vs. M_s (or V) can be used to estimate cake and medium resistances during centrifugal filtration, as is the case in vacuum and pressure filtration.

From equation (4.68) onwards it is assumed that all slurry fed to the centrifuge maintains a constant solids concentration, which is the same as that in the feed, whilst the cake is being formed. In reality, because the solids are usually denser than the liquid, the particles sediment in the centrifugal field and attain local velocities higher than that of the surrounding liquid phase. In a batch centrifuge, in which either a low permeability cake is formed or in which the solids are large and/or very dense, sedimentation can lead to the formation of a clear supernatant liquid layer which permeates through the formed cake. Cake formation by the combined effects of filtration and sedimentation have been modelled by Wakeman (1994). The model was able to predict the growth and decay of a clear liquid layer along with the cake growth and centrifugation rate. A more advanced approach to modelling the cake formation problem in centrifuges, accompanied by laboratory measurement procedure, has been proposed by Chan et al (2003).

4.6.2 Cake thickness dynamics (Alt, 1985)

Frequently it is important in the operation of a centrifuge to know the time when the inner radius of the liquid phase has reached a maximum height which is given by the geometry of the basket, particularly by the height of the lip ring through which the cake is discharged down a chute.

As equation (4.75) shows, the filtrate volume flow decreases with the rise of cake in the bowl as well as with the increase of the radius of the liquid phase, r_l. Both are connected to the feed rate and the solids ratio in the feed suspension. To understand the cake dynamics it is essential to develop equation (4.75) in a time dependent form.

From a mass balance in the bowl it may be shown that:

$$2\pi r h \frac{dr}{dt} = q_0 - q \tag{4.80}$$

where the left hand side represents the increase in the volume of the liquid and solid phases in the bowl, and the right hand side represents the feed volume flow q_0 and the filtrate flow q.

The relationship between the cake radius and the centrifuging time may be developed from the volumetric balance that equates the solids in the centrifuge feed to the increase of the cake volume in the basket:

$$\pi(1-\varepsilon_{av})\left(r_o^2 - r_c^2\right)h = qt\frac{V_s}{1-V_s} \tag{4.81}$$

where V_s is the solids volume fraction in the feed suspension. The product qt is the filtrate volume produced at time t. The cake radius is therefore:

$$r_c = \left(r_0^2 - \frac{V_s}{\pi h(1-\varepsilon_{av})(1-V_s)}qt\right)^{0.5} \tag{4.82}$$

Equation (4.75) becomes, after substitution of equation (4.80):

$$\frac{dr}{dt} = \frac{1}{2\pi rh}\left(q_0 - \frac{\pi h \rho_l \omega^2 \left(r_0^2 - r^2\right)}{\mu\left(\rho_s \alpha_{av}(1-\varepsilon_{av})\ln\left(\dfrac{r_0}{r_c}\right) + \dfrac{R}{r_0}\right)}\right) \tag{4.83}$$

Equations (4.82) and (4.83) can be used to calculate the movement of the cake and slurry surfaces respectively, for various process conditions.

Similar theoretical analyses are available on the performance characteristics of pusher centrifuges (Hallit, 1975; Zeitsch, 1977).

4.7 Filtration from suspensions of non–Newtonian liquids

In the previous sections it has been assumed that the liquid phase of the feed suspension to the filter (or the filtrate) exhibits Newtonian rheological properties, indicated by a linear and proportional dependence of the shear stress on the shear rate. In practice it is not uncommon for the liquid to display non-Newtonian characteristics. Constant pressure filtration of suspensions composed of "power law" non-Newtonian liquids has been studied by Kozicki (1990), and

Shirato *et al* (1977; 1980a, b) who showed that the specific resistance can vary significantly with the flow behaviour index for the liquid. A "power law" fluid obeys the equation:

$$\tau = K\dot{\gamma}^N \tag{4.84}$$

where τ is the shear stress, K the fluid consistency index, $\dot{\gamma}$ the shear rate and N the flow behaviour index. The working relationship for compressible cake filtration from a power law type fluid (equivalent to equation (4.18) for a Newtonian fluid) is:

$$\left(\frac{1}{A}\frac{dV}{dt}\right)^N = u^N = \frac{1}{K\alpha^*}\frac{dp_l}{dw} \tag{4.85}$$

where α^* is the local specific resistance when filtering from a power law fluid. If an average value of specific resistance can be defined in similar manner as for a Newtonian fluid, when the medium resistance term is neglected the "power law fluid equivalent" to equation (4.24) is:

$$\left(\frac{1}{A}\frac{dV}{dt}\right)^N = \frac{A\Delta p}{K\alpha_{av}^* cV} \tag{4.86}$$

where $c = \rho_l s/(1 - m_{av}s)$.

In contrast to the filtration of Newtonian slurries, the feed slurry concentration has relatively little effect on the rates of filtration, even for concentrated slurries of compressible materials. As the value of N decreases, the solids compressive pressure in the cake becomes greater. Consequently, a cake formed by filtration of a pseudoplastic fluid has a denser structure than that formed from a Newtonian fluid (Shirato *et al*, 1987). An alternative approach for incompressible filtrations, based on the Ergun equation and relating the friction factor to the Reynolds and Hedstrom numbers, has been proposed by Chase and Dachavijit (2003).

4.8 Process calculations

The preceeding sections also form the basis of many process calculations related to filtration, and a few example calculations are given in this section to illustrate use of the various equations and interactions between the concepts.

Example 4.5 Calculation of the filtration time when the pump curve provides for filtration approximated by constant rate followed by constant pressure operation.

A slurry is to be filtered in a pressure filter at a constant rate until the pressure reaches 300 kPa. This pressure is then kept constant for the remainder of the filtration. The constant rate period lasts for 1200 s, and half of the total filtrate volume is collected during this period. If the resistance of the filter medium (R) is low compared with that of the cake formed (R_c), estimate the total filtration time. Assume the compressibility of the cake to be negligible.

Solution

Let V_T be the total volume of filtrate collected.

During the constant pressure part of the filtration equation (4.43) is applicable:

$$\frac{t-t_i}{V-V_i} = \frac{K_1}{2}(V+V_i) + K_2$$

(A)

i.e. $\quad \dfrac{t-1200}{V_T - 0.5V_T} = \dfrac{K_1}{2}(V_T + 0.5V_T) + K_2$

where $K_1 = \alpha c \mu / A^2 \Delta p$ and $K_2 = \mu R'/A\Delta p$. The resistance R' is actually the resistance of the medium plus the cake deposited during the constant rate period, hence noting equation (4.7)

$$R' = R + R_c = R + \alpha w_{CR}$$

where w_{CR} is the mass of dry solids (per unit area) deposited during the constant rate period. Using equation (4.8),

$$w_{CR} = \frac{c}{A}\frac{V_T}{2}$$

and noting $R \ll R_c$ it follows that

$$R' \approx R_c = \frac{\alpha c}{A}\frac{V_T}{2}$$

and $K_2 = \mu R'/A\Delta p = (\alpha c \mu/A^2\Delta p)(V_T/2) = K_1(V_T/2)$. Substituting the results into equation (A) gives

$$\frac{t-1200}{0.5V_T} = 0.75K_1V_T + K_1\frac{V_T}{2} = 1.25K_1V_T$$

(B)

hence $t - 1200 = 0.625K_1V_T^2$

Using equation (4.49) for *the constant rate period*

$$\Delta p = \frac{\mu c \alpha}{A^2} \frac{V^2}{t} + \frac{\mu R}{A} \frac{V}{t}$$

where Δp is a function of the filtration time t and is written as $\Delta p(t)$. Rearrangement of this equation leads to

$$V^2 + \frac{AR}{\alpha c} V - \frac{A^2 \Delta p(t)}{\alpha c \mu} t = 0$$

Substituting $V = V_T/2$, $t = 1200$ and $\Delta p(t) = \Delta p$ gives

$$V_T^2 + \frac{2AR}{\alpha c} V_T - 4800 \frac{A^2 \Delta p}{\alpha c \mu} = 0 \qquad (C)$$

In equation (C), $\mu R/A\Delta p$ can be interpreted as K_2. Substituting the above results into equation (C) gives

$$V_T^2 + 2\frac{K_2}{K_1} V_T - \frac{4800}{K_1} = 0 \qquad (D)$$

During the constant rate period, since $R \ll R_c$ we may assume that $K_2 \ll K_1$ and equation (D) further simplifies to

$$V_T^2 = \frac{4800}{K_1} \qquad (E)$$

Combining equations (B) and (D) gives

$$t - 1200 = 0.625 \times 4800 = 3000$$

and the total filtration time (t) is estimated as 4200 seconds (1.17 h).

Note: If the cake were compressible, K_1 would not be constant, and K_2 in the constant pressure part of the filtration would also be dependent on the applied pressure.

Example 4.6 Calculation of the filtration time when a centrifugal pump is used to feed the filter.

Estimate the time taken to produce 40 tonnes of dry filter cake using a plate and frame filter press with a filtration area of 220 m². The feed slurry contains 15% by mass of solids. The filter cake resistance may be represented by

$$\alpha_{av} = \alpha_0 (1-n)\Delta p_c^n = 1.18 \times 10^8 \Delta p_c^{0.92}$$

where Δp is measured in kPa. The filter medium resistance is 3×10^{11} m^{-1}, the cake wetness (m_{av}) is 1.45, the liquid viscosity and density are 10^{-3} Pa s and 1100 kg m^{-3}, respectively. The centrifugal pump has the characteristics given in Table 4.13. Assume that a flow of < 0.005 m^3 s^{-1} is an unacceptably slow rate of filtration.

Table 4.13 The pump characteristics for the centrifugal pump used in Example 4.13

Pressure (kPa)	Flow rate (m³ s⁻¹)
360	0
300	0.022
250	0.03
200	0.035
150	0.037
100	0.0385
50	0.0395
0	0.04

Solution

Use equation (4.2) to calculate the pressure loss over the filter medium:

$$p_1 - p_0 = \frac{\mu R}{A}\frac{dV}{dt} = \frac{0.001 \times 3 \times 10^{11}}{220}\frac{dV}{dt} = 1.36 \times 10^6 \frac{dV}{dt} \qquad (A)$$

Plot this equation on the pump curve; the intersection of the two curves (Figure 4.22) indicates that 54 kPa must be exceeded to create flow through the filter medium. On the pump curve, prepare the possible range of flow rates and pressures as in Table 4.14.

Table 4.14 Calculation of the pressure and specific resistance variations with flow rate for Example 4.6

dV/dt (m³ s⁻¹)	p (kPa) (from the pump curve)	p – p₁ (kPa) (from the pump curve)	α_{av} (m kg⁻¹) (from equation (B))
0.0395	54	0	–
0.035	200	152	1.20×10^{10}
0.03	250	209	1.61×10^{10}
0.025	283	249	1.89×10^{10}
0.02	307	280	2.11×10^{10}
0.015	325	305	2.28×10^{10}
0.01	340	326	2.42×10^{10}
0.005	351	344	2.54×10^{10}

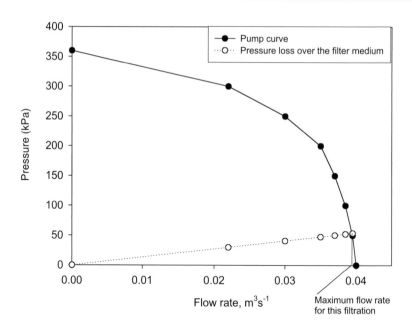

Figure 4.22 The pump curve and the pressure loss over the filter medium plotted for Example 4.6. The difference $(p - p_1)$ is the pressure loss over the filter cake (Δp_c), and $p \equiv \Delta p$ the pressure loss over the filter.

Now, $c = \dfrac{\rho s}{1 - ms}$ (see equation (4.13)) $= \dfrac{1100 \times 0.15}{1 - 1.45 \times 0.15} = 210.9 \ \text{kg m}^{-3}$

Using a rearranged form of equation (4.51):

$$\therefore \quad \Delta p = \frac{\alpha_{av} \times 0.001 \times 210.9}{220^2} V \frac{dV}{dt} + 1.36 \times 10^6 \frac{dV}{dt}$$

$$\therefore \quad V = \left(\frac{\Delta p}{dV\big/dt} - 1.36 \times 10^6 \right) \frac{2.29 \times 10^5}{\alpha_{av}}$$

(B)

Using Table 4.14 and equation (B), calculate the relationship between dt/dV and V as in Table 4.15.

The volume of filtrate that will be collected during the recovery of 40 tonnes of dry solids is

$$V = \frac{40 \times 10^3}{210.9} = 189.7 \ \text{m}^3$$

Using Table 4.15, plot dt/dV against V (Figure 4.23) and obtain the area under the curve to evaluate the integral in equation (4.53). From the area, the filtration time is 6045 s (1.68 hours).

Table 4.15 Calculation of the filtrate volume collected from the filter using equation (C)

$\dfrac{dV}{dt}$ $(m^3\ s^{-1})$	p (kPa)	$\dfrac{\Delta p}{dV/dt}$ $(Pa\ s\ m^{-3})$	$\dfrac{2.29\times10^5}{\alpha_{av}}$	V (m^3)	$\dfrac{dt}{dV}$ $(s\ m^{-3})$
0.0395	54	1.37×10^6	–	0	25.3
0.035	200	5.71×10^6	1.91×10^{-5}	83.1	28.6
0.03	250	8.33×10^6	1.42×10^{-5}	99.0	33.3
0.025	283	11.32×10^6	1.21×10^{-5}	120.5	40.0
0.02	307	15.35×10^6	1.09×10^{-5}	152.5	50.0
0.015	325	21.67×10^6	1.0×10^{-5}	203.1	66.7
0.01	340	34.00×10^6	9.46×10^{-6}	308.8	100.0
0.005	351	70.20×10^6	9.02×10^{-6}	620.9	200.0

Example 4.7 Estimation of the effect of a change of process conditions on filtration time.

A filter press fed with a suspension at constant pressure discharges 0.5 m³ of filtrate in 30 minutes and 1.5 m³ in the next 2.5 hours, at which time the press is full. An alteration to the process conditions upstream of the filter results in a change of particle size and shape, which causes the particle specific surface to change from 2.5×10^5 m^{-1}

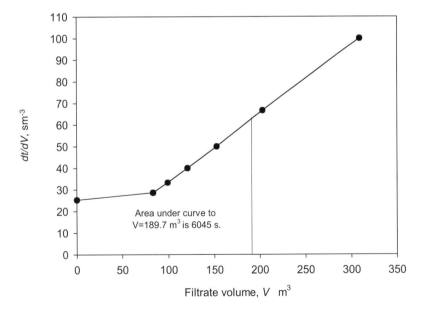

Figure 4.23 Obtaining the filtration time from Table 4.15.

to 3×10^5 m^{-1} and the porosity of the cakes formed to increase from 0.5 to 0.7. The density of the liquid and the solids is 1000 kg m^{-3} and 2500 kg m^{-3} respectively. The mass fraction of solids in the feed is 0.1. Estimate the time taken to fill the press with the new feed.

Solution

Using the constant pressure filtration equation (4.43):

$$\frac{t - t_i}{V - V_i} = \frac{K_1}{2}(V + V_i) + K_2$$

where $K_1 = \alpha_{av} c \mu / A^2 \Delta p$ and $K_2 = \mu R / A \Delta p$.

Between $t_i = 0$ and $t = 30 \times 60 = 1800$ s:

$$\frac{1800}{0.5} = \frac{K_1}{2} 0.5 + K_2 \qquad \text{(A)}$$

i.e. $3600 = 0.25 K_1 + K_2$

Between $t_i = 1800$ and $t = 3 \times 3600 = 10800$ s:

$$\frac{10800 - 1800}{2 - 0.5} = \frac{K_1}{2}(2 + 0.5) + K_2 \qquad \text{(B)}$$

i.e. $6000 = 1.25 K_1 + K_2$

Solve equations (A) and (B) simultaneously to give the filtration equation relevant *before* the change of process conditions, and to give $K_1 = 2400$ s m^{-6} and $K_2 = 3000$ s m^{-3}, then

$$\frac{t - t_i}{V - V_i} = 1200(V + V_i) + 3000 \qquad \text{(C)}$$

Define the total capacity of the press as ϕ m^3 (say), then the volume of solids held in the press is $(1 - \varepsilon_{av})\phi$. The volume of solids fed to the press is cV/ρ_s m^3, hence

$$\phi = \frac{cV}{(1 - \varepsilon_{av})\rho_s}$$

The total press capacity is unchanged, so

$$\phi|_b = \phi|_a$$

i.e. $$\frac{cV}{1 - \varepsilon_{av}}\bigg|_b = \frac{cV}{1 - \varepsilon_{av}}\bigg|_a$$

$$\therefore \qquad V_a = \frac{cV}{1 - \varepsilon_{av}}\bigg|_b \frac{1 - \varepsilon_{av}}{c}\bigg|_a$$

where the subscripts a and b indicate 'after' and 'before' the changes.
Now,

$$c_b = \frac{\rho s}{1 - m_b s}$$

$$m_b = 1 + \frac{0.5}{1 - 0.5} \frac{1000}{2500} = 1.4 \quad \text{equation (4.39)}$$

$$\therefore \quad c_b = \frac{1000 \times 0.1}{1 - 1.4 \times 0.1} = 116.3 \text{ kg m}^{-3}$$

Similarly,

$$m_a = 1 + \frac{0.7}{1 - 0.7} \frac{1000}{2500} = 1.93$$

$$c_a = \frac{1000 \times 0.1}{1 - 1.93 \times 0.1} = 123.9 \text{ kg m}^{-3}$$

$$\therefore \quad V_a = \frac{116.3 V_b}{1 - 0.5} \frac{1 - 0.7}{123.9} = 0.563 V_b$$

when the press is full.

K_1 is affected by the change in conditions, since both α_{av} and c are altered. Putting

$$K_1^* = \frac{K_1}{\alpha_{av} c}$$

K_1^* is unaffected by the changes. Using equation (4.31) with (1.2) gives:

$$\alpha_{av} = \frac{5 S_0^2 (1 - \varepsilon_{av})}{\rho_s \varepsilon_{av}^3}$$

$$\therefore \quad \alpha_a = \frac{5(3 \times 10^5)^2 0.3}{2500 \times 0.7^3} = 1.57 \times 10^8 \text{ m kg}^{-3}$$

$$\text{and } \alpha_b = \frac{5(2.5 \times 10^5)^2 0.5}{2500 \times 0.5^3} = 5.0 \times 10^8 \text{ m kg}^{-3}$$

Rewrite the filtration equation (C) for the changed slurry:

$$\frac{t - t_i}{V - V_i} = \frac{K_1^* (\alpha_{av} c)_a}{2} (V + V_i) + K_2$$

$$= \frac{K_1^*}{2} (\alpha_{av} c)_b \frac{(\alpha_{av} c)_a}{(\alpha_{av} c)_b} (V + V_i) + K_2$$

$$= \frac{K_1 \left(\alpha_{av} c \right)_a}{2 \left(\alpha_{av} c \right)_b} \left(V + V_i \right) + K_2$$

$$= \frac{2400}{2} \frac{1.57 \times 10^8 \times 123.9}{5 \times 10^8 \times 116.3} \left(V + V_i \right) + 3000$$

$$= 401.4 \left(V + V_i \right) + 3000$$

Putting $V_i = 0$ at $t_i = 0$, and $V = 0.563 \times 2 = 1.126$ m^3 at $t = t$ (that is, when the press is full):

$$t = 401.4 (1.126)^2 + 3000(1.126) = 3887 \text{ s}$$

that is, the press is full in 1.08 hours when using the new feed.

4.8.1 Optimisation of cycle times

Mechanization of pressure filters has led to shorter discharge times and this can have a major impact on the overall economy of the process. Generally, the rate of filtration is a maximum at the start and decreases progressively, and a decision must be taken as to when the filter cycle should be ended and the cake discharged. The average filtration rate must be calculated on the basis of total time and the overall production rate (Y) can be written as:

$$Y = \frac{\text{total filtrate (mass or volume) collected}}{\text{total cycle time}} \tag{4.87}$$

where the total cycle time includes the filter down time (t_{dt}), cake deliquoring time, cake washing time, and so on. If the filtration is at a constant pressure, equation (4.43) (setting $V_i = 0$) can be used and equation (4.87) becomes:

$$Y = \frac{V}{t + t_{dt}} = \frac{V}{0.5 K_1 V^2 + K_2 V + t_{dt}} \tag{4.88}$$

Differentiating this expression and putting $dY/dV = 0$ yields the volume of filtrate to be collected that corresponds to an optimum production rate:

$$\frac{dY}{dV} = \frac{\left(0.5 K_1 V^2 + K_2 V + t_{dt} \right) - V \left(K_1 V + K_2 \right)}{\left(0.5 K_1 V^2 + K_2 V + t_{dt} \right)^2} = 0 \tag{4.89}$$

That is, the optimum filtrate volume to be collected is

$$V_{opt} = \sqrt{\frac{2t_{dt}}{K_1}}$$

(4.90)

Substituting this into equation (4.43) gives an optimum time for cake formation during a constant pressure filtration:

$$t_{opt} = t_d + K_2 \sqrt{\frac{2t_{dt}}{K_1}}$$

(4.91)

If the filter medium resistance is neglected (that is, $K_2 = 0$) the cake form time is exactly equal to the total period for which the filter is carrying out its non-cake forming duties.

The clearance between leaves in a leaf filter or the frame thickness in a filter press may be insufficient to allow the handling of as much filtrate as V_{opt} in one cycle, hence it is useful to estimate the cake thickness L_{opt} corresponding to V_{opt}:

$$L_{opt} = \frac{cV_{opt}}{(1 - \varepsilon_{av}) \rho_s A}$$

(4.92)

Similar calculation steps can be followed to evaluate optimum conditions for constant rate and variable rate-variable pressure filtrations.

Example 4.8 Calculation of maximum plant capacity.

Calculate the maximum plant capacity for the filtration specified below. Assume a cake washing (see Chapter 6 for Cake Washing) pressure equal to the filtration pressure and a down time of 900 s. Take the volume of wash used to be 0.05 times the volume of filtrate produced. The required data are:

Filter area	32 m^2
Average specific resistance of the cake	5×10^{10} m kg^{-1}
Viscosity	0.001 Pa s
Filtration pressure	100 kPa
Medium resistance	10^{11} m^{-1}
Cake moisture, m_{av}	1.4
Feed solids concentration	300 kg m^{-3}
Solids density	2400 kg m^{-3}
Liquid density	1000 kg m^{-3}

Solution

Calculate the solids concentration according to equation (4.36):

$$s = \frac{\text{mass of solids}}{\text{mass of slurry}} = \frac{\dfrac{c'}{\rho_s}}{\dfrac{\rho}{\rho_s}\left(1-\dfrac{c'}{\rho_s}\right)+\dfrac{c'}{\rho_s}} = \frac{\dfrac{300}{2400}}{\dfrac{1000}{2400}\left(1-\dfrac{1000}{2400}\right)+\dfrac{300}{2400}} = 0.255$$

$$c = \frac{\rho s}{1-m_{av}s} = \frac{1000\times0.255}{1-1.4\times0.255} = 396.6 \ \text{kg m}^{-3}$$

Calculate the filtration parameters:

$$\frac{t-t_i}{V-V_i} = \frac{K_1}{2}(V+V_i)+K_2$$

$$K_1 = \frac{\alpha_{av}\mu c}{A^2\Delta p} = \frac{5\times10^{10}\times0.001\times396.6}{32^2\times10^5} = 194 \ \text{s m}^{-6}$$

$$K_2 = \frac{\mu R}{A\Delta p} = \frac{0.001\times10^{11}}{32\times10^5} = 31.25 \ \text{s m}^{-3}$$

Calculate the filtration time (putting $t_i = V_i = 0$):

$$t = \frac{194}{2}V^2 +31.25V = 97V^2 +31.25V$$

Calculate the washing time t_w (noting that in a filter press during washing, area = 0.5×filter area):

$$\frac{dV_w}{dt_w} = \frac{1}{4K_1V_w+2K_2} = \frac{1}{776V_w+62.5}$$

hence

$$t_w = \frac{776}{2}V_w^2 +62.5V_w$$

$$= \frac{776}{2}(0.05V)^2 +62.5(0.05V)$$

$$= 0.97V^2 +3.125V$$

Calculate the production rate:

$$t_{dt} = 900 \ \text{s}$$

Production rate, $Y = \dfrac{\text{total filtrate collected}}{\text{total cycle time}}$

$$= \frac{V}{t + t_w + t_{dt}}$$

$$= \frac{V}{97.97V^2 + 34.375V + 900}$$

The optimum occurs when $dY/dV = 0$, and it then follows that

$$V_{opt} = \sqrt{\frac{900}{97.97}} = 3.03 \ \text{m}^2$$

Calculate the cycle time and daily productivity:

Optimum cycle time, $t_c = t + t_w + t_{dt} = 97.97V_{opt}^2 + 34.375V_{opt} + 900$

$$\therefore \ t_c = 1904 \ \text{s}$$

Number of cycles per day $= \dfrac{24 \times 3600}{1904} = 45$

Hence, the optimum production capacity $= 3.03 \times 45 = 136 \ \text{m}^3 \ \text{day}^{-1}$, or $136 \times 396.6/1000 = 54$ tonne dry solids day^{-1}.

Example 4.9 Calculation of the most economic frame thickness.

Calculate the most economic frame thickness for the filtration in Example 4.8.

Solution

Using equation (4.92), and noting that filtration occurs on two surfaces in each frame of a plate and frame press:

$$L_{opt} = \frac{cV_{opt}}{(1-\varepsilon_{av})\rho_s \dfrac{A}{2}} = \frac{396.6 \times 3.03}{(1-\varepsilon_{av})2400 \dfrac{32}{2}} = \frac{0.0313}{(1-\varepsilon_{av})} \ \text{m}$$

Calculate the average porosity of the cake from its moisture content using equation (4.39) rearranged to:

$$\varepsilon_{av} = \frac{\rho_s(m-1)}{\rho + \rho_s(m-1)} = \frac{2400 \times 0.4}{1000 + 2400 \times 0.4} = 0.4898$$

Hence,

$$L_{opt} = \frac{0.0313}{1-0.4898} = 0.0613 \ \text{m}$$

The most economic frame thickness is ≈ 61 mm.

4.9 Advanced models of cake filtration

Although it is not the primary purpose of this book to present advanced models that have not gained acceptance for design purposes, a short discussion about some of these follows to place then into context with respect to the established models used in this chapter.

There are several other models in the literature that have been developed to model compressible cake formation during filtration. Smiles (1970) used the approach developed by Philip (1968), which was one of the first models based on the differential equations of continuity and momentum. A Lagrangian coordinate system and a similarity transformation were used to convert the governing equations to ordinary differential equations. Atsumi and Akiyama (1975) and Tosun (1986) criticised the boundary condition at the cake-slurry interface used by Smiles (1970), as it indicated no flux at this interface. Atsumi and Akiyama (1975) showed that the boundary condition at the cake-slurry interface may be valid for expression processes. Lagrangian material coordinates were also used by Kos (1975) who developed an expression for changes of the effective stress within a filter cake; as with Smiles (1970), the inertial terms in the solid and liquid phase force balance equations were neglected.

Atsumi and Akiyama (1975) used a moving boundary condition at the growing cake surface, and used a similarity transformation to solve the equations. The solution procedure forced the numerical model to be valid only for those filtrations in which the average porosity of the cake was time independent. Wakeman (1978) developed an Eulerian model with a moving boundary condition; by minimising the variance between porosity profiles through the cake as predicted by the model and as measured by experiment, a good fit between theory and experiment was obtained. Tosun (1986) developed an alternative moving boundary condition and used a solution procedure which eliminated the need to make a similarity transformation.

A common feature of these models, in spite of their advanced nature, is that they do not recognise interparticle interactions, and if they have any validity it is only for systems of large particles. In these cases, particles closer to the filter septum feel the presence of others as a result of hydrodynamic drag acting on all the particles farther from the septum. The formulation of these models does not include repulsive forces that act through the liquid phase, and as such none include a complete force balance for the solids phase. In some cases the model has been applied to "compressible" solid/liquid mixtures by use of empirical power law expressions that relate the permeability (or sometimes, specific resistance) of the cake to an applied pressure. Although these types of expressions can be helpful in design once the basic properties of the suspension filtration have been measured (indeed they are used in this book), they have no fundamental validity and arise only because of shortcomings in the original theories used to describe filtration.

More recent models developed by Auzerais *et al* (1988), Buscall and White (1987) and Landman *et al* (1991, 1993, 1995) attempted to overcome the theoretical deficiencies by using the concept of the particles being networked and defining a gel point at which the suspension becomes fully networked. This led to the definition of a compressive yield stress which is an explicit function of the volume fraction of solids in the solid-liquid mixture. It was further assumed that the yield stress was an implicit function of the strength of the interparticle bridging force and the shear history of the system during the flocculation process. The corollary is that the yield stress is zero at solid volume fractions below the gel point (Auzerais *et al* (1988) used the volume packing at close packing rather than the gel point). It is difficult to envisage the occurrence of a yield stress as a sharp demarkation identifying the onset of strength in a solid-liquid system. Like the empirical power law expressions used throughout this chapter, formulations involving the yield stress concept take no cognisance of the properties of the solid-liquid system and the stress also has to be obtained empirically.

As well as model development, it is important to have measurement techniques that realistically represent the filtration process and simultaneously enable the design parameters to be deduced from the model. Many measurement methods have been developed that extend current ones, but some newer ones are emerging. Some explain how cake permeability and compressibility can be obtained from cake filtration experiments (Green *et al*, 1998; Lu *et al*, 1998a, b; de Kretser *et al*, 2001; Usher *et al*, 2001; Chu *et al*, 2003; Couturier *et al*, 2003; Andersen *et al*, 2004; Scales *et al*, 2004). Care has to be taken if adopting these methods for industrial design, as many have not been evaluated in the practical context of design and scale-up but have relied on comparison with other models to establish their validity or on evaluation against specific feed types that may not be representative of many others (see the comment in Section 4.3.1.2 for an example of this problem.) In some cases the validity of the other models themselves have not been verified in a practical sense at pilot or full scale.

4.9.1 Constitutive properties

Theories relevant to both sedimentation (Chapter 3) and cake formation have been developed in the sedimentation literature (Fitch, 1966; Dixon, 1979; Tiller and Khatib, 1984; Buscall and White, 1987; Auzerais *et al*, 1988, 1990; Landman *et al*, 1988; Howells *et al*, 1990; Landman and White, 1992; Wakeman, 1986b; Garrido *et al*, 2003), the filtration literature (Ruth *et al*, 1933; Tiller, 1953 – 2003; Shirato *et al*, 1986; Wakeman, 1978, 1981d, 1986b; Willis, 1983; Smiles, 1986; Leu, 1986; Landman *et al*, 1991; Stamatakis and Tien, 1991; Holdich, 1993; Landman and Russel, 1993; Burger *et al*, 2001; Tien *et al*, 2001; Tien, 2002; Yim *et al*, 2003) and the expression literature (Willis, 1983; Shirato *et al*, 1986; Wakeman *et al*, 1991; Kapur *et al*, 2002). The classical works of Underwood (1926) and Ruth *et al* (1933) laid the basis for the theoretical and experimental developments of Tiller (1953 – 1984), Leu (1986), Shirato *et al* (1986), Tarleton (1998), Wakeman *et al* (1991), Tarleton *et al* (2001, 2003), Teoh *et al* (2001), and Johansson and Theliander (2003), and it is now well established that cake or sediment structure varies with filtration time. These works led the way to estimating the rate of liquid flow through compacts exhibiting compressible behaviour. They rely on measurement of the so-called constitutive equations that relate the specific resistance and the solids volume fraction or voids ratio of the cake to the local solids compressive pressure. The forms of equations used are generally power law expressions such as equations (4.25), (4.26) or (4.29) which can be written in general form as:

$$\alpha = \alpha_0 (a_1 + b_1 p_s)^{n_1} \tag{4.93}$$

and

$$\varepsilon = \varepsilon_0 (a_2 + b_2 p_s)^{m_1} \tag{4.94}$$

with a_1 or $a_2 = 0$ or 1. In the above and following equations $n_1 \ldots n_4$, m_1 and m_2 are constants.

Wakeman (1978, 1981d, 1986b) attempted an analysis of the formation and growth of compressible filter cakes based on the principles of continuity, Darcy's law and knowledge of the porosity (or solidosity) profile in the compact. The analysis showed good agreement between calculated and experimental solidosity profiles, and demonstrated the importance of the initial liquid pressure gradient in determining subsequent compact growth.

Buscall and White (1987) utilised the concept of the network possessing a compressive yield stress $P_y(\phi)$, and assumed this to be an implicit function of the strength of the interparticle bridging force and the previous shear history of the system during the flocculation process, and an explicit function of the local volume fraction ϕ of the solids. It was therefore expected that $P_y(\phi)$ would increase with ϕ due to the larger number of interconnections between particles and hence the greater strength of the networked structure, and that it would vanish below the volume fraction ϕ_g at which the suspension is completely networked (at the gel point). Power law expressions of the form:

$$P_y(\phi) = k \left(\left(\frac{\phi}{\phi_g} \right)^{n_2} - 1 \right)^{n_3} \tag{4.95}$$

with n_2 or n_3 equal to 1 were used to fit experimental $P_y(\phi)$ data. Auzerais et al (1988) used

$$P_y(\phi) = k \left(\frac{\phi^{n_4}}{\phi_{cp} - \phi} \right) \quad \text{for} \quad \phi \gg \phi_g \tag{4.96}$$

where ϕ_{cp} is the close packed volume fraction. In a series of papers Landman et al (1988 – 1995) developed the network model approach for sedimenting and filtering systems. This requires evaluation of a term $r(\phi)$, which is claimed to account for the local hydrodynamic interactions between particles which increases the drag on them. $r(\phi)$ can be regarded as a hindered settling factor or as a Darcy's law constant, and was represented by a simple power law expression (Landman and White, 1992):

$$r(\phi) = (1 - b\phi)^m \tag{4.97}$$

The model therefore requires the use of two equations which contain several unknown terms, and which can be evaluated only through comparison with experimental data. Experimental validation of the model has yet to be done. It is not yet clear if any of these constants are, for example, pressure dependent or whether the methodology is significantly better than that developed by Ruth et al (1933), which forms the basis of this chapter. Nonetheless, the effects of microstructural properties on the compressive yield stress has been examined (Channell et al, 2000) from a theoretical point of view.

Stamatakis and Tien (1991) used the equations of continuity, the Darcy equation and the so-called constitutive equations (4.93) and (4.94) to produce a formal numerical solution to the Tiller-Shirato model. Like Wakeman, they developed a moving boundary model to identify the location of the cake-slurry interface. As with the other models described above, there are unknown terms which can be evaluated only after filtration experiments have been carried out. However, unlike others, these authors described in detail their methodology for initiation of the numerical solution. For small times, they argued that the cake may be considered incompressible, and from Darcy's law calculated the cake thickness. The short time solidosity profile was then determined using an equation similar to (4.94). Experimental validation of this procedure was not attempted, and the approach does not recognise that equations (4.93) and (4.94) derive their pressure dependent form primarily because their purpose was to convert the incompressible Tiller-Shirato filtration equations in a convenient way to describe compressible cake filtrations (Tien and Bai, 2003).

4.9.2 Models incorporating interparticle interactions

Bowen and Jenner (1995) and Bowen *et al* (1996) rightly recognised the double layer interaction as a key constitutive parameter in filtration processes, both for deadend and for crossflow filtration. The micromechanics that are applied, however, does not allow for structure formation and as a consequence the theory is inadequate at low zeta potential, low solidosity or in shear fields. In Bowen and Jenner (1995), especially, a simple cell model for the double layer interaction is introduced, based on close packing of the particles. The solidosities used in the experimental verification, however, are of the order 0.002 at the initial stage, which is a doubtful application of a close packed cell. At low zeta potential the theory put forward is demonstratedly inapplicable, but the explanation offered by this fact (that conventional charge regulation theory needs revision at low ionic strengths) must be disputed as no allowance has been made for diffusive effects.

Koenders *et al* (1996–2000) developed an approach to analyse filtration data, taking account of the fundamental properties of the suspensions being separated. The theory was developed in two parts. Firstly, the initial deposition of solids onto the filter medium was analysed by taking account of fluctuations in the particle velocity near the medium. Secondly, the dynamics of cake formation from suspensions of interacting particles were modelled, taking account of (a) changes of the random fluctuations of particle velocities caused by local direction and magnitude changes of the fluid velocity, and (b) the skeletal structure stress in the cake, by using expressions for the two particle interactive potential and thereby avoiding the need to introduce empiricism. The approaches taken to the two stages were wholly consistent with one another. Simpler theories for constant pressure filtration of non-interacting particles conclude that the filtrate volume produced $\propto p^{0.5}$ (see, for example, equation (4.43)). Koenders and Wakeman showed that the filtrate volume produced during the initial stages of cake formation depends on the range of the zeta potential. For soft sphere interactions, the filtrate volume produced is $\propto p^2$ and for hard sphere interactions the filtrate volume $\propto p$. It was further shown that during cake formation the two particle interaction potential lies between the two limiting analytical expressions usually quoted in the colloid science literature. The value of this work for filter system design has not been substantiated, and although much work remains to be done to verify the generality of the models they have been used with success by Meeton (2000).

Antelmi *et al* (2001) investigated the phenomenon of cake collapse that appears to occur in filter cakes formed from aggregated particles using small-angle neutron scattering. They reported that the mechanisms that produce the collapse are very small relative motions of the particles, which leave the local particle coordination unchanged but allow large voids to be produced (see also Tarleton and Willmer, 1997b). These motions were inhibited by large, non-spherical particles where the friction forces between the particle surfaces are greater, thereby reducing the extent of collapse.

4.10 Conclusions

Notwithstanding their fundamental shortcomings, the equations and methods developed in Sections 4.1 to 4.8 in this chapter provide a good engineering approach for the design and sizing of filtration equipment. Critical evaluation of both the equations and the methods over many years has led to their widespread acceptance as valid for many industrial process engineering calculations, and it is with this purpose in mind that they are collated in a unified format in this chapter.

5 Cake deliquoring

Cake deliquoring is a term which is frequently applied in a rather loose way to either the desaturation of a filter cake by sucking or blowing gas (usually air) through it, or to the reduction of the cake moisture content by mechanical squeezing (compression dewatering). Chapter 2, Section 2.6, lays out the fundamental principles that lie behind the applications of cake deliquoring and these are developed further in this chapter. The terms deliquoring and dewatering are both commonly used to describe removal of liquid from the cake in this fashion, and are used interchangeably in this chapter.

From the practical point of view a theory or model (whether this be mathematical or empirical) must be able to quantify three factors:

- the moisture content of the cake being discharged from a filter;

- the time taken to reduce the cake moisture to a specified level; and

- the gas or air flow rate during the dewatering period (since this serves as a basis for the sizing of vacuum pumps or compressed air requirements).

During dewatering the wetting fluid, the residual filtrate, flows over the particle surfaces preventing contact of the non-wetting air with the solid. The shape of the flow channels met by each of the fluids is therefore quite different. However, it is generally accepted that Darcy's law, originally developed to describe the flow of a single phase fluid completely saturating the bed, can be extended to describe the flow of each fluid flowing simultaneously. Underlying this extension is the introduction of a relative permeability for each of the fluids in the bed, which is a function of the wetting fluid saturation. As dewatering proceeds the cake saturation is reduced until the irreducible level is eventually reached. Here filtrate flow ceases, and only air flows through the cake voids. Although this does not preclude any microscopic local movements of filtrate within the cake, the continuity of the flow has been broken and any further saturation reduction is likely to be due to evaporation rather then displacement. Most filter cake deliquoring operations terminate long before such a state is reached, but it has been proven necessary to correlate residual saturation data as a route to developing methods for the prediction of cake saturations.

There have been various other attempts at the interpretation of cake dewatering characteristics and the modelling of two-phase flow through filter cakes. Empirical methods have been refined by Silverblatt and Dahlstrom (1954) and Simons and Dahlstrom (1966); these methods require a lot of experimental data in order to develop the correlations, and thereby ensure that accurate data are available when calculating results for the solid/liquid system and the range of variables examined in the original tests. Such an approach lacks generality, and often turns out to be no more precise than the predictive methods described in detail below.

Control of filter cake behaviour towards the end of the dewatering process is important when subsequent *in situ* processing of the cake is to be carried out, particularly when a high pressure differential is used in the displacing gas phase. Side effects such as cake cracking tend to occur because the incompressible liquid in the pore of the cake is being replaced by a compressible gas and gas-liquid interfaces are being set up in the pores. These interfaces introduce additional interparticle stresses, the magnitude of which depend largely on the saturation or moisture content of the cake. The distribution of these interfaces at the microscopic level manifest themselves at the macroscopic level as cracks in the filter cake, sometimes also accompanied by cake shrinkage.

Once a cake has cracked a channel is formed. Any flowing fluid then passes preferentially through the crack rather than through the pores in the cake. If cracking occurs before the irreducible saturation is reached further dewatering is ineffective as the displacing gas passes through the formed channels. Similarly if the cake is to be washed after dewatering, any cracks must be sealed before washing is commenced to prevent wash liquor passing preferentially through the cracks.

5.1 Threshold pressure and irreducible saturation

The threshold pressure is a way of expressing the minimum pressure difference that must be applied across a cake to effect any dewatering whatsoever, whilst the irreducible saturation defines the lowest saturation which can be reached when dewatering a particular cake. The former can be predicted with reasonable confidence, but the latter is far more difficult to predict and is best measured in a simple laboratory test. However, both quantities are obtainable from the same experiment although the threshold pressure can be a troublesome measurement. The irreducible saturation can sometimes be inferred from the moisture content of a cake discharged from a test filter.

When historical data for a specified deliquoring problem are unavailable it is recommended that a capillary pressure curve be measured, as it is an experiment simply performed in the laboratory and can greatly improve the accuracy of later calculations. The necessary equipment is illustrated in Figure 5.1, and is essentially a filter at the bottom of a cylindrical vessel (a useful size is about 45 cm long by 10 cm in diameter and the septum should be of the same cloth as it is intended to use on the actual filter). The cylinder is filled with a known volume of suspension at a known solids concentration, which is then filtered to form a cake. The cake depth and volume of drained filtrate are recorded. A complete capillary pressure curve is obtained by successively

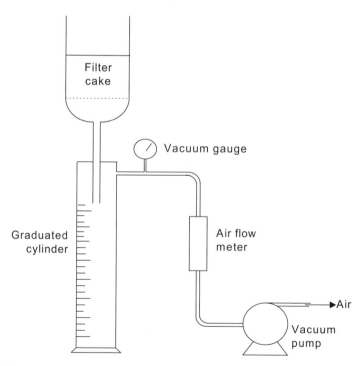

Figure 5.1 Apparatus for measuring threshold pressures and capillary pressure curves. The flow meter is not necessary for these purposes, but does give useful additional design information. An alternative form of the apparatus can be used to obtain data at higher pressures.

incrementing the driving force across the cake and recording corresponding decrements in the liquid volume remaining in the cake.

It is wise to measure the cake depth at each driving force as compressible cakes may remain almost saturated if their bulk volume is reduced but no flow of air occurs into the pores of the cake (although of course their moisture content is decreased). The amount of liquid in the cake is given by the difference between that added to the cylinder and that removed. The cake saturation is then calculated from the definition in equation (2.64):

i.e. from
$$S = \frac{\text{Volume of liquid in the cake}}{\varepsilon_{av} \, AL} \tag{5.1}$$

where (AL) is the bulk volume of the cake and ε_{av} is its average porosity, which may be calculated from

$$\varepsilon_{av} = 1 - \frac{W}{\rho_s AL} \tag{5.2}$$

where W is the dry weight of solids in the cake and ρ_s is the density of the solids. ε_{av} can also be obtained using equation (4.39).

If a complete capillary pressure curve determination is carried out, a curve similar to that shown in Figure 5.2 will be obtained. There is a

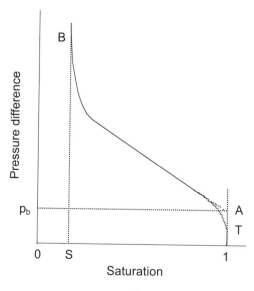

Figure 5.2 The typical form of a capillary pressure curve showing the modified threshold pressure (p_b) and the irreducible saturation (S_∞).

final finite saturation value beyond which no further reduction in liquid content of the cake is possible; this is the irreducible or infinite saturation, S_∞, and is shown on Figure 5.2. There is a minimum pressure required to give an initial reduction in saturation; this is the threshold pressure, given by point T on Figure 5.2. The point T is often difficult to identify, and it is easier to obtain a modified threshold pressure by back extrapolating the main portion of the curve as a straight line to the point A which also lies on the line $S = 1$. The threshold pressure value given by A is always slightly greater than that given by T, and can be predicted from (Wakeman, 1976b):

$$p_b = \frac{4.6(1-\varepsilon_{av})\sigma}{\varepsilon_{av}x} \tag{5.3}$$

where p_b is the threshold pressure, x is the mean particle size and σ is the liquid surface tension. Equation (5.3) can be used with confidence for threshold pressure determinations.

5.1.1 Irreducible saturation of vacuum or pressure deliquored cakes

Formulae exist which correlate irreducible saturations as a function of a capillary number, N_{cap}, expressing the ratio of the dewatering forces to the surface tension force retaining fluid in the cake. For vacuum or pressure drained cakes of a wide variety of more or less incompressible solids the irreducible saturation has been correlated as (Wakeman, 1976b):

$$S_\infty = 0.155\left(1+0.031N_{cap}^{-0.49}\right) \tag{5.4}$$

from which it can be shown that for $N_{cap} < 2.6 \times 10^{-5}$ displacement dewatering is unlikely to lead to any reduction of the saturation of the cake. The capillary number is defined by:

$$N_{cap} = \frac{\varepsilon_{av}^3 x^2(\rho_l gL+\Delta p)}{(1-\varepsilon_{av})^2 L\sigma} \tag{5.5}$$

5.1.2 Irreducible saturation of centrifuge cakes

When spinning a cake in a centrifuge, deliquoring is achieved as a result of body forces acting on the liquid volume in the cake voids. As in pressure (or vacuum) deliquoring where a minimum pressure must be applied in the gas phase to effect removal of the first drops of liquid from the cake, a minimum rotational speed of the centrifuge is required to effect initial deliquoring. In almost all practical cases this speed is very low and it is not necessary to be able to estimate its value. For centrifugal deliquoring, the irreducible saturation may be represented by equations of the form (Wakeman and Rushton, 1977):

$$
\begin{aligned}
S_\infty &= 0.0524 N_{cap}^{-0.19} \quad \text{for} \quad 10^{-5} \leq N_{cap} \leq 0.14 \\
&= 0.0139 N_{cap}^{-0.86} \quad \text{for} \quad 0.14 \leq N_{cap} \leq 10
\end{aligned}
\tag{5.6}
$$

where the capillary number is defined as

$$
N_{cap} = \frac{\varepsilon_{av}^3 \, x^2 \rho_l N^2 r}{(1 - \varepsilon_{av})^2 \sigma}
\tag{5.7}
$$

where r is the radius of rotation of the mid-point through the cake thickness. A sharp break in the correlation is noticeable when the ratio of the drainage force to the retentive forces (that is, the capillary number) is about 0.14; this probably coincides with a predominance of liquid being held in a pendular state (a state in which liquid is held around the points of contact of the particles) in the pores of a cake. This may also be interpreted as a breakdown of continuity in a majority of the liquid films throughout the cake voids.

Ambler (1988) reports alternative equations for the equilibrium saturation S_∞ and the instantaneous saturation S at any time t during the spin. In the following equations, the empirical constants k_1 and k_2 depend on the material being processed:

$$
S_\infty = k_1 \, x^{-0.5} \left(\frac{\omega^2 r_o}{g} \right)^{-0.5} \rho_l^{-0.25}
\tag{5.8}
$$

$$
S = \frac{k_2}{x} \left(\frac{\mu_l g}{\rho_l \omega^2 r_o} \right) \frac{(r_o - r_i)^a}{t^b} + S_\infty
\tag{5.9}
$$

where t is the centrifuging time. The exponent a on the cake thickness $(r_o - r_i)$ is reported in the range $0.5 < a < 1.0$ for relatively incompressible cakes, and the exponent b on the spin time takes values in the range $0.3 < b < 0.5$. Like the other models outlined in Chapter 2 (Section 2.6), the constants and exponents in the above equations cannot be determined *a priori*, limiting their usefulness as a predictive design tool.

Equations (5.4), (5.6) and (5.8), or similar ones, should be used with caution as the constants and exponents in the equations may have only limited validity; they have been applied successfully to deliquoring cakes composed of materials like coal fines, sandlike and coarse crystalline products.

5.2 Design charts for deliquoring

Design charts for deliquoring filter cakes in vacuum, pressure, centrifugal and gravity filters based on equations (2.64) to (2.86) have been developed (Wakeman, 1979a–c, 1982; Wakeman and Vince, 1986a,b). The equations are complex and their solution requires numerical integrations of partial differential equations; this is inconvenient for everyday use, and so the equations have been solved and the results presented graphically and then curve fitted.

An important parameter involved in the description of capillary pressure and relative permeability effects is the pore size distribution index λ (see equations (2.69) to (2.82)); strictly this requires measurement, but in order to create working design charts it is assumed to have a value of 5. To ease presentation of results from the solved equations, the dimensionless variables defined in Table 5.1 are used.

Table 5.1 Definition of the dimensionless terms used to calculate rates of deliquoring; these are written in terms of the average permeability k_{av} of the cake, which is related to the average specific cake resistance (equation (2.23)) by $k_{av} = 1/(\alpha_{av}\rho_s(1 - \varepsilon_{av}))$.

	Dimensionless deliquoring time, θ	Dimensionless gas flux, u_g^*	Dimensionless pressure, p^*
Pressure/vacuum	$\dfrac{k_{av}\, p_b}{\mu_l \varepsilon_{av}\left(1 - S_\infty\right)L^2} t_d$	$\dfrac{u_g \mu_g L}{k_{av} p_b}$	$\dfrac{p}{p_b}$
Centrifugal	$\dfrac{k_{av}\, p_b}{\mu_l \varepsilon_{av}\left(1 - S_\infty\right)r_o^2} t_d$	$-$	$\dfrac{\rho_l \omega^2 r_o^2}{p_b}$
Gravitational	$\dfrac{k_{av}\, p_b}{\mu_l \varepsilon_{av}\left(1 - S_\infty\right)L^2} t_d$	$-$	$\dfrac{\rho_l g L}{p_b}$

The output information required from the calculations are the variations of the cake saturation and the gas flow rate with deliquoring time; these enable calculation of cake moisture contents and blower or vacuum pump capacities, and are shown in Figures 5.3 and 5.4. In the cases of centrifugal deliquoring and gravitational drainage, no gas flow is imposed and the rates of deliquoring are shown in Figures 5.5 and 5.6.

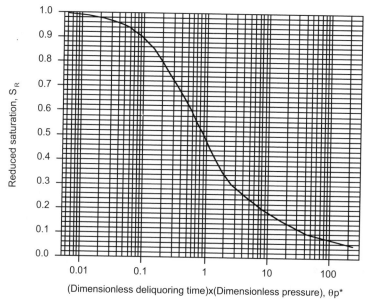

Figure 5.3 The reduced saturation of a filter cake as a function of a dimensionless deliquoring time during deliquoring by using vacuum or pressure applied in a gas phase.

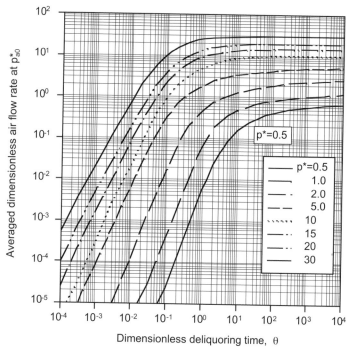

Figure 5.4 The dimensionless air flow rate $\bar{u}_g^*$ through a filter cake during deliquoring by using vacuum or pressure applied in a gas phase as a function of the dimensionless pressure p^* (note that the definition of dimensionless time on this plot is different from that on Figure 5.3); the curves are given for different dimensionless applied pressures.

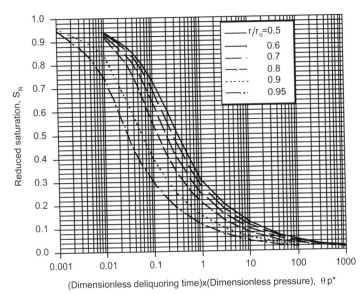

Figure 5.5 The reduced saturation of a centrifuge cake as a function of a dimensionless deliquoring time, the inner radius of the cake (r_i) in the basket and the radius of the basket (r_o).

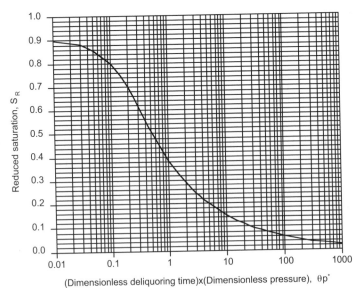

Figure 5.6 The reduced saturation of a filter cake or porous medium as a function of a dimensionless deliquoring time during gravity drainage.

Although there may be a small amount of 'windage' in a centrifige basket due to its rotation, it does not contribute to cake deliquoring – but it should be noted that some centrifuge designs can utilise a superimposed gas pressure (or vacuum) which does aid deliquoring. Equations which describe the graphical data are also given below to assist with design calculations.

5.2.1 Vacuum and pressure deliquoring

The variation of cake saturation with time is represented by a chart on which the reduced saturation is plotted against the product of the dimensionless deliquoring time and dimensionless pressure, defined by $S_R = \dfrac{S - S_\infty}{1 - S_\infty}$ and $\theta^* = \theta p^*$ respectively.

The curve in Figure 5.3 is described by the following equation

$$S_R = \frac{1}{1 + 1.08 \left(\theta p^*\right)^{0.88}} \qquad \text{for} \qquad 0.096 \leq \left(\theta p^*\right) \leq 1.915$$

$$= \frac{1}{1 + 1.46 \left(\theta p^*\right)^{0.48}} \qquad \text{for} \qquad 1.915 \leq \left(\theta p^*\right) \leq 204$$

(5.10)

The gas fluxes in Figure 5.4 are values averaged up to the dimensionless deliquoring time θ, that is

$$\overline{u}_g^* = \frac{1}{\theta} \int_0^\theta u_g^* \, d\theta$$

(5.11)

where u_g^* is the dimensionless value of the instantaneous gas flux (defined in Table 5.1) as would be determined using a rotameter or other similar instantaneous gas flow measuring device. The averaged gas flux provides the appropriate basis for sizing, for example, vacuum pumps. The curves in Figure 5.4 are described by the equation

$$\overline{u}_g^* = 10^{\left(a_0 + a_1 \log\theta + a_2 (\log\theta)^2 + a_3 (\log\theta)^3 + a_4 (\log\theta)^4\right)}$$

(5.12)

where the coefficients a_0 to a_4 are given in Table 5.2. Deliquoring is only possible when the applied pressure is greater than the threshold pressure, that is $p^* > 1$. A curve at $p^* = 0.5$ is included on Figure 5.4 as an aid to interpolation.

Table 5.2 The coefficients in equation (5.12) to describe gas flow rates during deliquoring by vacuum or pressure applied to the gas phase

p^*	a_0	a_1	a_2	a_3	a_4	Region of validity
0.5	-2.2828	1.9809	-0.3649	-0.0786	-0.0049	$10^{-4} \le \theta < 20$
	-1.6166	1.1793	-0.3787	0.0560	-0.0031	$20 \le \theta \le 10^6$
	0.1487	0	0	0	0	$\theta > 10^6$
1	-1.2305	1.6891	-0.6050	-0.2142	-0.0262	$10^{-4} \le \theta < 5$
	-0.8702	0.7782	-0.2668	0.0435	-0.0027	$5 \le \theta \le 10^6$
	0.1367	0	0	0	0	$\theta > 10^6$
2	-0.3275	0.5687	-1.1410	-0.3354	-0.0359	$10^{-4} \le \theta < 2$
	-0.2570	0.5046	-0.1602	0.0253	-0.0016	$2 \le \theta \le 10^6$
	0.4456	0	0	0	0	$\theta > 10^6$
5	0.3310	0.4431	-0.4369	-0.0126	-0.0058	$10^{-4} \le \theta < 0.8$
	0.2765	0.4395	-0.2264	0.0606	-0.0062	$0.8 \le \theta \le 10^4$
	0.7202	0	0	0	0	$\theta > 10^4$
10	0.4085	-0.6765	-1.3865	-0.3329	-0.0304	$10^{-4} \le \theta < 0.3$
	0.7071	0.4015	-0.2104	0.0503	-0.0046	$0.3 \le \theta \le 10^3$
	1.0039	0	0	0	0	$\theta > 10^3$
15	0.8288	-0.4022	-1.0965	-0.2551	-0.0227	$10^{-4} \le \theta < 0.18$
	0.9755	0.2690	-0.2139	0.1139	-0.0256	$0.18 \le \theta \le 10^2$
	1.1614	0	0	0	0	$\theta > 10^2$
20	0.7981	-0.8823	-1.3651	-0.3367	-0.0314	$10^{-4} \le \theta < 0.1$
	1.1400	0.2554	-0.1888	0.0724	-0.0108	$0.1 \le \theta \le 10^2$
	1.3010	0	0	0	0	$\theta > 10^2$
30	0.9539	-1.1129	-1.2848	-0.2757	-0.0225	$10^{-4} \le \theta < 0.05$
	1.4050	0.1420	-0.1481	0.0762	-0.0143	$0.05 \le \theta \le 10^2$
	1.4771	0	0	0	0	$\theta > 10^2$

5.2.2 Centrifugal deliquoring

Deliquoring curves for centrifuges are shown on Figure 5.5 as the reduced saturation of the cake plotted against the product (θp^*) at various values of r_i/r_o. The general equation describing these curves is

$$S_R = \frac{1}{1 + a_5 (\theta p^*)^{a_6}} \tag{5.13}$$

where the constants a_5 and a_6 are given in Table 5.3.

Table **5.3** The coefficients in equation (5.13) to describe deliquoring of centrifuge cakes

r_i/r_o	a_5	a_6	Region of validity
0.5	2.30	0.77	$0.01 \leq \theta p^* \leq 1$
	0.77	0.44	$1 < \theta p^* \leq 1000$
0.6	2.82	0.79	$0.01 \leq \theta p^* \leq 0.6$
	2.83	0.45	$0.6 < \theta p^* \leq 1000$
0.7	3.46	0.77	$0.01 \leq \theta p^* \leq 0.6$
	3.37	0.44	$0.6 < \theta p^* \leq 1000$
0.8	5.41	0.84	$0.01 \leq \theta p^* \leq 0.3$
	4.15	0.44	$0.3 < \theta p^* \leq 1000$
0.9	7.10	0.70	$0.001 \leq \theta p^* \leq 0.3$
	5.51	0.45	$0.3 < \theta p^* \leq 100$
0.95	14.09	0.76	$0.001 \leq \theta p^* \leq 0.1$
	7.64	0.45	$0.1 < \theta p^* \leq 100$

5.2.3 Gravitational drainage

Gravity drainage of packed beds and filter cakes is shown on Figure 5.6; the curve shown can be described by

$$S_R = \frac{1}{1+1.48(\theta p^*)^{0.67}} \quad \text{for} \quad 0.05 \leq (\theta p^*) \leq 4$$

$$= \frac{1}{1+1.99(\theta p^*)^{0.46}} \quad \text{for} \quad 4 < (\theta p^*) \leq 1000$$

(5.14)

5.2.4 Experimental validation of the design charts

Differences in final moisture contents for cakes formed on leaf filters with the calculations from the design charts for S_∞, and hence S, at deliquoring time t have been compared with industrial trials (Carleton and Mehta, 1983). These trials were done using large-scale drum, table and belt filters producing cakes of sand, gypsum, pigments, chalk and "impurities"; the results are reproduced in Figure 5.7. Excellent agreement was obtained between large scale measurements and predictions from the design charts for filter cakes with specific resistances in the range

$$5\times10^6 \text{ (sand)} < \alpha_{av} \text{ (m kg}^{-1}) < 3\times10^{11} \text{ (pigment)}.$$

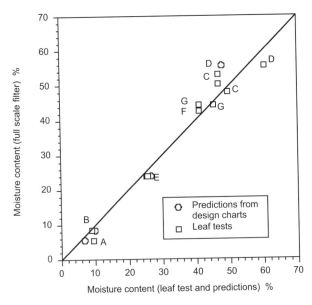

Figure 5.7 Comparison of cake moisture contents from full scale filters with leaf test data and with predictions from the design charts (Carleton and Mehta, 1983). (A – Sand, 50 mm cake, $\alpha_{av} = 5.9 \times 10^6$ m kg^{-1}, table filter; B – Sand, 70 mm cake, $\alpha_{av} = 1.4 \times 10^7$ m kg^{-1}, table filter; C – Gypsum in phosphoric acid, 30 to 46 mm cake, $\alpha_{av} = 1.9 \times 10^9$ m kg^{-1}, belt filter; D – 'Impurity', 46 mm cake, $\alpha_{av} = 5.5 \times 10^9$ m kg^{-1}, drum filter; E – Chalk, 5 mm cake, $\alpha_{av} = 2.1 \times 10^{11}$ m kg^{-1}, drum filter; F – Pigment, 4 mm cake, $\alpha_{av} = 2.1 \times 10^{11}$ m kg^{-1}, drum filter; G – Pigment, <1 mm cake, $\alpha_{av} = 2.7 \times 10^{11}$ m kg^{-1}, drum filter.)

The charts have also been used to predict the kinetics of filter cake deliquoring and the flow rate of air; Figures 5.8 and 5.9 show comparisons between predictions using the model and practical data when deliquoring sand cakes. Although predictions of the rate of desaturation of the cake are good, the design charts tend to overestimate air flow rates; in design this leads to overcapacity in the vacuum pumps. Similar comparisons are shown in Figure 5.10 and 5.11 for the moisture reduction kinetics of vacuum filtered froth concentrates (with a nominal size fraction –0.5+0 mm) from a coal preparation plant. These figures also indicate the extent to which the cake moisture content is reduced and the rate of deliquoring is increased as the applied pressure difference is increased. Although the gas rate predictions are better for thicker cakes, the design chart always errs on the side of predicting too high gas consumption – even though the tendency for channelling to occur is much greater with thinner cakes, which also leads to higher gas flow rates in practical situations. The likely reason for this probably lies in the assumption of a constant value of the pore size distribution index of $\lambda = 5$ (Table 2.4 shows the typical range over which λ can vary); whilst the effect of λ on cake

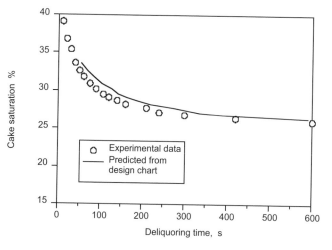

Figure 5.8
Comparison of experimental data for the kinetics of deliquoring HPF5 sand cake with predictions from the design chart, Figure 5.3 (Carleton and Salway, 1993).

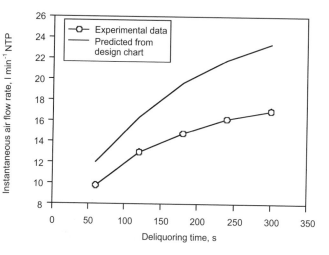

Figure 5.9
Comparison of experimental data for the air flow rate through the cake during deliquoring of HPF5 sand with predictions from the design chart, Figure 5.4 (Carleton and Salway, 1993).

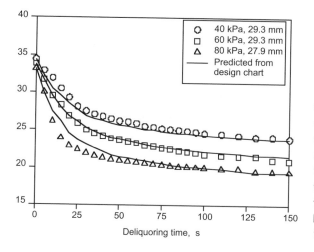

Figure 5.10
Comparison of experimental data for the moisture content reduction of flocculated fine coal filter cakes with predictions from the design chart, Figure 5.3 (Condie et al, 1996).

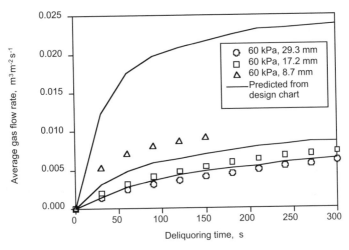

Figure 5.11 Comparison of experimental data for the air flow rate through flocculated fine coal filter cakes with predictions from the design chart, Figure 5.4 (Condie *et al*, 1996).

desaturation kinetics is small, its effect on gas consumption may be rather greater.

In Figure 5.12 deliquoring data obtained from an industrial pusher centrifuge (Wakeman and Mulhaupt, 1985) are compared with predictions from the design chart for centrifugal deliquoring; a characteristic of machine operation is the very short residence times of solids in the bowl, and hence the limited time available for deliquoring. Data are also

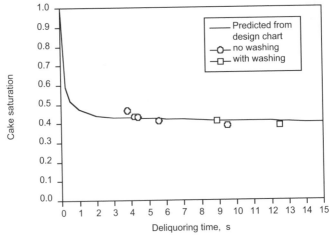

Figure 5.12 Comparison of cake deliquoring data from a pusher centrifuge ($r_o = 0.23$ m) with predictions from the design chart, Figure 5.5 (Wakeman and Vince, 1986b).

shown for centrifuge operation that includes cake washing. Models of centrifuge operation are particularly useful because there are so many inherent difficulties with obtaining reliable data from centrifuges unless specialist facilities are available.

5.3 Example calculations for deliquoring

To effect the deliquoring of a filter cake a pressure difference is applied in the gas phase (normally air), which is thereby caused to flow through the cake voids and displace part of the retained liquor. So long as this pressure difference exceeds the threshold pressure some saturation reduction of the cake will result. As liquid is drained from the cake the effective permeability of the cake to the liquid phase decreases, whilst simultaneously the effective permeability to the air phase increases. Flow of liquid from the cake ceases when the irreducible saturation is reached; if this saturation is envisaged to be always stagnant fluid during a dewatering operation, then for practical purposes it may be regarded as part of the immobile solids matrix and a reduced saturation can then be defined (see also equation (2.68)):

$$S_R = \frac{S - S_\infty}{1 - S_\infty} \tag{5.15}$$

$(S - S_\infty)$ is that part of the retained liquid which is potentially mobile at any time during the deliquoring, and $(1 - S_\infty)$ represents the total amount of potentially mobile liquid in the cake at the start of deliquoring.

The reduction of S_R (or S when S_∞ is known) as a function of the time of deliquoring can be calculated, and the increase in the air flow rate can also be determined. The results of primary interest are summarised in Figures 5.3 to 5.6. The lowering of the cake saturation during deliquoring is shown in Figures 5.3, 5.5 and 5.6, where the dimensionless time is defined in Table 5.1. The dimensionless pressure difference (see also Table 5.1 and Figure 5.13) for vacuum or pressure deliquoring is written as:

$$p_a^* = \left\{ \begin{array}{l} \text{dimensionless air} \\ \text{pressure at air inlet} \\ \text{surface of cake} \end{array} \right\} - \left\{ \begin{array}{l} \text{dimensionless air pressure} \\ \text{at air outlet surface of cake} \\ \text{(or filter medium)} \end{array} \right\}$$

$$= \left\{ \frac{p_a}{p_b} \right\}_{inlet} - \left\{ \frac{p_a}{p_b} \right\}_{outlet}$$

$$= p_{ai}^* - p_{ao}^* \tag{5.16}$$

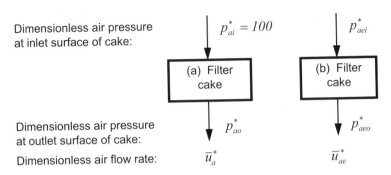

Figure 5.13 Definition of the pressure basis for (a) the design charts, and (b) the actual deliquoring problem.

In order to develop the vacuum and pressure design charts p_{ai}^* has been set to 100. In the equations in Table 5.1, k_{av} is the permeability of the cake to the liquid when its saturation is 100%; this can be estimated (see also equation (2.23)) provided the specific resistance and porosity of the cake have been measured, since:

$$k_{av} = \frac{1}{\alpha_{av} \rho_s (1 - \varepsilon_{av})} \qquad (5.17)$$

The curves in Figures 5.3, 5.5 and 5.6 all asymptote towards $S_R = 0$ (or $S = S_\infty$) as $t \to \infty$.

During deliquoring the air (or other gas) flow starts from zero when $S_R = S = 1$ and then progressively increases. The mean air flow rate passing through a cake subjected to deliquoring by vacuum or pressure is plotted against dimensionless time θ on Figure 5.4.

In general, the air rate through the cake, and thus the vacuum pump capacity, can be calculated from measurements of the total air flow after various dewatering periods or from a plot of the instantaneous air rate as a function of time (as measured by a rotameter, for example). The total air volume passing through the cake is then obtained by integration of the curve up to the requisite dewatering time. Vacuum pump capacity is conventionally based on the total filter cycle, rather than just the deliquoring time, and is measured at the pump inlet conditions. Thus, the gas volumes per unit area passing during each period of deliquoring in the filter operating cycle are totalled and divided by the cycle time to arrive at the design air rate. The average air rate during any period of deliquoring is obtained directly on a dimensionless basis using equations (5.11) or (5.12), or Figure 5.4. The dimensionless instantaneous air rate is defined in Table 5.1.

To facilitate generalisation, development of Figures 5.3 and 5.4 has assumed the pressure basis defined in Figure 5.13. This basis is not likely to correspond to that of the actual separation process under consideration, so $\bar{u}_a^*$ values read from the graphs must be corrected to the true absolute pressure conditions; the different bases of the charts and the actual dewatering are compared on Figure 5.13, from which pressure corrected mean dimensionless corrected air flow can be written as:

$$\bar{u}_{ae}^* = \bar{u}_a^* \frac{p_{ao}^*}{p_{aeo}^*} \left[\frac{p_{aeo}^{*^2} - p_{aei}^{*^2}}{p_{ao}^{*^2} - p_{ai}^{*^2}} \right] \tag{5.18}$$

5.3.1 Summary of deliquoring calculations for vacuum and pressure filters

Calculations for specific deliquoring problems are based on filter parameters as well as the cake, solid and fluid properties. The following data should be available:

(i)   vacuum or pressure levels;

(ii) time available on filter for dewatering, or, required moisture content of discharged cake;

(iii) typical specific cake resistance, α_{av};

(iv) typical cake porosity, ε_{av};

(v)   solids density, ρ_s;

(vi) expected cake thickness, L;

(vii) surface tension of liquid, σ;

(viii) liquid and air viscosities, μ_l and μ_a;

(ix) liquid density, ρ_l.

A typical calculation sequence would then take the form given below (Steps 2 and 3 are omitted if S_∞ has been obtained experimentally, and Step 4 can be omitted if p_b has been measured):

1. Calculate the effective mean particle size expressed from equation (2.37) as:

$$x = 13.4 \sqrt{\frac{1 - \varepsilon_{av}}{\alpha_{av} \rho_s \varepsilon_{av}^3}} \tag{5.19}$$

2. Calculate the capillary number from equation (5.5);

3. Calculate the irreducible saturation from equation (5.4);

4. Calculate the threshold pressure from equation (5.3), hence express

the expected pressures on the actual installation in dimensionless form and obtain the dimensionless pressure difference from:

$$p_a^* = p_{aei}^* - p_{aeo}^*$$

5. Calculate the cake permeability from equation (5.17). In most problems either the dewatering time or a required cake moisture content will be known.
   (a) If the time available for dewatering, t_d, is known, calculate the dimensionless time, θ, from Table 5.1. Or,
   (b) if the moisture content at discharge is specified and t_d is to be determined, calculate S and S_R from Step 9;

6. If θ is known read S_R from Figure 5.3 (p^* is known from Step 4 since $p_a^* = p^*$), or, if S_R is known, read θp^*. Alternatively use equation (5.10) to calculate S_R or θp^*;

7. Read $\bar{u}_a^*$ from Figure 5.4. Alternatively use equation (5.12) together with Table 5.2;

8. Correct $\bar{u}_a^*$ for inlet/exit pressures:

$$\bar{u}_{ae}^* = \left\{ \begin{array}{l} \text{Predicted average} \\ \text{dimensionless} \\ \text{air flow} \end{array} \right\} = \left\{ \begin{array}{l} \bar{u}_a^* \text{ from} \\ \text{Step 7.} \end{array} \right\} \left\{ \begin{array}{l} \text{Inlet/exit pressure} \\ \text{correction from} \\ \text{equation (5.18)} \end{array} \right\} \quad (5.20)$$

To allow for evacuation of the filter drainage passages, or for filling of equipment when compressed air is used, the predicted air rate should be multiplied by 1.1,

$$\bar{u}_a^* \big|_{\text{corrected}} = 1.1 \times \bar{u}_{ae}^* \quad (5.21)$$

9. The cake saturation and moisture content at discharge are given by (the calculation sequence within Step 9 is reversed if moisture content is specified and t_d is unknown):

$$S = S_R(1 - S_\infty) + S_\infty \quad (5.22)$$

$$M' = S \frac{\varepsilon_{av}}{1 - \varepsilon_{av}} \frac{\rho}{\rho_s} 100 \quad \% \quad (5.23)$$

where M' has the units (mass liquid/mass solid). A preferred term for moisture content is:

$$M = \frac{M'}{100 + M'} 100 \quad \% \quad (5.24)$$

which has the units (mass liquid/(mass of liquid + mass of solid));

10. Calculate air rate on a total cycle basis (for vacuum installations) using Table 5.1 in the form:

$$u_a = \frac{k_{av}\, p_b \bar{u}_a^* \big|_{\text{corrected}}}{\mu_a L} \tag{5.25}$$

The air rate on total cycle basis $= u_a \times \dfrac{\text{deliquoring time}}{\text{cycle time}}$ (5.26)

measured at the operating vacuum (p_v N m^{-2}) of the filter;

11. Correct design air rate for pressure loss through ancillary vessels and/or pipework systems (allowing a loss of 3.3 kN m^{-2}):

$$\left\{\begin{matrix}\text{Design} \\ \text{air rate}\end{matrix}\right\} = \left\{\begin{matrix}\text{Air rate on total cycle} \\ \text{basis from Step 10.}\end{matrix}\right\} \frac{p_{aei}^* - \dfrac{p_v}{p_b}}{p_{aei}^* - \dfrac{p_v + 3300}{p_b}}$$

measured at (p_v+3300) N m^{-2} vacuum.

Or, for pressure systems,

$$\left\{\begin{matrix}\text{Design} \\ \text{air rate}\end{matrix}\right\} = \left\{\begin{matrix}u_a \text{ from} \\ \text{Step 10.}\end{matrix}\right\} \frac{p_{aeo}^*}{p_{aeo}^* + \dfrac{\Delta p + 3300}{p_b}}$$

measured at ($p_{aeo}^* p_b + \Delta p + 3300$) N m^{-2} absolute pressure where Δp is the pressure difference over the cake during deliquoring.

Variations in barometric pressure have been taken into account by assigning correct pressure values in Step 4.

Example 5.1 Calculations for rotary vacuum filters.
It is proposed that an existing rotary drum filter be used on a new separation. The total cycle time is 90 s and 45% of the cycle is available for cake dewatering; calculate the cake moisture content at discharge and the vacuum pump flow capacity requirements. The following data are available:

Barometric pressure	1.013×10^5 N m^{-2}
Pressure difference over cake (p_v)	65 kN m^{-2}
Cake thickness (L)	0.1 m
Average specific cake resistance (α_{av})	10^9 m kg^{-1}
Average cake porosity (ε_{av})	0.38
Solids density (ρ_s)	2200 kg m^{-3}
Surface tension of filtrate (σ)	0.07 N m^{-1}
Viscosity of filtrate (μ_l)	10^{-3} Pa s
Density of filtrate (ρ_l)	1100 kg m^{-3}

Solution
Following the procedure laid out above:

1. $x = 13.4\sqrt{\dfrac{0.62}{10^9 \times 2200 \times 0.38^3}} = 3.04 \times 10^{-5}$ m

2. $N_{cap} = \dfrac{0.38^3 \times (3.04 \times 10^{-5})^2 (1.1 \times 10^3 \times 9.81 \times 0.1 + 65 \times 10^3)}{0.62^2 \times 0.1 \times 0.07} = 1.25 \times 10^{-3}$

3. $S_\infty = 0.155(1 + 0.031(1.25 \times 10^{-3})^{-0.49}) = 0.282$

4. $p_b = \dfrac{4.6 \times 0.62 \times 0.07}{0.38 \times 3.04 \times 10^{-5}} = 1.73 \times 10^4$ N m^{-2}

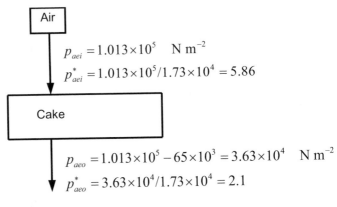

$$\boxed{\text{Air}}$$

$p_{aei} = 1.013 \times 10^5$ N m^{-2}

$p_{aei}^* = 1.013 \times 10^5 / 1.73 \times 10^4 = 5.86$

$$\boxed{\text{Cake}}$$

$p_{aeo} = 1.013 \times 10^5 - 65 \times 10^3 = 3.63 \times 10^4$ N m^{-2}

$p_{aeo}^* = 3.63 \times 10^4 / 1.73 \times 10^4 = 2.1$

$p_a^* = 5.86 - 2.1 = 3.76$

5. $k_{av} = \dfrac{1}{10^9 \times 2200 \times 0.62} = 7.33 \times 10^{-13}$ m^2

$t_d = 90 \times 0.45 = 40.5$ s

$\theta = \dfrac{7.33 \times 10^{-13} \times 1.73 \times 10^4 \times 40.5}{10^{-3} \times 0.1^2 \times 0.38 \times 0.718} = 0.188$

$\theta p_a^* = 0.188 \times 3.76 = 0.707$

6. From Figure 5.3, $S_R \approx 0.56$. Alternatively, using equation (5.10):

$S_R = \dfrac{1}{1 + 1.08(0.707)^{0.88}} = 0.557$

7. From Figure 5.4, by linear interpolation between the curves for $p^* = 2$ and $p^* = 5$:

$$\frac{\bar{u}_a^* - 0.7}{3.76 - 5} = \frac{0.07 - 0.7}{2 - 5}$$

$$\therefore \bar{u}_a^* = 0.44 \quad \text{measured at } p_{ao}^*$$

Alternatively, using equation (5.12) and Table 5.2:

$$\bar{u}_a^* = 10^{-0.3275+0.5687(-0.7258)-1.141(-0.7258)^2-0.3354(-0.7258)^3-0.0359(-0.7258)^4}$$

$$= 0.0598 \quad \text{at} \quad p_a^* = 2$$

$$\bar{u}_a^* = 10^{0.331+0.4431(-0.7258)-0.4369(-0.7258)^2-0.0126(-0.7258)^3-0.0058(-0.7258)^4}$$

$$= 0.606 \quad \text{at} \quad p_a^* = 5$$

$$\therefore \quad \bar{u}_a^* = 0.606 - \frac{0.606 - 0.0598}{5 - 2}(5 - 3.76) = 0.380 \quad \text{measured at } p_{ao}^*.$$

8. $p_{ao}^* = 100 - 3.76 = 96.24$ (taking note of Figure 5.13). Using equation (5.18):

$$\frac{\bar{u}_{ae}^*}{\bar{u}_a^*} = \frac{96.24}{2.1}\left[\frac{2.1^2 - 5.86^2}{96.24^2 - 100^2}\right] = 1.86$$

$$\therefore \left.\bar{u}_a^*\right|_{\text{predicted}} = 0.38 \times 1.86 = 0.707 \quad \text{measured at } p_{aeo}^*$$

and $\left.\bar{u}_a^*\right|_{\text{corrected}} = 1.1 \times 0.707 = 0.778 \quad \text{measured at } p_{aeo}^*$

9. $\quad S = 0.557(1 - 0.282) + 0.282 = 0.682$

$$\therefore \quad M' = 0.682\frac{0.38}{0.62}\frac{1100}{2200}100 = 21.0\%$$

Hence, the moisture content of the discharged cake is

$$M = \frac{21.0}{100 + 21.0}100 = 17.4\%$$

10. $u_a = \dfrac{7.33 \times 10^{-13} \times 1.73 \times 10^4 \times 0.778}{1.81 \times 10^{-5} \times 0.1}$

$= 5.45 \times 10^{-3} \text{ m}^3 \text{ m}^{-2} \text{ s}^{-1}$ measured at 65 kN m^{-2} vacuum

$\therefore$ air rate on total cycle basis $= 5.45 \times 10^{-3} \dfrac{40.5}{90}$

$= 2.45 \times 10^{-3} \text{ m}^3 \text{ m}^{-2} \text{ s}^{-1}$

measured at 65 kN m^{-2} vacuum.

11. The design air rate

$= 2.45 \times 10^{-3} \dfrac{5.86 - \dfrac{65 \times 10^3}{1.73 \times 10^4}}{5.86 - \dfrac{65 \times 10^3 + 3300}{1.73 \times 10^4}} = 2.69 \times 10^{-3} \text{ m}^3 \text{ m}^{-2} \text{ s}^{-1}$

measured at 68.3 kN m^{-2} (i.e. $65 + 3.3$ kN m^{-2}) vacuum.

Example 5.2 Calculations for pressure filters.

A single, horizontal leaf Nutsche filter is to be used to filter small batches of a product by applying air pressure to the aqueous slurry. If the operating temperature is 40°C, calculate the air flow rate and the time taken to reach a moisture content (M) of 20%. The following data apply:

Pressure difference over cake (Δp)	4×10^5 N m^{-2}
Cake thickness (L)	0.3 m
Average specific cake resistance (α_{av})	10^{10} m kg^{-1}
Average cake porosity (ε_{av})	0.43
Solids density (ρ_s)	1700 kg m^{-3}
Surface tension of filtrate (σ)	0.0696 N m^{-1}
Viscosity of filtrate (μ_l)	0.65×10^{-3} Pa s
Density of filtrate (ρ_l)	992 kg m^{-3}

The downstream side of the leaf can be assumed to be at atmospheric pressure (1.013×10^5 N m^{-2}).

Solution
Following the procedure set out:

1. $x = 13.4 \sqrt{\dfrac{0.57}{10^{10} \times 1700 \times 0.43^3}} = 8.7 \times 10^{-6}$ m

2. $N_{cap} = \dfrac{0.43^3 \times (8.7 \times 10^{-6})^2 (992 \times 9.81 \times 0.3 + 4 \times 10^5)}{0.57^2 \times 0.3 \times 0.0696} = 3.6 \times 10^{-4}$

3. $S_\infty = 0.155(1 + 0.031[3.6 \times 10^{-4}]^{-0.49}) = 0.389$

4. $p_b = \dfrac{4.6 \times 0.57 \times 0.0696}{0.43 \times 8.7 \times 10^{-6}} = 4.88 \times 10^4 \quad \text{N m}^{-2}$

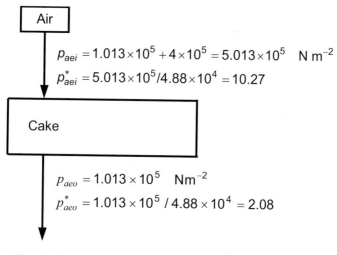

Air

$p_{aei} = 1.013 \times 10^5 + 4 \times 10^5 = 5.013 \times 10^5 \quad \text{N m}^{-2}$

$p_{aei}^* = 5.013 \times 10^5 / 4.88 \times 10^4 = 10.27$

Cake

$p_{aeo} = 1.013 \times 10^5 \quad \text{N m}^{-2}$

$p_{aeo}^* = 1.013 \times 10^5 / 4.88 \times 10^4 = 2.08$

$p_a^* = 10.27 - 2.08 = 8.19$

5. $k_{av} = \dfrac{1}{10^{10} \times 1700 \times 0.57} = 1.03 \times 10^{-13} \quad \text{m}^2$

The time available for dewatering is unknown. The dimensionless time is

$$\theta = \frac{1.03 \times 10^{-13} \times 4.88 \times 10^4 \times t}{0.65 \times 10^{-3} \times 0.3^2 \times 0.43 \times 0.611} = 3.27 \times 10^{-4} t$$

6. The cake moisture content is specified as 20%, hence

$$M = 20 = \frac{M'}{100 + M'} 100$$

$\therefore\ M' = 25\%$

$$= S \frac{0.43}{0.57} \frac{992}{1700} 100$$

$\therefore\ S = 0.568$

and $S_R = \dfrac{0.568 - 0.389}{1 - 0.389} = 0.293$

7. From Figure 5.3:

$\theta p_a^* \approx 2.9$

$$\therefore \quad \theta = \frac{2.9}{8.19} = 0.354$$

and $\quad t = \dfrac{0.354}{3.27 \times 10^{-4}} = 1083 \text{ s}$

i.e. the dewatering time is 1083 seconds (18 minutes).

8. From Figure 5.4, by linear interpolation between curves for $p^* = 5$ and $p^* = 10$:

$$\frac{\bar{u}_a^* - 3}{8.19 - 10} = \frac{1.1 - 3}{5 - 10}$$

$\therefore \quad \bar{u}_a^* = 2.31 \quad$ measured at p_{ao}^*.

Alternatively, using equation (5.12) and Table 5.2:

$\bar{u}_a^* = 10^{0.331+0.4431(-0.451)-0.4369(-0.451)^2-0.0126(-0.451)^3-0.0058(-0.451)^4}$
$\quad = 1.104 \quad$ at $\quad p_a^* = 5$

$\bar{u}_a^* = 10^{0.7071+0.4015(-0.451)-0.2104(-0.451)^2+0.0503(-0.451)^3-0.0046(-0.451)^4}$
$\quad = 3.009 \quad$ at $\quad p_a^* = 10$

$\therefore \quad \bar{u}_a^* = 3.009 + \dfrac{3.009-1.104}{10-5}(8.19-10) = 2.281 \quad$ measured at p_{ao}^*.

9. $\quad p_{ao}^* = 100 - \Delta p_a^* = 100 - 8.19 = 91.81$(taking note of Figure 5.13). Using equation (5.18):

$$\frac{\bar{u}_{ae}^*}{\bar{u}_a^*} = \frac{91.81}{2.08}\left[\frac{2.08^2-10.27^2}{91.81^2-100^2}\right] = 2.84$$

$\bar{u}_a^*\big|_{\text{predicted}} = 2.3 \times 2.84 = 6.53 \quad$ measured at p_{aeo}^*

$\therefore \quad \bar{u}_a^*\big|_{\text{corrected}} = 1.1 \times 6.53 = 7.18 \quad$ measured at p_{aeo}^*

10. $\quad u_a = \dfrac{1.03 \times 10^{-13} \times 4.88 \times 10^4 \times 7.18}{1.91 \times 10^{-5} \times 0.3} = 6.3 \times 10^{-3} \quad \text{m}^3\text{m}^{-2}\text{s}^{-1}$

measured at 1.013×10^5 N m^{-2} and 40°C.

11. Design air rate $= 6.3 \times 10^{-3} \dfrac{2.08}{2.08 + \dfrac{4 \times 10^5 + 3300}{4.88 \times 10^4}}$

$= 1.27 \times 10^{-3}$ m^3 m^{-2} s^{-1} measured at 505 kN m^{-2} (i.e. 501.3 + 3.3 kN m^{-2}) and 40°C.

5.3.2 Effects of particle size, cake thickness and pressure difference

Particle size, cake thickness and the applied pressure difference all affect both the time to deliquor a cake to a specified moisture content and the average gas flow. The relative effects of these variables are demonstrated by defining a "base case", and a change in any one of the variables can then be compared with this case. For this purpose the following base case is defined:

Cake porosity	0.4
Interfacial tension	0.07 N m^{-1}
Mean size of particle in the cake	10 μm
Cake thickness	20 mm
Pressure difference	300 kPa
Filtrate density	1000 kg m^{-3}
Filtrate viscosity	10^{-3} Pa s
Air viscosity	10^{-5} Pa s
True density of the solids	2000 kg m^{-3}

It is also assumed that the filtrate is at atmospheric pressure (101.3 kPa) and that a discharged filter cake with a 12% moisture is required. The expected deliquoring time and average gas flux can be calculated by following Example 5.2, and the results are shown on Figure 5.14. The procedure outlined here can be used to evaluate changes in process or operating variables on the characteristics of specific deliquoring problems; the general trends are shown in Figure 5.14 (Wakeman, 1997).

The effect of doubling the mean particle size shortens the time required for deliquoring, but extends considerably the gas rate through the cake due to a reduced breakthrough pressure and an increased cake permeability; doubling the applied pressure difference also reduces the deliquoring time and increases the air rate, but the effect is not as great as changing the particle size. Increasing the cake thickness has the reverse effect, and leads to longer deliquoring times but reduced gas rates.

5.3.3 Summary of deliquoring calculations for centrifugal filters

Calculations for specific centrifugal deliquoring problems are based on the centrifuge parameters as well as the cake, solid and fluid properties. The following data should be available to define these:

(i) rotational speed of the centrifuge, N;

(ii) diameter/radius of the centrifuge bowl or basket, r_o;

(iii) time available in centrifuge cycle for dewatering, or, required moisture content of discharged cake;

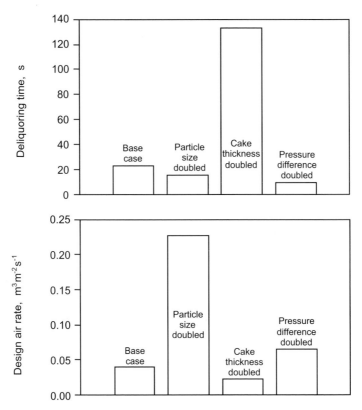

Figure 5.14 Trend effects of particle size, cake thickness and pressure difference changes on deliquoring times and gas flow rates.

(iv) typical specific cake resistance, α_{av} (which can be obtained from a pressure leaf test);

(v) typical cake porosity, ε_{av};

(vi) solids density, ρ_s;

(vii) surface tension of liquid, σ;

(viii) liquid viscosity, μ_l;

(ix) liquid density, ρ_l.

A typical calculation sequence would then take the form given below (Steps 2 and 3 are omitted if S_∞ has been obtained experimentally):

1. Calculate the effective particle size from equation (5.19);

2. Calculate the capillary number from equation (5.7);

3. Calculate the irreducible saturation from equation (5.6);

4. Calculate the cake permeability from equation (5.17). In most problems either the deliquoring time or a required cake moisture content will be known. (a) If the time available for deliquoring, t_d, is known, calculate the dimensionless time, θ, from Table 5.1. Or, (b) if the moisture content at discharge is specified and t_d is to be determined, calculate S and S_R from Step 6;

5. If θp^* is known read S_R from Figure 5.5, or, if S_R is known, read θp^*. Alternatively use equation (5.13) together with Table 5.3;

6. The cake saturation and moisture content at discharge are given by (the calculation sequence within Step 6 is reversed if moisture content is specified and t_d is unknown):

$$S = S_R(1 - S_\infty) + S_\infty \tag{5.22}$$

$$M' = S\frac{\varepsilon_{av}}{1 - \varepsilon_{av}}\frac{\rho}{\rho_s}100 \quad \% \tag{5.23}$$

where M' has the units (mass liquid/mass solid). A preferred term for moisture content is:

$$M = \frac{M'}{100 + M'}100 \quad \% \tag{5.24}$$

which has the units (mass liquid/(mass of liquid + mass of solid)).

Example 5.3 Calculations for centrifugal filters.

A batch-type centrifugal filter is to be used to dewater a suspension of inorganic crystals. The following data are available:

g-factor (F_c)	400
Inside radius of drum (r_o)	0.4 m
Inner radius of centrifuge cake (r_i)	0.36 m
Average specific cake resistance (α_{av})	10^7 m kg^{-1}
Average cake porosity (ε_{av})	0.4
Solids density (ρ_s)	2300 kg m^{-3}
Surface tension of liquid (σ)	0.07 N m^{-1}
Liquid viscosity (μ_l)	10^{-1} Pa s
Liquid density (ρ_l)	1000 kg m^{-3}

Estimate the spin time of the centrifuge required to reduce the cake moisture content to 10%.

Solution

The driving force is produced by the hydrostatic pressure resulting from the liquid on top of the cake and also from the liquid present inside the cake. In this example it is assumed that there is no liquid layer lying over the cake surface. The g-factor is defined by:

$$F_c = \frac{r_o \omega^2}{g}$$ (5.27)

In this example, $F_c = 400 = \dfrac{0.4 \omega^2}{9.81}$

∴ angular speed of the centrifuge drum, $\omega = 99$ s^{-1}

i.e. the rotational speed of the drum is $N = \dfrac{\omega}{2\pi} = 15.75$ rps (945 rpm).

Following the procedure set out above:

1. Using equation (5.19) to calculate the particle size:

$$x = 13.4 \sqrt{\frac{0.6}{10^7 \times 2300 \times 0.4^3}} = 271 \times 10^{-6} \text{ m}$$

2. Using equation (5.7) to calculate the capillary number:

$$N_{cap} = \frac{0.4^3 \times \left(271 \times 10^{-6}\right)^2 \times 1000 \times 15.75^2 \times 0.4}{0.6^2 \times 0.07} = 0.0185$$

3. Using equation (5.6) to calculate the irreducible saturation of the cake:

$$S_\infty = 0.0524(0.0185)^{-0.19} = 0.112$$

4. The cake permeability is

$$k_{av} = \frac{1}{\alpha_{av} \, \rho_s \left(1 - \varepsilon_{av}\right)} = \frac{1}{10^7 \times 2300 \times 0.6} = 7.25 \times 10^{-11} \text{ m}^2$$

The moisture content of the discharged cake is required to be 10%, hence its saturation is

$$S = \frac{1 - \varepsilon}{\varepsilon} \frac{\rho_s}{\rho_l} \frac{\dfrac{M}{100}}{1 - \dfrac{M}{100}} = \frac{0.6}{0.4} \frac{2300}{1000} \frac{0.1}{1 - 0.1} = 0.383$$

and $S_R = \dfrac{0.383 - 0.112}{1 - 0.112} = 0.305$

5. From Figure 5.5, using $\dfrac{r_i}{r_o} = \dfrac{0.36}{0.4} = 0.9$, $\theta p^* \approx 0.2$.

A more precise value can be estimated from equation (5.13) and Table 5.3:

$$S_R = 0.305 = \frac{1}{1 + 7.10\left(\theta p^*\right)^{0.70}} \qquad i.e. \ \theta p^* = 0.197$$

Using the equations in Table 5.1:

$$\theta p^* = \frac{k_{av} p_b}{\mu_l \varepsilon_{av}\left(1 - S_\infty\right) r_o^2} \frac{\rho_l \omega^2 r_o^2}{p_b} t_d$$

$$= \frac{k_{av} \rho_l \omega^2}{\mu_l \varepsilon_{av}\left(1 - S_\infty\right)} t_d = \frac{7.25 \times 10^{-11} \times 1000 \times 99^2}{10^{-1} \times 0.4 \times (1 - 0.112)} t_d = 0.02 t_d$$

where the cake permeability k_{av} is calculated from equation (5.17).

The spin time required for the dewatering is 0.197/0.02 i.e. 10 seconds.

The effects of changes in the g-factor, the particle size and the solids loading, with the base case operation given by Example 5.3, are shown on Figure 5.15. Doubling the g-factor reduces the spin time to reach the same residual moisture content in the cake; the effect of reducing the particle size is more complex, as particle size affects the specific resistance of the cake, the capillary number and hence the irreducible saturation, leading to a significant lengthening of the spin time; an increase of the solids loading primarily affects the inner radial position

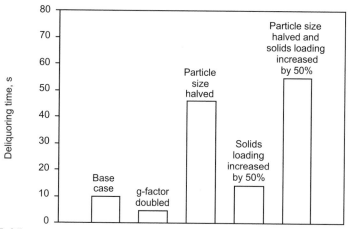

Figure 5.15 Trend effects of g-factor, particle size and solids loading on deliquoring times in a centrifuge.

of the cake surface and hence the cake thickness, which has a small effect on the capillary number and a less significant effect on the irreducible saturation of the cake.

5.4 Cake cracking

Control of filter cake behaviour towards the end of the deliquoring process is important when subsequent in situ processing of the cake is to be carried out on the filter. Side effects such as cake cracking occur because, firstly, the pore fluid is a mixture of gas (commonly air) and liquid and, secondly, the presence of gas-liquid interfaces introduces additional interparticle pressures which vary with changes in cake moisture content. The distribution of these interfaces is reflected in the tendency of the cake to crack, as is the extent to which cake shrinkage occurs during deliquoring.

Cracking and the formation of channels is most marked with compressible filter cakes. When cake washing follows deliquoring in the filter cycle, the result of these phenomena is that much of the cake is only partially washed since wash liquor passes preferentially through the cracks. The formation of cracks has been noted to occur in three stages (Wakeman, 1975b): a pinhole type fracture occurs initially, which generally leads to a "star" type cracking of the cake, which in turn propagates such that the large proportions of the cake surface are subject to cracking. This sequence of events can often be observed on operating filters, such as on rotary vacuum drum filters where the cake is exposed. The increasing extent of cake cracking with higher deliquoring pressures is illustrated in Figures 5.16a–c, where the typical result of more severe cracking is observed at higher pressures.

5.4.1 Stresses in partially saturated filter cakes

A draining filter cake is a moist agglomerate of particles which cohere because of mobile liquid surfaces (capillary and contact forces). The strength of the cake is greatly affected by its saturation by the liquid. Depending on the proportions of the voids filled with liquid the cake will fall into one of four categories:

(i) $S = 1$ – The voids are completely filled by liquid, that is, the cake is saturated;

(ii) $1 > S > 0.8$ – Some liquid has been drained from the cake and most of the undrained liquid exists in a capillary state in the voids;

Figure 5.16a
Cracking of a TiO_2 filter cake deliquored at a pressure of 100 kPa (Filtration pressure 100 kPa; 12 mm formed cake thickness).

Figure 5.16b
Cracking of a TiO_2 filter cake deliquored at a pressure of 300 kPa (Filtration pressure 300 kPa; 15 mm formed cake thickness).

Figure 5.16c
Cracking of a TiO_2 filter cake deliquored at a pressure of 500 kPa (Filtration pressure 500 kPa; 13 mm formed cake thickness).

(iii) $0.8 > S > 0.3$ – Both liquid filled regions and liquid bridges contribute to the cake strength in this so-called funicular region;

(iv) $S < 0.3$ – When there is little liquid there exists separate liquid bridges between the particles, often referred to as the pendular (or conate) state.

Deliquoring starts at $S = 1$ where the cake has a fairly low tensile or compressive strength, and as $S \to 0.9$ the cake strength increases – it is likely that the maximum strength occurs close to the point where the threshold pressure can be measured. At lower saturations the cake strength decreases, and goes to zero as $S \to 0$.

It is interesting to consider how this affects the stress distribution inside the filter cake. The saturation distributions through three types of cakes are shown on Figure 5.17. A low suction potential cake (for example, one composed of large, non-interacting particles as might be deliquored on a centrifuge) has a more or less uniform saturation in the gas inflow region and here the saturation is at the irreducible level; closer to the filter medium the saturation increases if the medium offers a significant resistance to the fluid flow. A high suction potential cake is saturated in the region close to the medium, and the saturation falls from unity to a value greater than the irreducible level at the gas inflow face. Moderate or intermediate suction potential cakes show no saturation gradients close to the filter medium (where $S = 1$) or close to the gas inflow surface (where $S = S_\infty$), but between these regions there is a gradient in the saturation. Wherever a gradient exists there is a greater likelihood of cake cracking to be observed (Wakeman, 1974).

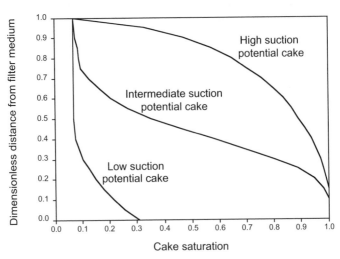

Figure 5.17 Curves showing typical saturation distributions through the thickness of filter cakes. Deliquoring stresses occur in regions where there is a saturation gradient.

It is observed in practice that cracking tends to occur with finer particle filter cakes, particularly if higher pressure differences are used for deliquoring. These correspond to higher suction potential cakes in the above categorisation, where the deliquoring stresses are concentrated into regions of the cake closest to the gas inflow surface. From a practical point of view, when handling cakes with moderate to high suction potentials incremental application of the deliquoring pressure helps to prevent cake cracking since the stress boundary moves deeper into the cake. Assessment of the tendency of a cake to crack is important for the correct location of wash liquor sprays or weirs on vacuum filters to ensure wetting of the cake before it cracks due to deliquoring. This is preferential to resorting to the use of a wash blanket or a drag blanket after cracking has occurred as such accessories can be troublesome in use.

5.5 Conclusions

This chapter sets out models for deliquoring that have been developed from fundamental principles. It has also been used to assess process engineering aspects of practical problems. Most parameters of practical importance are included in the model, making it particularly useful for predicting effects of process changes on the rate of deliquoring and on the moisture content of discharged filter cakes. To enable the calculations to be done either manually or be set up as a computer program, the deliquoring and air flow rate charts are presented in graphical and tabulated forms for the convenience of the reader. More accurate information is obtained from the model if experimental data are available for the irreducible saturation S_∞; as correlations for this parameter are often unreliable, it is recommended that this data is measured whenever possible.

6 Cake washing

Cake washing is the process of removing solutes from the voids in a filter cake by the application of a wash liquor or solvent which is miscible with the retained filtrate; soluble impurities are flushed out of the voids to leave them filled by a more pure liquid. The reasons for washing are normally either to remove liquid impurities from a valuable solid product, or to increase the recovery of a liquid product from the cake. Ideally, washing would be continued until the concentration of solute in the cake has fallen to some acceptable level, or until the maximum allowable degree of dilution of the collected filtrate has been reached. On batch filters washing can be continued until a specified concentration of solute in the cake has been obtained, but on continuous filters the time allowable for washing is frequently limited by the time essential to complete other stages (cake formation, compression, deliquoring and discharge) in the filter cycle.

Residual liquor is usually removed by one of three methods: (i) displacement washing where wash liquor is passed through the cake once, (ii) countercurrent washing where all or part of the wash liquor (sometimes mixed with filtrate) emanating from a downstream section of cake is passed through a section of the cake farther upstream, or (iii) reslurry washing where the cake is discharged from the filter and then mixed with wash liquid in a separate vessel before being filtered again (sometimes this whole operation can be carried out on the filter). There are occasionally advantages to be gained from combining two of these methods.

The preferred way to describe washing is through the use of a washing curve, which can be represented by several alternative graphs (see Section 6.1). The most basic representation is a plot of the solute concentration in the washings, normalised by the solute concentration in the liquor retained in the cake prior to washing, that is:

$$\frac{\text{solute concentration in washings from the filter cake}}{\text{solute concentration in the cake liquor before washings starts}}$$

plotted against a 'wash ratio' term defined as:

$$\text{Wash ratio} = \frac{\text{Volume of wash liquid used}}{\text{Volume of liquor in the cake at the start of washing}} \quad (6.1)$$

$$= \frac{\text{Volume flow rate of wash liquor} \times \text{Washing time}}{\text{Volume of liquor in the cake at the start of washing}}$$

In the ideal case for a displacement washing only one void volume of wash liquor is required for 100% recovery of the residual liquor. Curve 2 on Figure 6.1 is the ideal washing curve, where 100% of the solute in the cake is displaced by one wash ratio of liquid, that is, a pistonlike displacement of liquor from the cake occurs. This ideal represents a limiting situation which can never be reached in practice. In practical washings only 30–86% of the retained filtrate is removed by this one-displacement wash and the percentage of residual liquor removed by one pore volume of wash liquor (or one wash ratio in the case of an initially saturated cake) is a measure of the washing efficiency.

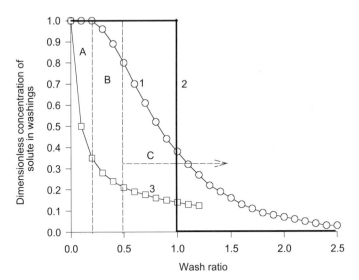

Figure 6.1 Plots of the dimensionless concentration of solute in the wash effluent (ϕ) from the cake versus the wash ratio (or alternatively wash volume used or washing time) are referred to as washing curves; curve 1 shows a typical washing curve, curve 2 shows the ideal washing curve and curve 3 shows an extreme which is obtained by washing highly deliquored cakes in, for example, a centrifuge. The initial part of curve 3 does not show any displacement washing as the pores of the cake are almost empty before washing starts. On curve 1, in region A retained liquor is displaced from the pores in the cake, region C is controlled by mixing and mass transfer processes, and B is an intermediate region.

A typical displacement washing curve, see Figure 6.1 curve 1, can be divided into two distinct regions representing different modes of solute removal from the pores of the filter cake. The initial stage (A) involves the direct displacement of the retained fluid by the wash liquor, and the final one (C) arises from a variety of local mass transfer processes that take place in the pores of the cake (see Chapter 2, Section 2.7). There is an intermediate stage (B) when some pores are subject to displacement and others to the mass transfer mechanisms.

Extreme washing curve shapes arise when cakes have been dewatered almost to their irreducible saturation level, as, for example, may be the case with centrifuge cakes (curve 3 on Figure 6.1). The mass transfer mechanism is then operable from the onset of washing and there is little or no hydraulic displacement of residual filtrate.

Intermediate curves with shapes similar to curve 1 are often found, their exact form being dependent on the cake structure and permeability and wash liquid flow characteristics. Any models which purport to determine washing characteristics must therefore be capable of predicting a wide variety of washing curve shapes and must be flexible in their applicability. More detailed analyses of the interpretations of differing forms of washing curves are available (Wakeman, 1975b, 1981a).

6.1 The wash liquid and washing curves

The effectiveness of washing depends on the rate of permeation of the wash liquid into the pores of the filter cake as well as mass transfer processes. The pore structure of the voids, of which the pore size distribution is most important, in the filter cake affects the wash liquid flow and hence also the effectiveness of washing. Wash liquid will travel more rapidly through the larger pores and breakthrough of wash liquid occurs through these larger pores prior to the use of one void volume. In addition, velocity gradients normally exist across a pore and this causes breakthrough to occur even sooner. These effects are often called "fingering" and if any choice in the use of wash liquid is available these effects can be reduced. Examination of the stability criterion suggested by Helfferich (1962), which is

$$\left(\mu_w - \mu_l\right)\frac{v}{k} > \left(\rho_w - \rho_l\right)g$$

This suggests that to minimise fingering, a wash liquid having a high viscosity and low density relative to that of the retained filtrate should be chosen. Very often the wash liquid will have a lower density than that of the filtrate or mother liquor, since it will have very little or no dissolved species in it, or it may be a low density solvent. The other criterion, that of having a higher viscosity, is not very likely in practice. Inferences implicit in the above inequality are not always borne out in practice as other criteria are more important.

The effect of increasing the temperature of the wash liquid to decrease its viscosity will favour the creation of fingers, but the increased rate of mass transfer may outweigh the disadvantages of fingering.

The use of solvents as wash liquids may be necessary in some applications in order to remove some dissolved species or to replace one liquid by another in the cake pores and if this is so then special arrangements have to be made to recover the solvents, especially if they are volatile.

Obviously the wash liquid should not cause precipitation of some other dissolved species present in the liquid phase. This could cause problems with blocking the pores of the filter cake itself, or those of the filter medium or even blocking internal pipework in the filter.

6.1.1 Measurement of washing curves

Laboratory testing in order to obtain washing data is concerned with evaluation of two main parameters:

1. How much of the soluble species will be removed by a given volume of wash liquid;

2. How long will the wash liquid take to effect such a removal.

Both displacement and reslurry washing can be investigated on the small scale. The conditions of the test work should replicate as far as is practicable the actual conditions under which the filter will operate; for example, if washing is to be carried out in a centrifuge, then measurement of data in a vacuum filter apparatus is not likely to be representative of the plant situation. In like manner, in reslurry washing the degree of agitation and the liquid-solid ratio should be similar to that which will pertain in the full scale operation. In addition, the temperature, pH, and composition of the wash liquid and slurry or cake being washed must correspond to the conditions likely to be met in practice, otherwise misleading results may be obtained.

If the washing is to be done on a vacuum filter, then an apparatus similar to that shown in Figure 4.17 may be used. The main difference is that a means of feeding wash liquid to the surface of the cake needs to be added; this is often a spray or a weir over which the wash liquid flows. Whichever is used, feeding the wash liquid to the cake should not cause any disruption of the surface. After cake formation in the way described in Section 4.4, during washing a steady flow rate of wash liquid is required to keep a thin (in the order of 1 mm deep) pool of wash liquid covering the surface of the cake. The cake should not be allowed to deliquor – this is sometimes observed when the cake surface appearance changes from being shiny or reflective to being mat or dull. After passing a fixed volume of wash liquid through the cake the experiment is stopped, the cake removed from the filter, and the amount of the soluble species remaining in the cake is measured. By repeating the experiment several times but using a different volume of wash liquid each time, the variation of the amount of the soluble species in the cake can be plotted as a function of the amount of wash liquid used. The amount of wash liquid to use in an experiment can be estimated from a knowledge of the amount of residual liquid in the cake just before washing is started (equation (4.38) *et seq* in Section 4.2.7 can help to estimate this amount if it is not measured). The wash liquid amount should be varied from about 0.5 up to 3.5 times the residual amount.

In reslurry washing, apart from the amount of wash liquid to be mixed with the cake or thickened slurry, the main parameter to be investigated is the time taken to remove the solubles from the solid phase if they are adsorbed. If there are no adsorbed solubles, then washing by reslurrying is only a matter of diluting the liquid retained in the cake or thickened slurry. The degree of agitation is important when the solubles are adsorbed, and should be similar to that likely to be experienced at full scale. The agitation stage is then followed by re-filtration, when it should be noted that the filtration characteristics measured already might have changed.

The aim of experimental washing curve determinations is to gain a knowledge of the amount of wash liquor and the washing time required to remove a given quantity of solute from a filter cake. It is usually necessary to calculate the amount of solute to be removed from a specification of the allowable solute concentration remaining in the cake at the end of the wash. The results can be plotted in a variety of ways:

(i) as the dimensionless instantaneous concentration of solute in the wash effluent plotted against the wash ratio (as in Figure 6.1),

(ii) as the dimensionless concentration of solute in the wash effluent collected in a washings receiver vessel plotted against wash ratio,

(iii) as the percentage of solute remaining in the cake plotted against wash ratio, and

(iv) as the percentage of solute removed from the cake plotted against wash ratio.

The usefulness of the wash ratio term for scale-up purposes is evident by inclusion of the wash flow rate and the washing time. However, in all but the most carefully performed experiments it is difficult to obtain the value of the wash ratio with a high degree of accuracy, and it is almost impossible to obtain an accurate value from an operating filter. The porosity of the cake is important in this context as it not only affects the liquor volume in the cake at the start of washing, but it also affects the volume flow rate of wash through the cake. When the porosity is very low, the washing rate is very low but the washing effectiveness is very high.

The maximum quantity of wash water that can be used in a given operation is dictated by plant flowsheet considerations, the shape of the washing effectiveness curve (washing curve) and the required solute content of the washed cake. On continuous filters it is frequently the case that the maximum quantity of wash water should be equivalent to a wash ratio of 1.5 to 2.5. Where high solute removals are required a two stage washing with intermediate reslurrying is often necessary; these stages may involve counter-current flow of wash water, or fresh wash may be used on both washing stages. On batch filters higher wash ratios can be employed.

There are several simple and reliable methods for monitoring washing performance. Usually the instantaneous concentration of solute (ϕ) emerging from the cake is monitored or, alternatively, the wash liquor effluent from the cake is collected and its bulk concentration (ϕ_{av}) determined; these are indicated in Figure 6.2. The former gives a more accurate measure of the washing curve, which may be important towards the end of the wash when the solute concentrations can be very low.

From many industrial filters the wash effluent is collected in a receiver vessel, making it difficult to determine effects of small changes in washing effectiveness as a result of, for example, operational changes to the process. In these instances samples of cake are often collected and analysed for their solute content. For these reasons it is necessary to be able to convert between the instantaneous concentration of solute in the wash effluent, the concentration in the mixed effluent and the amount remaining in the cake.

Whenever washing data are being obtained from plant or laboratory experiments it is important to measure also the solute concentration in the cake, even though the solute concentration in the washing may have

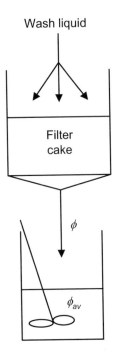

Wash liquid

Filter cake

ϕ

ϕ_{av}

Figure 6.2 The variation of the instantaneous concentration of solute (ϕ) with time or wash volume is measured continuously in the wash effluent or by analysing samples of the effluent; the time variation of the concentration in a mixed effluent (ϕ_{av}) is obtained by continuously collecting and mixing the wash effluent.

been measured carefully. The reason for this is that porous particles or poor flow distributions in the cake lead to hold-up of solute during washing that is not always detectable from measurements of solute concentration in the washings, and also to establish a solute mass balance over the washing operation.

6.1.2 Representations of washing data

As pointed out in the above, washing data can be represented in a number of ways. The first is a plot of solute concentration in the washing from the cake against washing time on a dimensionless basis i.e. a plot of ϕ/ϕ_0 versus the wash ratio W. Sometimes pore volumes are reported in the literature rather than wash ratio, but the term is seldom used in practice because of the problems that are usually associated with determining the porosity of the filter cake. The basic terms are related and defined (noting the definition in equation (6.1)) as:

$$\text{Wash ratio} = W = \frac{Qt}{\varepsilon_{av} ALS} \tag{6.2}$$

$$= \frac{\text{volume of wash liquid used}}{\text{void volume of the cake} \times \text{cake saturation at start of washing}}$$

$$= \frac{\text{number of pore volumes of wash liquid used}}{S}$$

where Q is the volume flow rate of wash liquid actually passing through the cake, t is the time of washing, ε_{av}, A and L are the cake average porosity, surface area and thickness, and S is the cake saturation at the start of washing.

An alternative representation of the data is to plot the fraction of solute removed (F) from the filter cake against the wash ratio, where the fractional solute removal (sometimes expressed as a percentage) is given by:

$$F = \int_0^W \frac{\phi - \phi_w}{\phi_0 - \phi_w} \, dW \tag{6.3}$$

where ϕ_w is the solute concentration in the wash liquor feed (usually $\phi_w = 0$). The general shape of a fractional recovery curve is shown in Example 6.1.

The percentage of solute remaining in the cake (R) may also be plotted against the wash ratio, where R is defined by:

$$R = 1 - F = 1 - \int_0^W \frac{\phi - \phi_w}{\phi_0 - \phi_w} \, dW \tag{6.4}$$

The value of F should approach 1 asymptotically (or R should approach 0), and is obtainable by graphical integration of the washing curve.

6.1.3 "Washing efficiency"

Although not often used, it can be convenient in practice to refer to "wash efficiency"; this is generally taken to mean the fraction of solute removed by one wash ratio, i.e.

$$E = F \quad \text{at} \quad W = 1$$

$$\therefore \quad E = \int_0^1 \frac{\phi - \phi_w}{\phi_0 - \phi_w} \, dW \tag{6.5}$$

Clearly this is not strictly an efficiency, but it does express the effectiveness of washing any particular cake. An ideal washing would be obtained when $E = 1$, that is, all solutes are removed from the cake by one wash ratio.

Example 6.1 The analysis of test data.
The data from a washing test using solute-free wash liquor ($\phi_w = 0$) were measured to be:

ϕ/ϕ_0	1.0	1.0	0.9	0.8	0.4	0.1	0.06	0.04	0.01
W	0	0.4	0.68	0.75	1.0	1.25	1.375	1.5	2.7

Determine the fractional recovery of solute curve for this washing.

Solution

W	$\displaystyle\int_{W_1}^{W_2} \frac{\phi}{\phi_0}\, dW$	$\displaystyle F = \int_{0}^{W} \frac{\phi}{\phi_0}\, dW$
0	0	0
0.4	1×0.4 = 0.4	0.40
0.68	0.95×0.28 = 0.266	0.666
0.75	0.85×0.07 = 0.0595	0.725
1.0	0.6×0.25 = 0.15	0.875
1.25	0.25×0.25 = 0.0625	0.938
1.375	0.08×0.125 = 0.01	0.948
1.5	0.05×0.125 = 0.0062	0.954
2.7	0.025×1.2 = 0.03	0.984

The fractional recovery of solute curve is a plot of F versus W and is shown on Figure 6.3.

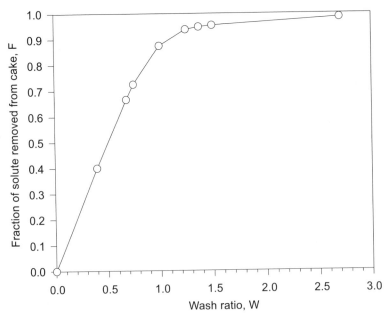

Figure 6.3 The fractional recovery of solute removed from a filter cake plotted against the wash ratio for the washing data given in Example 6.1.

After analysing the data from a washing experiment, then the mass balance on the amount of solute should be checked to ensure the integrity of the data – the mass of solute in the cake before washing must be equal to the amount of solute in the cake after washing plus the amount removed by the wash liquid. Or, alternatively, $F \leq 1$. This is also a convenient way to ensure that the experiment has been carried out correctly.

6.2 Cake washing models for design

One of the foremost problems in modelling a washing operation is the selection of a theory or concept which gives a reasonable interpretation of experimental data, and which can then be expressed in a general predictive way so that changes in process conditions can be accounted for and/or new designs assessed. The so-called dispersion model gives a good description of washing under a wide variety of conditions and has been developed for practical use. Equation (2.88) with boundary condition (2.94) describes cake washing; solutions to this equation are shown in Figure 6.4. The data for Figure 6.4 are given in Table 6.1.

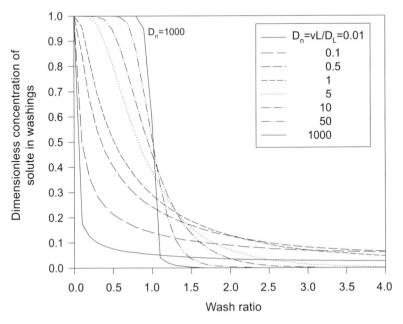

Figure 6.4 Washing curves plotted as the dimensionless instantaneous concentration of solute in the wash effluent $(\phi - \phi_w/\phi_0 - \phi_w)$ where ϕ_0 is the solute concentration in the wash effluent at $t \rightarrow 0$, i.e. the solute concentration in the liquor retained in the cake prior to washing, plotted against the wash ratio (W). The parameter on the plot is the dispersion number (D_n).

Table 6.1 The variation of the instantaneous concentration of solute in the wash effluent during washing of a saturated cake; this is the solute concentration of wash discharged from the cake before any mixing of the wash effluent is allowed to occur. The same data are plotted on Figure 6.4

Wash Ratio W	Instantaneous concentration of solute in wash effluent, $\phi^* = (\phi-\phi_w)/(\phi_0-\phi_w)$ Dispersion number, $D_n = vL/D_L$											
	0.01	0.05	0.1	0.5	1	5	10	50	100	500	1000	10000
0.0	1.000	1.000	1.000	1.000	1.000	1.000	1.000	1.000	1.000	1.000	1.000	1.000
0.1	0.176	0.369	0.500	0.855	0.959	1.000	1.000	1.000	1.000	1.000	1.000	1.000
0.2	0.123	0.259	0.355	0.666	0.821	0.997	1.000	1.000	1.000	1.000	1.000	1.000
0.3	0.100	0.208	0.285	0.546	0.690	0.966	0.997	1.000	1.000	1.000	1.000	1.000
0.4	0.086	0.179	0.245	0.463	0.592	0.898	0.986	1.000	1.000	1.000	1.000	1.000
0.5	0.076	0.159	0.214	0.406	0.513	0.807	0.945	1.000	1.000	1.000	1.000	1.000
0.6	0.069	0.144	0.193	0.358	0.449	0.706	0.870	0.995	1.000	1.000	1.000	1.000
0.7	0.064	0.131	0.176	0.321	0.400	0.610	0.766	0.968	0.993	1.000	1.000	1.000
0.8	0.061	0.121	0.161	0.289	0.357	0.524	0.657	0.866	0.945	1.000	1.000	1.000
0.9	0.056	0.114	0.150	0.263	0.321	0.445	0.548	0.700	0.766	0.931	0.950	0.960
1.0	0.054	0.107	0.141	0.243	0.290	0.383	0.455	0.519	0.531	0.583	0.600	0.620
1.1	0.051	0.100	0.132	0.224	0.263	0.324	0.366	0.334	0.291	0.141	0.040	0.020
1.2	0.048	0.095	0.124	0.207	0.241	0.271	0.286	0.197	0.139	0.061	0.020	0.010
1.3	0.046	0.090	0.117	0.192	0.220	0.226	0.224	0.110	0.063	0.037	0.010	0.000
1.4	0.044	0.086	0.111	0.179	0.201	0.188	0.172	0.062	0.034	0.022	0.007	0.000
1.5	0.042	0.083	0.107	0.169	0.186	0.159	0.137	0.035	0.020	0.015	0.003	0.000
1.6	0.041	0.080	0.103	0.158	0.172	0.132	0.103	0.021	0.013	0.010	0.001	
1.7	0.039	0.076	0.098	0.148	0.159	0.112	0.080	0.012	0.010	0.007	0.000	
1.8	0.038	0.074	0.095	0.139	0.147	0.096	0.061	0.008	0.006	0.004	0.000	
1.9	0.036	0.072	0.092	0.133	0.138	0.082	0.047	0.006	0.004	0.002	0.000	
2.0	0.035	0.069	0.089	0.126	0.128	0.070	0.035	0.004	0.002	0.000		
2.1	0.034	0.067	0.086	0.119	0.121	0.062	0.028	0.003	0.001	0.000		
2.2	0.033	0.065	0.083	0.114	0.111	0.054	0.021	0.002	0.000	0.000		
2.3	0.033	0.063	0.081	0.109	0.106	0.043	0.015	0.001	0.000			
2.4	0.032	0.061	0.079	0.104	0.100	0.037	0.012	0.000	0.000			
2.5	0.031	0.059	0.076	0.098	0.093	0.031	0.008	0.000				
2.6	0.031	0.058	0.074	0.094	0.086	0.027	0.006	0.000				
2.7	0.030	0.057	0.073	0.090	0.081	0.024	0.005					
2.8	0.030	0.056	0.071	0.087	0.077	0.022	0.004					
2.9	0.029	0.055	0.070	0.084	0.073	0.016	0.003					
3.0	0.029	0.054	0.069	0.081	0.069	0.013	0.002					
3.1	0.029	0.053	0.068	0.079	0.066	0.011	0.001					
3.2	0.028	0.052	0.067	0.077	0.063	0.010	0.000					
3.3	0.028	0.051	0.066	0.075	0.061	0.009	0.000					
3.4	0.028	0.050	0.065	0.073	0.059	0.007	0.000					
3.5	0.027	0.049	0.064	0.071	0.057	0.006						
3.6	0.027	0.049	0.063	0.069	0.054	0.005						
3.7	0.027	0.048	0.063	0.068	0.052	0.004						
3.8	0.027	0.047	0.062	0.067	0.050	0.003						
3.9	0.027	0.047	0.062	0.066	0.048	0.002						
4.0	0.026	0.046	0.061	0.065	0.046	0.001						

On Figure 6.4 the dispersion number ($D_n = vL/D_L$) has a significant effect on the shape of the washing curve, and hence on the effectiveness of washing. As this number is increased, the shape of the curve approaches that which would be associated with an ideal washing situation (Curve 2 on Figure 6.1). The dispersion number is properly interpreted in terms of the Reynolds number (Re) for wash liquor flow through the cake and the Schmidt number (Sc) describing the ratio of the molecular diffusivity of momentum to the molecular diffusivity of mass, i.e.

$$D_n = \frac{vL}{D_L} = \frac{\rho vx}{\mu} \frac{\mu}{\rho D} \frac{L}{x} \frac{D}{D_L} = Re\,Sc\,\frac{L}{x}\,\frac{D}{D_L} \tag{6.6}$$

The ratio of the molecular diffusion coefficient (D) to the axial dispersion coefficient (D_L) is dependent on the tortuosity of the cake, an increase in tortuosity generally corresponding to a decrease in the dispersion coefficient. Tortuosity factors of $\sqrt{2}$ have been found generally suitable, hence:

$$\frac{D_L}{D} = \frac{1}{\sqrt{2}} = 0.707 \tag{6.7}$$

for ReSc < 1. The dispersion coefficient for larger values of the ReSc product has been correlated for bed thicknesses greater than about 10 cm by:

$$\frac{D_L}{D} = 0.707 + 1.75\,Re\,Sc \tag{6.8}$$

and for L < 10 cm by

$$\frac{D_L}{D} = 0.707 + 55.5(Re\,Sc)^{0.96} \tag{6.9}$$

Higher value dispersion numbers (vL/D_L) are associated with more effective washings, which are achieved with small values of the axial dispersion coefficient or thicker filter cakes. Equations (6.8) and (6.9) can be rearranged into $D_L \propto vx$, that is smaller dispersion coefficients are obtained with cakes composed of smaller particles and lower wash velocities. (At any applied pressure difference across the cake, lower wash velocities would naturally be associated with finer particle cakes due to the lower permeability of the cake). Hence it follows that more effective washing should be found with thick filter cakes of fine particles.

In terms of the potential usefulness of a mathematical model, the dispersion model has probably come closest to being of practical use. Its validity has been demonstrated with experimental data for a wide range of suspension types; these include fibre pulps (Sherman, 1964; Gren, 1972), fine chemicals and precipitates (Jemaa et al, 1974; Wakeman and Attwood, 1988), lime mud (Eriksson and Theliander, 1994; Eriksson et al, 1996) and clays (Michaels et al, 1967).

Values of diffusion coefficients are available in books concerned with mass transfer. Cake washing calculations usually use literature values of the binary molecular diffusion coefficient in a pure solvent to estimate the axial diffusion coefficient. However, it should be recognised that the presence of many solutes in complex solutions can alter the pure liquid diffusion coefficients quite markedly, particularly when the solutes show strong thermodynamic interactions or when their concentrations are high. Although it is always preferable to use a reasonably precise value for the diffusion coefficient in calculations, an approximate value is available since the infinite dilution diffusion coefficient of ions in water and the diffusion coefficients of many liquids are of the order 10^{-9} m^2 s^{-1}. The precision of the value used is of lesser importance when estimating washing performance on industrial equipment, since large correction factors (see Section 6.3.1) are applied to the calculated dispersion numbers. For example, using equation (6.14) calculated dispersion numbers of 80, 800 and 8000 lead to practical values for use in filter sizing calculations of 6.4, 9.5 and 12.6.

When integrating Figure 6.4 (or the data in Table 6.1) into a design procedure it is convenient to have equations that give a good representation of the data; Table 6.2 presents sets of equations which can be used for this purpose.

Table 6.2 Curvefit coefficients for W vs ϕ^* relationships at various $D_n = vL/D_L$ values, representing the curves on Figure 6.4

W vs ϕ^* relation		Region of validity
$D_n = 0.01$	$\phi^*=1$	$0 \leq W < 0.1$
	$\phi^*=0.0524W^{-0.5311}$	$W \geq 0.1$
$D_n = 0.05$	$\phi^*=1$	$0 \leq W < 0.1$
	$\phi^*=0.1036W^{-0.5848}$	$W \geq 0.1$
$D_n = 0.1$	$\phi^*=1$	$0 \leq W < 0.1$
	$\phi^*=0.1364W^{-0.6037}$	$W \geq 0.1$
$D_n = 0.5$	$\phi^*=1$	$0 \leq W < 0.1$
	$\phi^*=1.0754-2.6090W+3.5187W^2-2.3065W^3+0.5687W^4$	$0.1 \leq W < 1.6$
	$\phi^*=0.2481W^{-0.9950}$	$W \geq 1.6$
$D_n = 1$	$\phi^*=1$	$0 \leq W < 0.1$
	$\phi^*=1.1433-1.9882W+1.8761W^2-0.9206W^3+0.1798W^4$	$0.1 \leq W <1.7$
	$\phi^*=0.3515W^{-1.4654}$	$W \geq 1.7$
$D_n = 5$	$\phi^*=1$	$0 \leq W \leq 0.1$
	$\phi^*=1.0583+0.0795W-1.7285W^2+1.2410W^3-0.2603W^4$	$0.1 < W < 2$
	$\phi^*=2.1739\exp(-1.7383W)$	$2 \leq W < 4.2$
	$\phi^*=0$	$W \geq 4.2$
$D_n = 10$	$\phi^*=1$	$0 \leq W \leq 0.2$
	$\phi^*=0.6630+2.3569W-4.9493W^2+2.9684W^3-0.5826W^4$	$0.2 < W < 1.7$
	$\phi^*=11.5698\exp(-2.9575W)$	$1.7 \leq W < 3.3$
	$\phi^*=0$	$W \geq 3.3$
$D_n = 50$	$\phi^*=1$	$0 \leq W \leq 0.5$
	$\phi^*=-5.3263+26.272W-37.56W^2+21.47W^3-4.3437W^4$	$0.5 < W < 1.4$
	$\phi^*=44.08\exp(-4.667W)$	$1.4 \leq W < 2.4$
	$\phi^*=0$	$W \geq 2.4$
$D_n = 100$	$\phi^*=1$	$0 \leq W \leq 0.6$
	$\phi^*=-10.01+39.21W-46.46W^2+19.90W^3-2.122W^4$	$0.6 < W < 1.1$
	$\phi^*=0.5789W^{-8.0948}$	$1.1 \leq W < 2.2$
	$\phi^*=0$	$W \geq 2.2$
$D_n = 500$	$\phi^*=1$	$0 \leq W \leq 0.8$
	$\phi^*=-16.77+53.01W-49.87W^2+14.23W^3$	$0.8 < W < 1.1$
	$\phi^*=0.3095W^{-7.5097}$	$1.1 \leq W < 2.1$
	$\phi^*=0$	$W \geq 2.1$
$D_n = 1000$	$\phi^*=1$	$0 \leq W \leq 0.8$
	$\phi^*=1.0583+0.0795W-1.7285W^2+1.2410W^3$	$0.8 < W < 1.1$
	$\phi^*=94.13\exp(6.9979W)$	$1.1 \leq W < 1.8$
	$\phi^*=0$	$W \geq 1.8$
$D_n = 10000$	$\phi^*=1$	$0 \leq W \leq 0.8$
	$\phi^*=-9.7886+26.83W-17.72W^2+1.3069W^3$	$0.8 < W < 1.1$
	$\phi^*=0.0427W^{-7.9662}$	$1.1 \leq W \leq 1.4$
	$\phi^*=0$	$W > 1.4$

Based on equation (6.3) a chart can be developed for the calculation of the fractional removal of solute from saturated filter cakes. The set of curves on Figure 6.4 shows the variation of solute concentration with the usage of wash liquor, when the cake is fully saturated prior to washing. Figure 6.5 illustrates a corresponding family of curves for the fraction of soluble material recovered from the cake.

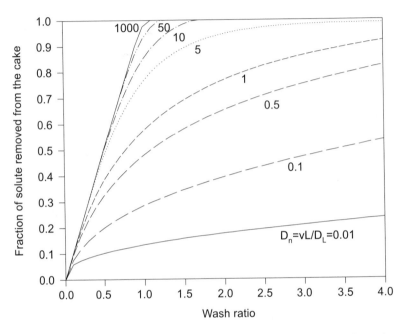

Figure 6.5 The variation of the fraction of solute removed (*F*) from a saturated filter cake with wash ratio (*W*) and dispersion number (D_n), obtained by integration of the corresponding curves in Figure 6.4.

Wash efficiency, *E*, defined as the percentage of solute removed from the cake by one wash ratio, has been adopted by practitioners as a useful tool for comparing different washings. The curves on Figure 6.5, or the corresponding data in Table 6.3, indicate a wide range of wash efficiencies varying from a low value up to almost 100%. These calculated high efficiencies can be obtained only in carefully controlled laboratory washing experiments, when the model boundary conditions can be matched. Due to the inability to match laboratory conditions, washing operations on industrial filters generally show lower efficiencies of between 30% and 86%.

Table 6.3 The variation of the fraction of solute recovered from an initially saturated cake; the same data are plotted on Figure 6.5

Wash Ratio W	Fractional removal of solute from the cake, $F = \int_0^W \dfrac{\phi - \phi_w}{\phi_0 - \phi_w} dW$											
	Dispersion number, $D_n = vL/D_L$											
	0.01	0.05	0.1	0.5	1	5	10	50	100	500	1000	10000
0.0	0.000	0.000	0.000	0.000	0.000	0.000	0.000	0.000	0.000	0.000	0.000	0.000
0.1	0.059	0.068	0.075	0.093	0.098	0.100	0.100	0.100	0.100	0.100	0.100	0.100
0.2	0.074	0.100	0.118	0.169	0.187	0.200	0.200	0.200	0.200	0.200	0.200	0.200
0.3	0.085	0.123	0.150	0.229	0.262	0.298	0.300	0.300	0.300	0.300	0.300	0.300
0.4	0.094	0.143	0.176	0.280	0.327	0.391	0.399	0.400	0.400	0.400	0.400	0.400
0.5	0.102	0.159	0.199	0.323	0.382	0.476	0.496	0.500	0.500	0.500	0.500	0.500
0.6	0.110	0.175	0.220	0.361	0.430	0.552	0.586	0.600	0.600	0.600	0.600	0.600
0.7	0.116	0.188	0.238	0.395	0.472	0.618	0.668	0.698	0.700	0.700	0.700	0.700
0.8	0.122	0.201	0.255	0.426	0.510	0.675	0.739	0.790	0.797	0.800	0.800	0.800
0.9	0.128	0.213	0.270	0.454	0.544	0.723	0.799	0.868	0.882	0.897	0.898	0.898
1.0	0.134	0.224	0.285	0.479	0.575	0.764	0.850	0.929	0.947	0.972	0.975	0.977
1.1	0.139	0.234	0.299	0.502	0.602	0.800	0.891	0.971	0.988	1.000	1.000	1.000
1.2	0.144	0.244	0.311	0.524	0.628	0.830	0.923	0.998	1.000	1.000	1.000	1.000
1.3	0.149	0.253	0.323	0.544	0.651	0.854	0.949	1.000	1.000	1.000	1.000	1.000
1.4	0.153	0.262	0.335	0.562	0.672	0.875	0.969	1.000	1.000			
1.5	0.158	0.270	0.346	0.580	0.691	0.892	0.984	1.000				
1.6	0.162	0.279	0.356	0.596	0.709	0.907	0.996					
1.7	0.166	0.286	0.366	0.611	0.725	0.919	1.000					
1.8	0.169	0.294	0.376	0.626	0.741	0.930	1.000					
1.9	0.173	0.301	0.385	0.639	0.755	0.938	1.000					
2.0	0.177	0.308	0.394	0.652	0.768	0.946						
2.1	0.180	0.315	0.403	0.664	0.781	0.953						
2.2	0.184	0.322	0.412	0.676	0.792	0.959						
2.3	0.187	0.328	0.420	0.687	0.803	0.963						
2.4	0.190	0.334	0.428	0.698	0.813	0.967						
2.5	0.193	0.340	0.435	0.708	0.823	0.971						
2.6	0.196	0.346	0.443	0.718	0.832	0.974						
2.7	0.199	0.352	0.450	0.727	0.840	0.976						
2.8	0.202	0.357	0.458	0.736	0.848	0.978						
2.9	0.205	0.363	0.465	0.744	0.856	0.980						
3.0	0.208	0.368	0.472	0.752	0.863	0.982						
3.1	0.211	0.374	0.478	0.760	0.870	0.983						
3.2	0.214	0.379	0.485	0.768	0.876	0.984						
3.3	0.217	0.384	0.492	0.776	0.882	0.985						
3.4	0.220	0.389	0.498	0.783	0.888	0.986						
3.5	0.222	0.394	0.505	0.790	0.894	0.986						
3.6	0.225	0.399	0.511	0.797	0.900	0.987						
3.7	0.228	0.404	0.517	0.804	0.905	0.987						
3.8	0.231	0.409	0.524	0.811	0.910	0.988						
3.9	0.233	0.413	0.530	0.818	0.915	0.988						
4.0	0.236	0.418	0.536	0.824	0.920	0.988						

6.3 Continuous washing processes

Continuous washing can be carried out on either batch or continuous filters, and is used to describe that process when wash effluent is continuously mixed during the washing operation. This is indicated schematically in Figure 6.2 for a batch filter and Figure 6.6 for a continuous filter.

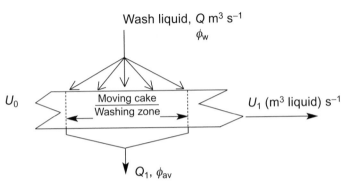

Figure 6.6 Definition of terms used to describe washing on a continuous filter. When the cake does not consolidate during washing and it enters and leaves the washing zone saturated, $U_0 = U_1$ and $Q = Q_1$ (Wakeman, 1997).

Wash liquor applied to a batch filter results in liquor of more or less equal solute strength, ϕ, coming from the entire outlet surface of the cake, all parts of the cake having been subject to washing for the same time period. However, when the wash effluent is collected in a single receiver, liquid of varying strength is mixed to give an average solute concentration ϕ_{av}, that is:

$$\phi_{av} = \frac{1}{t}\int_0^t \phi \, dt \qquad (6.10)$$

where t is the duration of the wash. When wash liquor is applied to a continuous filter with cake moving through the washing zone of length δ at a velocity ψ, then each elemental length of cake $\Delta\delta$ is subjected to washing for a time $t = \Delta\delta/\psi$. The wash effluent from each element is mixed on the downstream side of the cake, and its solute strength is again given by equation (6.10) but where $t = \delta/\psi$. The curves on Figure 6.4 describe the variation of ϕ with time (or W), and for most filter applications must be integrated according to equation (6.10); these curves also describe laboratory washing data, when it is more convenient to measure ϕ rather than ϕ_{av} (which is rarely the case on an industrial filter). Equation (6.10) can be written in terms of the wash ratio W and dimensionless concentrations and related to the fractional removal of solute from the cake F:

$$\phi^*_{av} = \frac{\phi_{av} - \phi_w}{\phi_0 - \phi_w} = \frac{1}{W} \int_0^W \frac{\phi - \phi_w}{\phi_0 - \phi_w} dW = \frac{F}{W} \tag{6.11}$$

Whereas on a batch filter equation (6.2) is useful (with $S = 1$), that is

$$W = \frac{Qt}{\varepsilon_{av} AL} \tag{6.12}$$

where $(\varepsilon_{av} AL)$ is the pore volume in the cake, on a continuous filter

$$W = \frac{Q}{U} \tag{6.13}$$

where U is the flow rate of liquid moving with the cake solids.

The variation of the solute concentration ϕ_{av} in the mixed wash effluent with wash ratio W and dispersion number is shown in Figure 6.7 (and corresponding data are given in Table 6.4). These curves are obtained by suitable integration of the curves in Figure 6.4.

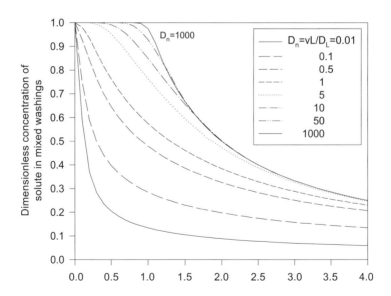

Figure 6.7 Variation of the solute concentration in the wash effluent from a continuous filter (or continuously mixed washings from a batch filter) (ϕ^*_{av}) with wash ratio (W) and dispersion number (D_n). Data for the curves are given in Table 6.4. As with washing on batch filters, the most effective washings are obtained with high dispersion numbers, or thicker cakes of finer particles.

Table 6.4 The variation of the average concentration of solute in a mixed wash effluent during washing of an initially saturated cake; the same data are plotted on Figure 6.7

Wash Ratio W	Average solute concentration in a mixed wash effluent, $\phi_{av}^* = (\phi_{av}-\phi_w)/(\phi_0-\phi_w)$ Dispersion number, $D_n = vL/D_L$											
	0.01	0.05	0.1	0.5	1	5	10	50	100	500	1000	10000
0.0	1.000	1.000	1.000	1.000	1.000	1.000	1.000	1.000	1.000	1.000	1.000	1.000
0.1	0.588	0.685	0.750	0.928	0.980	1.000	1.000	1.000	1.000	1.000	1.000	1.000
0.2	0.369	0.499	0.589	0.844	0.935	1.000	1.000	1.000	1.000	1.000	1.000	1.000
0.3	0.283	0.411	0.499	0.765	0.875	0.993	1.000	1.000	1.000	1.000	1.000	1.000
0.4	0.236	0.356	0.441	0.700	0.816	0.978	0.998	1.000	1.000	1.000	1.000	1.000
0.5	0.205	0.319	0.398	0.647	0.764	0.953	0.991	1.000	1.000	1.000	1.000	1.000
0.6	0.183	0.291	0.366	0.602	0.717	0.920	0.977	1.000	1.000	1.000	1.000	1.000
0.7	0.166	0.269	0.340	0.565	0.675	0.883	0.954	0.997	1.000	1.000	1.000	1.000
0.8	0.153	0.251	0.319	0.532	0.638	0.843	0.924	0.987	0.996	1.000	1.000	1.000
0.9	0.143	0.236	0.300	0.504	0.605	0.803	0.888	0.964	0.980	0.996	0.997	0.998
1.0	0.134	0.224	0.285	0.479	0.575	0.764	0.850	0.929	0.947	0.972	0.975	0.977
1.1	0.126	0.213	0.271	0.457	0.548	0.727	0.810	0.883	0.898	0.917	0.915	0.917
1.2	0.120	0.203	0.259	0.436	0.523	0.691	0.769	0.832	0.841	0.842	0.836	0.835
1.3	0.114	0.195	0.249	0.418	0.500	0.657	0.730	0.780	0.777	0.773	0.770	0.770
1.4	0.109	0.187	0.239	0.402	0.480	0.625	0.692	0.720	0.718	0.716	0.715	0.714
1.5	0.105	0.180	0.230	0.386	0.461	0.595	0.656	0.670	0.668	0.668	0.667	0.667
1.6	0.101	0.174	0.223	0.372	0.443	0.567	0.623	0.627	0.626	0.626	0.625	0.625
1.7	0.097	0.168	0.215	0.360	0.427	0.541	0.591	0.589	0.589	0.589	0.588	0.588
1.8	0.094	0.163	0.209	0.348	0.412	0.516	0.559	0.556	0.556	0.556	0.556	0.556
1.9	0.091	0.158	0.203	0.336	0.397	0.494	0.529	0.527	0.527	0.526	0.526	0.526
2.0	0.088	0.154	0.197	0.326	0.384	0.473	0.502	0.500	0.500	0.500	0.500	0.500
2.1	0.086	0.150	0.192	0.316	0.372	0.454	0.478	0.476	0.476	0.476	0.476	0.476
2.2	0.083	0.146	0.187	0.307	0.360	0.436	0.456	0.455	0.455	0.455	0.455	0.455
2.3	0.081	0.143	0.183	0.299	0.349	0.419	0.436	0.435	0.435	0.435	0.435	0.435
2.4	0.079	0.139	0.178	0.291	0.339	0.403	0.417	0.417	0.417	0.417	0.417	0.417
2.5	0.077	0.136	0.174	0.283	0.329	0.388	0.400	0.400	0.400	0.400	0.400	0.400
2.6	0.076	0.133	0.170	0.276	0.320	0.374	0.385	0.385	0.385	0.385	0.385	0.385
2.7	0.074	0.130	0.167	0.269	0.311	0.362	0.371	0.370	0.370	0.370	0.370	0.370
2.8	0.072	0.128	0.163	0.263	0.303	0.349	0.357	0.357	0.357	0.357	0.357	0.357
2.9	0.071	0.125	0.160	0.257	0.295	0.338	0.345	0.345	0.345	0.345	0.345	0.345
3.0	0.069	0.123	0.157	0.251	0.288	0.327	0.333	0.333	0.333	0.333	0.333	0.333
3.1	0.068	0.121	0.154	0.245	0.281	0.317	0.323	0.323	0.323	0.323	0.323	0.323
3.2	0.067	0.118	0.152	0.240	0.274	0.307	0.313	0.313	0.313	0.313	0.313	0.313
3.3	0.066	0.116	0.149	0.235	0.267	0.298	0.303	0.303	0.303	0.303	0.303	0.303
3.4	0.065	0.114	0.147	0.230	0.261	0.290	0.294	0.294	0.294	0.294	0.294	0.294
3.5	0.064	0.113	0.144	0.226	0.255	0.282	0.286	0.286	0.286	0.286	0.286	0.286
3.6	0.063	0.111	0.142	0.222	0.250	0.274	0.278	0.278	0.278	0.278	0.278	0.278
3.7	0.062	0.109	0.140	0.217	0.245	0.267	0.270	0.270	0.270	0.270	0.270	0.270
3.8	0.061	0.108	0.138	0.213	0.239	0.260	0.263	0.263	0.263	0.263	0.263	0.263
3.9	0.060	0.106	0.136	0.210	0.235	0.253	0.256	0.256	0.256	0.256	0.256	0.256
4.0	0.059	0.105	0.134	0.206	0.230	0.247	0.250	0.250	0.250	0.250	0.250	0.250

The fraction of recoverable solute removed from the cake is defined similarly for both batch and continuous washing operations and is given by equation (6.3) (or Figure 6.5 and Table 6.3).

6.3.1 Correction factors for full scale filters

Washing efficiencies on continuous vacuum filters actually vary from a low of about 35% to a high of 86% for various applications (Dahlstrom, 1978). Dahlstrom has also suggested that a correction factor be used with empirical equations to take into account full scale deviations, caused by uneven cake thicknesses across the filtration area and maldistributions of wash water across the cake. The correction equates to lowering the laboratory wash efficiency by approximately 10%, the actual value being dependent on the type of filter and method of wash fluid application.

Comparison of laboratory data with full-scale experiences has suggested that a correction factor of this sort is necessary, and a correction must also be applied to allow for deviations from the theoretical model conditions which cannot be met in full-scale practice. Rather than correct the calculated efficiency, it is more convenient to correct the dispersion number (vL/D_L) for laboratory washing data before it is applied to full-scale applications. The eventual result is, of course, tantamount to a correction factor on the washing efficiency. Figure 6.8 shows an empirical relationship, based on an analysis of data from

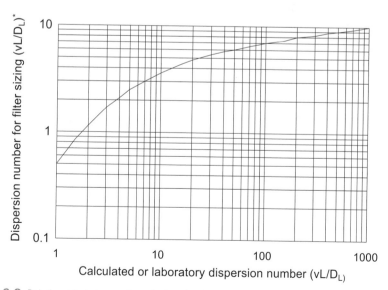

Figure 6.8 Relationship between the calculated or laboratory dispersion number and the corrected value to be used in the estimation of washing behaviour on rotary drum filters (Purchas and Wakeman, 1986).

rotary drum filters, between the calculated or laboratory dispersion number (vL/D_L) and the dispersion number appropriate for rotary drum filter sizing $(vL/D_L)^*$. Similar analyses are currently not available for other types of filters. The relationship on Figure 6.8 is described by (Rushton *et al*, 1996):

$$\left(\frac{vL}{D_L}\right)^* = 0.5 + 1.35 \ln\left(\frac{vL}{D_L}\right) \tag{6.14}$$

When the quantity of wash liquor required to accomplish a washing has been estimated, in order to estimate the total volume of liquor to be delivered to rotary drum filters another factor should be applied to account for splashing and run-off from the cake surface, that is, to account for the wash liquid which does not actually pass through the cake. The fraction of wash water going through the cake has been found to be between 64 and 100% (Koudstaal, 1976), but this must inevitably be largely affected by the care exercised during the filter operation. The assumption that 70% of the wash liquor actually passes through the cake is probably conservative.

In the case of plate and frame presses, the calculated wash ratio should be increased to account for the wash volume required to fill the feed channels in the press. This may double the amount of wash liquid that needs to be supplied to the filter, but the extent of the increase varies from one installation to another and may be estimated from a knowledge of the pipework volume that needs to be filled (including the feed channels inside the filter).

Example 6.2 Calculation for washing on a rotary vacuum filter.
Calculate the washing time and the number of wash ratios required, and assess the possibility of recovering 95% of the solute from the cake on a rotary drum filter on which 29% of the cycle is available for washing. The following data are available:

Pressure difference (Δp):	26 kPa
Cake thickness (L):	0.05 m
Average specific cake resistance (α_{av}):	10^9 m kg^{-1}
Average cake porosity (ε_{av}):	0.48
Solids density (ρ_s):	2400 kg m^{-3}
Viscosity of wash liquor (μ_w):	10^{-3} Pa s
Diffusivity of solute (D):	10^{-9} m^2 s^{-1}
Medium resistance (R):	10^9 m^{-1}
Drum speed (D_s):	0.004 revs s^{-1}

Solution

An estimate of the effective particle size is calculable from equation (5.19):

$$x = 13.4 \sqrt{\frac{0.52}{10^9 \times 2400 \times 0.48^3}} = 1.9 \times 10^{-5} \quad m$$

The superficial velocity of the wash flowing through the cake is given by Darcy's Law:

$$u = \text{superficial velocity of wash liquor} = \frac{\Delta p}{\mu_w \left[\alpha_{av} \rho_s (1 - \varepsilon_{av}) L + R \right]} \quad (6.15)$$

$$= \frac{26 \times 10^3}{10^{-3} \left[10^9 \times 2400 \times 0.52 \times 0.05 + 10^9 \right]}$$

$$= 4.1 \times 10^{-4} \quad m \, s^{-1}$$

The pore velocity (v) is related to the superficial velocity by

$$v = \frac{u}{\varepsilon_{av}}$$

$$= \frac{4.1 \times 10^{-4}}{0.48} = 8.5 \times 10^{-4} \quad m \, s^{-1}$$

$$(6.16)$$

Now, $ReSc = \dfrac{vx}{D} = \dfrac{8.5 \times 10^{-4} \times 1.9 \times 10^{-5}}{10^{-9}} = 16$

Hence, using equation (6.9): $\dfrac{D_L}{D} = 0.707 + 55.5(16)^{0.96} = 795$

and using equation (6.6): $\dfrac{vL}{D_L} = 16 \dfrac{0.05}{1.9 \times 10^{-5}} \dfrac{1}{795} = 53$

From Figure 6.8 (or equation (6.14)) an appropriate dispersion number to be used for filter sizing would be:

$$\left(\frac{vL}{D_L} \right)^* = 0.5 + 1.35 \ln(53) = 6$$

and hence, using Table 6.3, to obtain a solute recovery of $F = 95\%$, about 2.1 wash ratios are needed (using $vL/D_L = 5$ on the Table).

Allowing for wash liquor losses from the drum surface, up to 2.1/0.7 (i.e. 3.0) wash ratios should be fed to the filter.

To determine if the calculated results are a possible proposition on the filter operating at the given conditions, calculate the washing time from equation (6.12):

$$t = \frac{W\varepsilon_{av}AL}{Q} = \frac{W\varepsilon_{av}L}{u} = \frac{WL}{v} = \frac{2.1\times0.05}{8.5\times10^{-4}} = 124 \quad s$$

The washing time available in the operating cycle of the filter is $0.29D_s^{-1} = 0.29/0.004$, i.e. 73 seconds, so rather less than 95% of solute will be recovered at these operating conditions.

If the washing time required to draw the necessary wash ratio through the cake were longer than the time available in the operating wash cycle on the drum, a filter design would be based on the washing stage as the controlling factor, or else a lower fractional recovery would need to be accepted. A further technical alternative to this solution is considered in the Section on multiple washing filter installations.

Example 6.3 Calculation of washing in a plate and frame press.

A cake is formed in a plate and frame press containing 10 frames, the internal dimensions of each frame measure 0.6 m square and 0.05 m thick. Estimate the wash liquid consumption and the time taken to reduce the cake solute content to 5% of its initial value during a "through washing" operation. (The term through washing is used to describe feeding wash liquor to one side of the cake and allowing it to pass through the entire thickness of the cake – noting that cake formation takes place on two surfaces in each chamber of the press). So if a cake of thickness L is formed on an area A in the filter press, then washing is carried out on a cake with thickness $2L$ and with an area $A/2$.)

The following data are available:

Pressure difference (Δp):	400 kPa
Average specific cake resistance (α_{av}):	10^{12} m kg^{-1}
Average cake porosity (ε_{av}):	0.4
Solids density (ρ_s):	3000 kg m^{-3}
Viscosity of wash liquor (μ_w):	10^{-3} Pa s
Diffusivity of solute (D):	10^{-9} m^2 s^{-1}
Medium resistance (R):	10^9 m^{-1}

Solution

Following the calculation sequence in the previous example, the effective particle size is:

$$x = 13.4\sqrt{\frac{0.6}{10^{12} \times 3000 \times 0.4^3}} = 7.5 \times 10^{-7} \quad m$$

Assuming the frames to be filled by cake, the superficial wash velocity is:

$$u = \frac{400 \times 10^3}{10^{-3}[10^{12} \times 3000 \times 0.6 \times 0.05 + 10^9]} = 4.44 \times 10^{-6} \quad ms^{-1}$$

The pore velocity of wash liquid is:

$$v = \frac{4.44 \times 10^{-6}}{0.4} = 1.11 \times 10^{-5} \quad ms^{-1}$$

Hence, $ReSc = \dfrac{1.11 \times 10^{-5} \times 7.5 \times 10^{-7}}{10^{-9}} = 8.3 \times 10^{-3}$

Since $ReSc < 1$, $\dfrac{D_L}{D} = 0.707$

and $\dfrac{vL}{D_L} = 8.3 \times 10^{-3}\dfrac{0.05}{7.5 \times 10^{-7}}\dfrac{1}{0.707} = 783$

So, from Table 6.3 the fractional recovery at $vL/D_L = 783$ can be interpolated as:

$$W = 0.9 + (1 - 0.9)\frac{0.95 - 0.897}{0.972 - 0.897} = 0.971 \quad when \quad vL/D_L = 500$$

$$W = 0.9 + (1 - 0.9)\frac{0.95 - 0.898}{0.972 - 0.898} = 0.967 \quad when \quad vL/D_L = 1000$$

In this case, 0.97 wash ratios are required to reduce the retained solute content of the cake to 5% of its initial value (i.e. $F = 1 - 0.05 = 0.95$).

The washing time $= \dfrac{WL}{v} = \dfrac{0.97 \times 0.05}{1.11 \times 10^{-5}} = 4369 \quad s \quad$ (i.e.1.21 hours)

Now, the wash ratio should be increased to account for filling of pipes with wash liquor. This may well double the wash ratio, that is, 2.0 wash ratios of wash liquid should be fed to the press and then, consumption of wash liquid per frame $= W\varepsilon_{av}AL = 2.0 \times 0.4 \times 0.6^2 \times 0.05 = 0.014$ m^3 and the total wash liquid consumption $= 0.014 \times 10 = 0.14$ m^3.

Long wash times are not uncommon on pressure filters where fine particle cakes are being processed, even though only a small number of wash ratios are passed through the cake – the long time arises from the high specific resistance (or low permeability) of the cake, which is mainly due to small particle sizes and/or low cake porosity.

6.3.2 Washing deliquored filter cakes

When a filter cake has been deliquored prior to washing, mass transfer processes in any region of the cake voids are ineffective until the wash liquor has resaturated that part of the cake. If air is trapped in the voids solute transfer does not occur across the gaseous phase that separates the residual liquor from the wash liquid, thereby leaving a microscopic portion of the cake poorly washed. The fundamentals of this problem were outlined in Chapter 2 (Section 2.7). After solving the equations that describe simultaneous resaturation and mass transfer in the cake, washing curves can be obtained that represent the dimensionless concentration of solute in the washings in terms of the wash ratio, the cake saturation at the start of washing and the dispersion number.

This information can be presented in a simplified form to facilitate estimation of the increase or decrease in the number of wash ratios or the volume of wash liquor required when the cake has been deliquored. For practical purposes a corrected wash ratio (W_{corr}) may be estimated from:

$$W_{corr} = W_{S=1} + 15.1(1-S)\exp(-1.56\phi^*) - 7.4(1-S^2)\exp(-1.72\phi^*) \quad (6.17)$$

where $W_{S=1}$ and ϕ^* are the wash ratio and instantaneous concentration of solute determined using Figure 6.4 (or Table 6.1). If the wash ratio has been calculated using Figures 6.5 or 6.7, then the corresponding value of ϕ^* for substitution into equation (6.17) can be obtained from Figure 6.4.

Example 6.4 Calculation of the effect of deliquoring on cake washing.
Using the data in Example 6.2 it was shown that 2.1 wash ratios would be needed to wash the cake if it were saturated at the start of washing if the required solute recovery was to be achieved (in a washing time of 124 s). Calculate the effect of deliquoring the cake to saturations of 0.2, 0.4, 0.6 or 0.8 before commencement of the wash.

Solution

From Table 6.1, the instantaneous concentration of washings coming from the cake in the final stages of washing is $\phi^* = 0.062$ when $vL/D_L = 5$.

The corrected wash ratios can now be calculated from equation (6.17). For example, at a saturation of 0.2 the corrected wash ratio is:

$$W_{corr} = 2.1 + 15.1(1-0.2)\exp(-1.56\times0.062)$$
$$-7.4(1-0.2^2)\exp(-1.72\times0.062) = 6.68$$

The volume consumption of wash liquor required is calculated using equation (6.2):

$$\frac{Qt}{A} = W\varepsilon_{av}LS = 6.68\times0.48\times0.05\times0.2 = 0.0321 \quad m^3\,m^{-2}$$

This calculation is repeated for each saturation and the results summarised as:

	Cake saturation at the start of washing				
	0.2	0.4	0.6	0.8	1.0
Wash ratio, W_{corr}	6.68	4.74	6.52	2.45	2.1
Wash liquor consumption, m³ m⁻²	0.032	0.046	0.094	0.047	0.050

This illustrates that in order to significantly reduce wash liquor consumption the cake saturation must be reduced to below about 0.5 or be close to unity, indicating that it is not always beneficial to deliquor the cake prior to washing. The implications of cake cracking were noted in Section 5.4, which must obviously be avoided if effective washing of the cake is to be achieved.

6.3.3 Cake washability

"Washability" is a qualitative descriptive term which is sometimes used, but its precise meaning has tended to be vague. It can be useful to know which parameters (or combination of parameters) are likely to give a cake which is more readily washed, that is, a cake with superior washability characteristics. From the aforegoing it is apparent that the dispersion number, vL/D_L, controls the washability of a cake; on Figures 6.4 and 6.5 it can be seen that more effective washing is obtained with greater values of vL/D_L.

To express the washability concept in more acceptable engineering terminology, the pore velocity v is obtained from equation (6.15) as

$$v = \frac{u}{\varepsilon_{av}} = \frac{\Delta p}{\varepsilon_{av}\mu[\alpha_{av}\rho_s(1-\varepsilon_{av})L+R]} \tag{6.18}$$

If it is assumed that the cloth resistance is much lower than the cake resistance, that is $R<<[\alpha_{av}\rho_s(1-\varepsilon_{av})L]$, then for fine particle cakes ReSc <1 and $D_L \approx D$ and

$$\frac{vL}{D_L} \approx \frac{\Delta p}{\alpha_{av}\, \mu\, \rho_s \varepsilon_{av}(1 - \varepsilon_{av})D} \tag{6.19}$$

The negligible cloth resistance assumption is not always valid; when it is not the ratio of cloth to cake resistances appears in the denominator of equation (6.18) and vL/D_L is decreased, leading to a less readily washed system.

Since the molecular diffusivity D does not vary greatly for mass transfer in liquid phases, better washing should be expected when a low viscosity wash liquid is used with a high pressure difference across the filter, on which there exists a cake with as low a specific resistance as possible (this does not contradict Section 6.1 where a high viscosity wash liquid is required to minimise fingering, as mass transfer is not taken into account in that section). Clearly, such an advantageous combination of parameters would be rare in the practical situation, but α_{av}, μ and ε_{av} can sometimes be adjusted by, for example, the use of flocculants, the addition of reagents to alter pH, or the use of a warm wash feed (Wakeman, 1981b).

For coarser particle cakes $D_L/D \rightarrow ReSc \rightarrow vx/D$ and $vL/D_L \rightarrow L/x$. That is, the ratio of cake thickness to particle size largely determines the effectiveness of washing and better washing conditions can be expected with thicker filter cakes.

6.4 Reslurry washing

Some filter cakes have a high inherent resistance to flow of wash liquor which leads to prolonged wash times, others are prone to cracking which causes channelling of wash liquor. Cake cracking as a result of dewatering stresses sets up channels through the cake that makes displacement washing ineffective. These situations can make displacement washing an unrealistic option in practice.

A technical alternative to overcome these difficulties is to reslurry the filter cake using wash liquor as the reslurrying liquid, and then to re-filter the resulting suspension. This may involve additional process vessels for mixing and redispersing the cake solids, which is usually the case for most designs of filter. However, sometimes the operation may be carried out *in situ* on the filter, for example when using some designs of enclosed pressure vessel filters. The desirability of reslurrying depends largely on the nature of the cake solids; for example, when the slurry has a reduced filtration rate after mechanical action (e.g. agitation) on the cake during the reslurry process, then reslurrying may be impracticable. Reslurrying can often produce solids products with a lower solute content than those obtained from displacement washing operations, but this is achieved using larger wash ratios (that is, larger volumes of wash liquor). Also, more complex plant flowsheets are usually necessary.

6.4.1 Multiple reslurry systems

To determine the quantity of wash liquor required in a reslurry process and the solute concentration in the filtrate after the reslurry, consider the multiple stage reslurry system shown in Figure 6.9. A cake containing a volume of residual filtrate V_r with a solute concentration ϕ_0 is mixed with a volume V_w/m of wash liquor containing an initial impurity concentration ϕ_w. There are m reslurry stages, and each stage is fed with the same quantity of wash liquor, so the total volume of wash liquor fed to the reslurry system is V_w. Filtration is carried out between each reslurry stage.

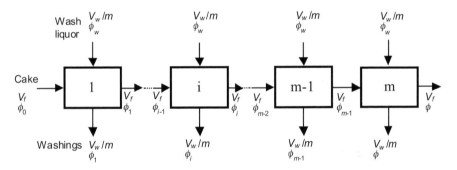

Figure 6.9 A multiple reslurry system comprising m stages. Filtration is carried out under the same conditions at each stage, and the resultant cake is mixed with fresh wash liquor.

Consider Figure 6.9, and the solute material balance for the ith stage is

$$V_r\phi_{i-1} + \frac{V_w}{m}\phi_w = \left(V_r + \frac{V_w}{m}\right)\phi_i \tag{6.20}$$

which rearranges to

$$\frac{\phi_i - \phi_w}{\phi_{i-1} - \phi_w} = \left(1 + \frac{V_w}{mV_r}\right)^{-1} \tag{6.21}$$

The solute concentration in the liquid discharged from the final stage is obtained by noting that

$$\frac{\phi-\phi_w}{\phi_0-\phi_w}=\frac{\phi_1-\phi_w}{\phi_0-\phi_w}\times\frac{\phi_2-\phi_w}{\phi_1-\phi_w}\times\cdots\times\frac{\phi_i-\phi_w}{\phi_{i-1}-\phi_w}\times\cdots\times\frac{\phi-\phi_w}{\phi_{m-1}-\phi_w}\qquad(6.22)$$

Substitution of equation (6.21) into equation (6.22) gives an expression in the usual form of the concentration ratio for m reslurries:

$$\frac{\phi-\phi_w}{\phi_0-\phi_w}=\left(1+\frac{V_w}{mV_r}\right)^{-m}\qquad(6.23)$$

The ratio V_w/V_r is the overall wash ratio used in the reslurry system, and $V_w/(mV_r)$ is the wash ratio used on each stage in the system.

Example 6.5 Calculation of the number of reslurry stages.
A suspension is filtered using a sequence of filters, and the cake from each filter is discharged into separate reslurry vessels. The same volume of solute free wash water ($\phi_w=0$) is fed to each reslurry vessel and the wash ratio at each stage (i.e. $V_w/(mV_r)$) is limited to 1.5 in order to re-suspend the cake solids. If the solute retained with the cake solids is to be 0.1% of the concentration in the feed solution, calculate the number of reslurry stages required.

Solution

Figure 6.9 depicts this situation, where each stage comprises of a filter and a reslurry vessel.

The solute remaining in the cake is $R=1-F=0.001$ (equation (6.4)).

We need to calculate the solute concentration leaving the reslurry system with the cake solids, ϕ. The solute entering the first reslurry vessel with the cake solids is $V_r\phi_0$, the solute discharging with the solids from the final filter is $V_r\phi$ (assuming the moisture content of the cake discharged from each filter is the same), hence the solute removed from the cake by reslurrying is $V_r(\phi_0-\phi)$. The fractional removal of solute is

$$F=\frac{V_r\left(\phi_0-\phi\right)}{V_r\phi_0}=1-\frac{\phi}{\phi_0}=0.999$$

$$\therefore\quad\frac{\phi}{\phi_0}=0.001$$

Now substitute the values into equation (6.23) to calculate the number of reslurry stages:

$$0.001=\left(1+1.5\right)^{-m}$$

$$\therefore\quad m=-\frac{\log 0.001}{\log 2.5}=7.5$$

i.e. 8 reslurry stages would be needed to achieve the specified washing.

If the particles are porous or aggregated and solute (solution) is retained within the internal structure of each particle, the solute material balance for the ith stage becomes:

$$V_r \phi_{i-1} + p_{i-1} w + \frac{V_w}{m} \phi_w = \left(V_r + \frac{V_w}{m} \right) \phi_i + p_i w \qquad (6.24)$$

when the solute is linearly distributed between the solid and liquid phases according to the equation

$$p_i = \frac{k}{\rho_f} \phi_i \qquad (6.25)$$

where p_i is the amount of solute adsorbed onto the cake solids. k is the distribution coefficient which has units consistent with those of p_i and ϕ_i (e.g. k has the units (kg solute/kg solid)/(kg solute/kg liquid) when p_i has units (kg solute/kg solid) and ϕ_i has units (kg solute/m^3 liquid)). Rearrangement of equations (6.22) and (6.24) gives an equation relating the solute concentration in the liquid phase to the volume of reslurry liquid used and the number of reslurry stages:

$$\frac{\phi - \phi_w}{\phi_0 - \phi_w} = \left(1 + \frac{V_w}{mV_r \left(1 + \frac{kw}{\rho_f V_r} \right)} \right)^{-m} \qquad (6.26)$$

where w is the dry weight of solids in the cake and ρ_f is the density of the liquid forming the suspension. Derivation of this equation has assumed that sufficient time is allowed for the distribution of solute between the solids and their surrounding liquid to reach an equilibrium state during the reslurry. Alternative expressions may be derived when the solute distribution between the solid and liquid phases is described by forms of equation that differ from (6.25).

In the case of non-porous, non-aggregated solids $k = 0$ (and equation (6.23) results). If the filtration and deliquoring pressure (or vacuum) are different at each filtration stage, then the value of V_r may be different at each stage and the solute concentration in the cake leaving the final stage is given by repeated application of equation (6.26) with $m = 1$.

6.4.2 Countercurrent multiple reslurry systems

It is sometimes possible to reduce reslurry liquor consumption by employing a countercurrent reslurry system, where the mth reslurry is supplied with fresh wash liquor which is subsequently filtered by the $(m+1)$th filtration stage (this may be a separate filter in the case of a multiple filter installation, or it may be a repeated reslurry/filtration step in the case of a single filter). The filtrate from this filtration is used in the $(m-1)$th reslurry, which is fed to the mth filtration, and so on.

For a washing system of this type, when each cake contains the same quantity of residual liquid immediately before it is reslurried, the expression describing the solute concentration in the final wash effluent is:

$$
\frac{\phi - \phi_w}{\phi_0 - \phi_w} = \left\{ 1 + \sum_{i=1}^{m} \left(\frac{V_w}{V_r} \right)^i \right\}^{-1}
\tag{6.27}
$$

and, for solids obeying the solute distribution equation (6.25):

$$
\frac{\phi - \phi_w}{\phi_0 - \phi_w} = \left\{ 1 + \sum_{i=1}^{m} \left(\frac{V_w}{V_r \left(1 + \dfrac{kw}{\rho_f V_r} \right)} \right)^i \right\}^{-1}
\tag{6.28}
$$

Reslurry systems are compared with countercurrent systems in Table 6.5. The data show that countercurrent flowsheets require less wash liquor to reduce the solute concentration to any given value, or fewer washing stages may be required for any particular washing. A further advantage when the solute is adsorbed to the solids is that, if necessary, the solids residence time in the reslurry can be extended to allow solute transfer into the liquid phase – the relative sizing of the filters and intermediate reslurry vessels must be given careful consideration in these cases.

Equations (6.20) to (6.28) assume that the solute is completely mixed in the wash liquor, and equations (6.24) to (6.26) and (6.28) further assume that the equilibrium distribution of solute between the solid and liquid phases has been reached. When complete mixing by relurrying is not the only means of washing, the procedures outlined in Section 6.5 need to be followed.

Table 6.5 The variation of solute concentration in the liquids from multiple reslurry systems; the columns in boldface type are for countercurrent systems (equation (6.27)), and the other columns are for a reslurry system (equation (6.23))

V_w/V_f	Concentration of solute in final liquid from the reslurry system, $(\phi-\phi_w)/(\phi_0-\phi_w)$					
	$m = 2$		$m = 3$		$m = 4$	
0	1	**1**	1	**1**	1	**1**
1	0.444	**0.333**	0.422	**0.250**	0.410	**0.200**
2	0.250	**0.143**	0.215	**0.067**	0.197	**0.034**
3	0.160	**0.077**	0.125	**0.025**	0.107	**0.008**
4	0.111	**0.048**	0.079	**0.012**	0.062	**0.003**
5	0.082	**0.032**	0.053	**0.006**	0.039	**0.001**
6	0.062	**0.023**	0.037	**0.004**	0.026	**0.0006**
7	0.049	**0.018**	0.027	**0.002**	0.017	**0.0004**
8	0.040	**0.014**	0.020	**0.002**	0.012	**0.0002**
9	0.033	**0.011**	0.016	**0.001**	0.009	**0.0001**
10	0.028	**0.009**	0.012	**0.0009**	0.007	**0.00009**
15	0.014	**0.004**	0.005	**0.0003**	0.002	**0.00002**
20	0.008	**0.002**	0.002	**0.0001**	0.0008	–

Example 6.6 Calculation of the number of stages in a countercurrent reslurry system.

Using the data in Example 6.5, calculate the number of countercurrent reslurry stages that would be needed to achieve the washing if the same overall wash ratio were used.

Solution

In this case we need to use equation (6.26). With $\phi/\phi_0 = 0.001$ and an overall wash ratio of $V_w/V_r = 1.5\times8 = 12$:

$$0.001 = \cfrac{1}{1+\left(\cfrac{V_w}{V_r}\right)^1+\left(\cfrac{V_w}{V_r}\right)^2+\left(\cfrac{V_w}{V_r}\right)^3+\ldots\ldots}$$

where as many terms are included in the denominator of this expression as are necessary to make the right hand side equal to the left hand side. The calculated values are tabulated below:

Number of stages	Number of terms in the denominator	Value of r.h.s.
1	2	0.077
2	3	0.0064
3	4	0.00053

From the table, 3 stages are required to achieve the washing.

Comparison of this result with that from Example 6.4 illustrates how fewer washing stages, and hence a smaller wash volume, are required when a countercurrent washing system is used.

6.5 Multiple washing filter installations

The adaptation of batch washing data to continuous washing systems is described in Section 6.3, and an important further development of this is its use with multiple washing systems. Countercurrent washing is used to make better use of the available wash liquor when "difficult to wash" cakes are being handled. The difficulties may arise when the time available in the filter cycle for washing is limited, when the wash flow rate through the cake is very low, or when the particles are porous or tend to adsorb solutes strongly. In these cases the time needed by the washing stage can be unacceptably long, or the solute concentration in the wash liquor is low after a single pass through the cake. The technique is generally limited to filtration systems employing rotary drum, horizontal belt, or pan filters. In the case of drum filters intermediate repulping or reslurry vessels and liquor recycles are necessary, whereas with belt or pan filters the additional mixing, pumping and filtration area requirements are largely eliminated.

Any countercurrent system could be chosen to illustrate the calculation procedures which have evolved (Wakeman, 1981c; Tomiak, 1994). The necessary information for multiple washing calculations is summarised in this section. Other methods have appeared in the literature which ignore the kinetics of washing (for example Hermia and Taeymans (1978), Hermia (1981) and Hermia and Letesson (1982)). These assume perfect mixing of wash liquor and retained filtrate within the pores of the cake, and hence the solute concentration in the liquor retained in the cake leaving the washing zone is identical to that in the wash effluent; this assumption enables a McCabe-Thiele type of construction to determine the number of washing stages. In practice, this assumption and its consequences are rarely satisfied as the time for washing is limited by the mode of operation of the filter; the washing kinetics must always be taken into account and combined with material balances.

The basic equations for multiple washing can be derived with reference to Figure 6.6. As cake moves through the washing zone, solute is removed to give a wash effluent with an instantaneous concentration, ϕ, which varies with wash time, t. The solute from any one washing zone is mixed and the concentration of solute removed over the entire zone is $(\phi_{av} - \phi_w)$. Considering the ith washing stage in a sequence of stages, the average concentration of the effluent over the total wash time $(t = \delta/\psi$) is (using equation (6.11)):

$$
\frac{\phi_{i,av} - \phi_w}{\phi_0 - \phi_w} = \frac{1}{t}\int_0^t \frac{\phi - \phi_w}{\phi_0 - \phi_w}\,dt
$$

$$
= \frac{1}{W}\int_0^W \frac{\phi - \phi_w}{\phi_0 - \phi_w}\,dW \tag{6.29}
$$

$= X_i$ when applied to the ith washing stage

The fraction of recoverable solute removed from the cake in the ith washing zone is:

$$
\overline{F}_i = \frac{\phi_0 - \phi_{c,i}}{\phi_0 - \phi_w} \tag{6.30}
$$

when the cake is saturated as it enters the washing zone, and noting that for any solute to be removed from the cake $\phi_w < \phi_0$. The solute concentration in the cake liquid leaving the ith washing zone $(\phi_{c,i})$ is obtained from a solute material balance as:

$$
\phi_{c,i} = \phi_0 + \frac{Q}{U}(\phi_w - \phi_{i,av}) = \phi_0 - W(\phi_w - \phi_{i,av}) \tag{6.31}
$$

Substitution into equation (6.30) and using equation (6.11) gives:

$$
\overline{F} = W\frac{\phi_{i,av} - \phi_w}{\phi_0 - \phi_w} = \int_0^W \frac{\phi - \phi_w}{\phi_0 - \phi_w}\,dW \tag{6.32}
$$

$= \overline{F}_i$ when applied to the ith washing stage

X_i and $\overline{F}_i$ can now be used to superimpose the washing kinetics on the material balance calculations; numerical values of these factors are available from Tables 6.3 and 6.4 (or from experimental washing data). If too few wash ratios are used (too low a wash liquor flow or too short a wash time) such that $X_i = 1$, then the wash effluent strength becomes identical to the solute concentration in the cake entering the washing zone, i.e. $\phi_{av} = \phi_0$. Under these conditions a single washing stage will always achieve as much washing as a multiple washing system, and the latter is therefore not a realistic technical alternative.

6.5.1 Rotary vacuum filter flow sheets

There are several possible alternative flow sheets utilising continuous countercurrent washing on rotary vacuum filters. One of the better, in

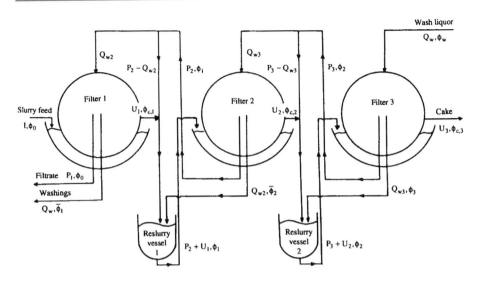

Figure 6.10 A countercurrent washing system using rotary vacuum filters with cake reslurrying, with part of the filtrate used as wash liquor on the preceding filter. I, P, Q and U are the liquid flow rates, and ϕ the solute concentration in the liquid phase.

so far as less wash liquor is usually needed to obtain a given cake purity, is shown in Figure 6.10. The filtrate from the ith filter is divided into two streams; one is used to wash the cake on the $(i-1)$th filter, whilst the other mixed with the wash effluent from the ith filter and then the mixture is used to reslurry the cake discharging from the $(i-1)$th filter. If we assume the process conditions to be the same on each filter, then for practical purposes the only variable through the flow sheet is solute concentration. Furthermore, the numerical value of the dimensionless solute concentration in the wash effluent is the same on each filter, and so equation (6.29) can be expressed in the following way for this flow sheet:

$$X_1 = \frac{\phi_{1,av} - \phi_1}{\phi_0 - \phi_1}$$

$$= X_2 = \frac{\phi_{2,av} - \phi_2}{\phi_1 - \phi_2} \tag{6.33}$$

$$= X_3 = \frac{\phi_{3,av} - \phi_w}{\phi_2 - \phi_w}$$

and the same fraction of recoverable solute is removed at each filter. That is, equation (6.32) is interpreted for this flow sheet as:

$$\bar{F}_1 = \frac{\phi_0 - \phi_{c,1}}{\phi_0 - \phi_1}$$

$$= \bar{F}_2 = \frac{\phi_1 - \phi_{c,2}}{\phi_1 - \phi_2} \qquad (6.34)$$

$$= \bar{F}_3 = \frac{\phi_2 - \phi_{c,3}}{\phi_2 - \phi_w}$$

where $\phi_{c,1}$, $\phi_{c,2}$ and $\phi_{c,3}$ are the solute concentrations in the cake liquid leaving the first, second and third washings.

Also, we can write:

$$I = P_2 + U_1 = P_3 + U_2$$
$$U = U_1 = U_2 = U_3 \qquad (6.35)$$
$$Q_w = Q_{w,3} = Q_{w,2} = Q_{w,1}$$

for the liquid flow rates when the operating conditions are the same on each filter. On any one filter a maximum of only one concentration value is known (ϕ_0 on the first filter and ϕ_w on the final one); but on the final filter the discharged cake purity is sometimes known (when it is not it can be estimated), hence using $\bar{F}_3$ and X_3 a solute balance can be obtained over the last filter. Let us assume the discharged cake purity ($\phi_{c,3}$) is unknown, as will often be the case; then the fraction of the solute input to the filter train removed by the combined effects of washing and filtration can be written as:

$$F^* = 1 - \frac{U}{I}\frac{\phi_{c,3}}{\phi_0} \qquad (6.36)$$

Sometimes it will be more convenient to define the fraction of solute entering the washing sequence which is removed by washing processes, that is:

$$F^{**} = 1 - \frac{U\phi_{c,3}}{U\phi_0} = 1 - \frac{\phi_{c,3}}{\phi_0} \qquad (6.37)$$

It can be observed that $F^* > F^{**} > \bar{F}_i$. Whichever equation is used, $\phi_{c,3}/\phi_0$ can be determined if a value for F^* or F^{**} is estimated in the first instance. If ϕ_w is known, ϕ_2/ϕ_0 then follows from equation (6.34), and $\phi_{3,av}/\phi_0$ from equation (6.33).

A solute balance over reslurry vessel 2 gives (noting $W = Q_w/U$):

$$\frac{\phi_{c,2}}{\phi_0} = (1+W)\frac{\phi_2}{\phi_0} - W\frac{\phi_{3,av}}{\phi_0} \tag{6.38}$$

whence ϕ_1/ϕ_0 is obtainable from equation (6.34), $\phi_{2,av}/\phi_0$ from (6.33), $\phi_{c,1}/\phi_0$ from (5.34), and $\phi_{1,av}/\phi_0$ from (6.33). Finally, the overall solute balance must be checked to see if a correct value for F^* or F^{**} was assumed in the first place. This is written as:

$$\frac{\phi_0}{\phi_0} = W\left(\frac{\phi_{1,av}}{\phi_0} - \frac{\phi_w}{\phi_0}\right) + \frac{\phi_{c,3}}{\phi_0} \tag{6.39}$$

If ϕ_0/ϕ_0 calculated from equation (6.39) $\neq 1$, then an incorrect value of F^* or F^{**} has been assumed; a new value must then be used and the calculation repeated.

Alternatively, a preferable procedure of elimination of the variables can be adopted as shown in the following example.

Example 6.7 Solute concentration calculations for 3-Stage RVF installation.

For the filtration conditions given in Example 6.2, calculate the solute material balance around the combined displacement and reslurry washing installation shown in Figure 6.10. Estimate the percentage removal of solute which should be obtainable on the multiple filters when a solute free wash liquor is fed to the filter train ($\phi_w = 0$).

Solution

The operating conditions of the filter in Example 6.2 show that the wash ratio is limited to $2.1 \times \dfrac{73}{124} = 1.2$. Assuming filtration conditions to be the same on each filter, then 1.2 wash ratios can be applied to each filter. Also, it was calculated in Example 6.2 that the dispersion number $\left(\dfrac{vL}{D_L}\right)^* = 6$. Values of the fractional removal of solute from the cake (F) and the average solute concentration in a mixed wash effluent (X) can be interpolated from Tables 6.3 and 6.4 respectively as:

$$\frac{F - 0.83}{0.923 - 0.83} = \frac{6 - 5}{10 - 5}$$

$$\therefore \quad F = 0.83 + (0.923 - 0.83)\frac{6 - 5}{10 - 5} = 0.85$$

and

$$\frac{X-0.691}{0.769-0.691}=\frac{6-5}{10-5}$$

$$\therefore \quad X = 0.691+(0.769-0.691)\frac{6-5}{10-5}=0.71$$

Since the operating conditions of each filter are the same,

$F = \bar{F}_1 = \bar{F}_2 = \bar{F}_3 = 0.85$ (the definition of F is given by equation (6.34))

and

$X = X_1 = X_2 = X_3 = 0.71$ (the definition of X is given by equation (6.33))

Using a technique of eliminating the unknown concentrations, express all concentrations in terms of the solute concentration in the fluid discharged with the cake, $\phi_{c,3}/\phi_0$. From equation (6.34):

$$\frac{\phi_2}{\phi_0}=\frac{1}{1-\bar{F}_3}\left(\frac{\phi_{c,3}}{\phi_0}-\bar{F}_3\frac{\phi_w}{\phi_0}\right)=\frac{1}{1-0.85}\left(\frac{\phi_{c,3}}{\phi_0}-0.85\times0\right)=6.67\frac{\phi_{c,3}}{\phi_0}$$

From equation (6.33):

$$\frac{\phi_{3,av}}{\phi_0}=\frac{\phi_w}{\phi_0}+X_3\left(\frac{\phi_2}{\phi_0}-\frac{\phi_w}{\phi_0}\right)=0+0.71\left(6.67\frac{\phi_{c,3}}{\phi_0}-0\right)=4.74\frac{\phi_{c,3}}{\phi_0}$$

From equation (6.38):

$$\frac{\phi_{c,2}}{\phi_0}=(1+W)\frac{\phi_2}{\phi_0}-W\frac{\phi_{3,av}}{\phi_0}=(1+1.2)6.67\frac{\phi_{c,3}}{\phi_0}-1.2\times4.74\frac{\phi_{c,3}}{\phi_0}=8.99\frac{\phi_{c,3}}{\phi_0}$$

From equation (6.34):

$$\frac{\phi_1}{\phi_0}=\frac{1}{1-\bar{F}_2}\left(\frac{\phi_{c,2}}{\phi_0}-\bar{F}_2\frac{\phi_2}{\phi_0}\right)=\frac{1}{1-0.85}\left(8.99\frac{\phi_{c,3}}{\phi_0}-0.85\times6.67\frac{\phi_{c,3}}{\phi_0}\right)=22.14\frac{\phi_{c,3}}{\phi_0}$$

From equation (6.33):

$$\frac{\phi_{2,av}}{\phi_0}=\frac{\phi_2}{\phi_0}+X_2\left(\frac{\phi_1}{\phi_0}-\frac{\phi_2}{\phi_0}\right)=9.67\frac{\phi_{c,3}}{\phi_0}+0.71\left(22.14\frac{\phi_{c,3}}{\phi_0}-6.67\frac{\phi_{c,3}}{\phi_0}\right)=20.65\frac{\phi_{c,3}}{\phi_0}$$

From equation (6.34):

$$\frac{\phi_{c,1}}{\phi_0}=\frac{\phi_0}{\phi_0}-\bar{F}_1\left(\frac{\phi_0}{\phi_0}-\frac{\phi_1}{\phi_0}\right)=1-0.85\left(1-22.14\frac{\phi_{c,3}}{\phi_0}\right)=0.15+18.82\frac{\phi_{c,3}}{\phi_0}$$

From equation (6.33):

$$\frac{\phi_{1,av}}{\phi_0} = \frac{\phi_1}{\phi_0} + X_1\left(\frac{\phi_0}{\phi_0} - \frac{\phi_1}{\phi_0}\right) = 22.14\frac{\phi_{c,3}}{\phi_0} + 0.71\left(1 - 22.14\frac{\phi_{c,3}}{\phi_0}\right) = 0.71 + 6.42\frac{\phi_{c,3}}{\phi_0}$$

From equation (6.39):

$$\frac{\phi_0}{\phi_0} = W\left(\frac{\phi_{1,av}}{\phi_0} - \frac{\phi_w}{\phi_0}\right) + \frac{\phi_{c,3}}{\phi_0} = 1.2\left(0.71 + 6.42\frac{\phi_{c,3}}{\phi_0} - 0\right) + \frac{\phi_{c,3}}{\phi_0} = 0.85 + 7.70\frac{\phi_{c,3}}{\phi_0}$$

Now, knowing that $\phi_0/\phi_0 = 1$, working from this last equation it follows that:

$$\frac{\phi_{c,3}}{\phi_0} = \frac{1 - 0.85}{7.70} = 0.019 \qquad \frac{\phi_2}{\phi_0} = 0.127$$

$$\frac{\phi_{3,av}}{\phi_0} = 0.090 \qquad \frac{\phi_{c,2}}{\phi_0} = 0.171$$

$$\frac{\phi_1}{\phi_0} = 0.421 \qquad \frac{\phi_{2,av}}{\phi_0} = 0.392$$

$$\frac{\phi_{c,1}}{\phi_0} = 0.508 \qquad \frac{\phi_{1,av}}{\phi_0} = 0.832$$

and from equation (6.37):

The fraction of cake solutes removed by the washing is $F^{**} = 0.981$.

It should be noted that in a flowsheet such as Figure 6.10, sufficient liquid must be available at each reslurry vessel to produce a slurry which can be pumped to the next filter. Solute material balance calculations must be done in conjunction with solid and liquid mass flow calculations (taking account of the suspension rheology) to ensure that pumpable slurries exist at each stage.

The example given above refers to only one possible arrangement of the washes applied to a series of filters and calculations; similar calculation procedures can be applied to other filter types and other plant flowsheets. Other possible cases where countercurrent reslurry washing is practised on vacuum drum filters include (i) mixing of the filtrate and washings, with the mixture used as wash liquor on the preceding filter, and (ii) keeping the filtrate and washings separate, with the washings used as wash liquor on the preceding filter. Figure 6.11 shows the general case where each filter is operating under different conditions,

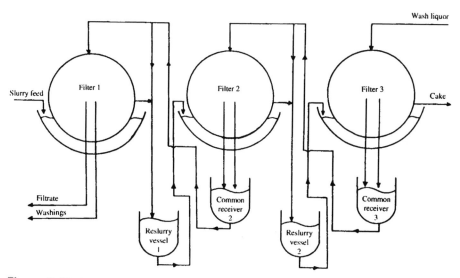

Figure 6.11 A 3-stage countercurrent washing system for rotary vacuum filters: the filtrate and washings are mixed and used as wash liquor on the preceding filter.

with the filtrate (mother liquor) and washings mixed before part is used as wash on the preceding filter and part is used to reslurry the cake from the preceding filter. In the case shown in Figure 6.12 filtrate (mother liquor) and washings are kept separate; the filtrate is used to reslurry the cake from the preceding filter and the washings used to wash the cake on the preceding filter.

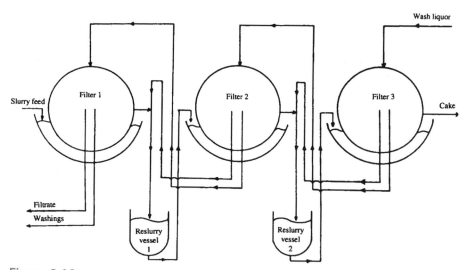

Figure 6.12 A 3-stage countercurrent washing system for rotary vacuum filters: filtrate and washings are kept separate and the washings used as wash liquor on the preceding filter. Note the fewer vessels required compared with the flowsheet in Figure 6.11.

6.5.2 Countercurrent washing on belt filters

For multiple washing operations belt filters can be a more attractive proposition in that extra equipment is not needed for reslurrying and mixing, hence pumping and filtration area requirements are not as large. At the same time there are fewer possible alternative flow sheets to consider (the same is true of tilting pan filters which offer somewhat similar advantages to belt filters as far as washing is concerned). A simple countercurrent washing scheme for belt filters is shown in Figure 6.13, where the wash liquor is applied to the final stage, and the wash effluent from each stage is used as wash liquor on the preceding one.

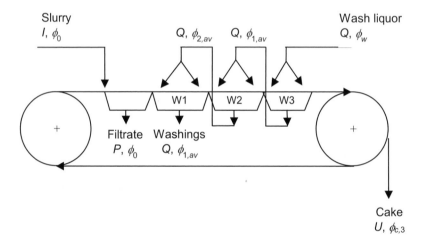

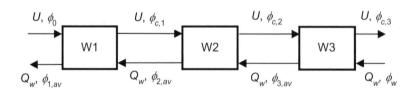

Figure 6.13 Countercurrent washing on a belt filter with three washing zones (W1, W2 and W3); the lower diagram indicates the flow of liquid (U) associated with the cake in relation to the flow of wash liquor (Q_w) and the solute concentration (ϕ) in each stream.

The approach to the analysis of these systems is similar to that used in Section 6.5.1. Consider Figure 6.13, and use equations (6.29) and (6.32) to give:

$$X_1 = \frac{\phi_{1,av} - \phi_{2,av}}{\phi_0 - \phi_{2,av}}$$

$$= X_2 = \frac{\phi_{2,av} - \phi_{3,av}}{\phi_{c,1} - \phi_{3,av}} \tag{6.40}$$

$$= X_3 = \frac{\phi_{3,av} - \phi_w}{\phi_{c,2} - \phi_w}$$

and

$$\bar{F}_1 = \frac{\phi_0 - \phi_{c,1}}{\phi_0 - \phi_{2,av}}$$

$$= \bar{F}_2 = \frac{\phi_{c,1} - \phi_{c,2}}{\phi_{c,1} - \phi_{3,av}} \tag{6.41}$$

$$= \bar{F}_3 = \frac{\phi_{c,2} - \phi_{c,3}}{\phi_{c,2} - \phi_w}$$

Equations (6.40) and (6.41) can be solved for specific problems in a similar manner to the previous example by writing the solute concentrations of the various streams in terms of the cake liquid discharge concentration $\phi_{c,3}/\phi_0$. $\phi_{c,3}/\phi_0$ is obtained from the overall solute balance

$$\frac{\phi_0}{\phi_0} = 1 = W\left(\frac{\phi_{1,av}}{\phi_0} - \frac{\phi_w}{\phi_0}\right) + \frac{\phi_{c,3}}{\phi_0} \tag{6.42}$$

and fraction of solute recovered can be calculated from equation (6.37).

Example 6.8 Solute mass balances with countercurrent washing on belt filters.

For the belt filter shown in Figure 6.13 with three washing zones, calculate the solute material balances along the filter and the fraction of solute removed by the washing. Solute free wash liquor ($\phi_w = 0$) is used for the wash, and a wash ratio of 0.6 is applied at each washing zone. Assume $vL/D_L = 10$.

Solution

Since the same wash ratio is used in each zone, $\bar{F}_1 = \bar{F}_2 = \bar{F}_3 = \bar{F}$ (say) and $X_1 = X_2 = X_3 = X$ (say). From Table 6.3, $\bar{F} = 0.586$ and from Table 6.4, $X = 0.977$.

Express all concentrations in terms of the solute concentration in the fluid discharged with the cake; from equation (6.41):

$$\frac{\phi_{c,2}}{\phi_0} = \frac{1}{1-\overline{F}}\left(\frac{\phi_{c,3}}{\phi_0} - \overline{F}\frac{\phi_w}{\phi_0}\right) = \frac{1}{0.414}\left(\frac{\phi_{c,3}}{\phi_0} - 0.586 \times 0\right) = 2.415\frac{\phi_{c,3}}{\phi_0}$$

From equation (6.40):

$$\frac{\phi_{3,av}}{\phi_0} = \frac{\phi_w}{\phi_0} + X\left(\frac{\phi_{c,2}}{\phi_0} - \frac{\phi_w}{\phi_0}\right) = 0 + 0.977\left(2.415\frac{\phi_{c,3}}{\phi_0} - 0\right) = 2.359\frac{\phi_{c,3}}{\phi_0}$$

From equation (6.41):

$$\frac{\phi_{c,1}}{\phi_0} = \frac{1}{1-\overline{F}}\left(\frac{\phi_{c,2}}{\phi_0} - \overline{F}\frac{\phi_{3,av}}{\phi_0}\right) = \frac{1}{0.414}\left(2.415\frac{\phi_{c,3}}{\phi_0} - 0.586 \times 2.359\frac{\phi_{c,3}}{\phi_0}\right) = 2.494\frac{\phi_{c,3}}{\phi_0}$$

From equation (6.40):

$$\frac{\phi_{2,av}}{\phi_0} = \frac{\phi_{3,av}}{\phi_0} + X\left(\frac{\phi_{c,1}}{\phi_0} - \frac{\phi_{3,av}}{\phi_0}\right) = 2.359\frac{\phi_{c,3}}{\phi_0} + 0.977\left(2.494\frac{\phi_{c,3}}{\phi_0} - 2.359\frac{\phi_{c,3}}{\phi_0}\right) = 2.491\frac{\phi_{c,3}}{\phi_0}$$

From equation (6.40):

$$\frac{\phi_{1,av}}{\phi_0} = \frac{\phi_{2,av}}{\phi_0} + X\left(\frac{\phi_0}{\phi_0} - \frac{\phi_{2,av}}{\phi_0}\right) = 2.491\frac{\phi_{c,3}}{\phi_0} + 0.977\left(1 - 2.491\frac{\phi_{c,3}}{\phi_0}\right) = 0.977 + 0.057\frac{\phi_{c,3}}{\phi_0}$$

From equation (6.42):

$$\frac{\phi_{c,3}}{\phi_0} = 1 - W\left(\frac{\phi_{1,av}}{\phi_0} - \frac{\phi_w}{\phi_0}\right) = 1 - 0.6\left(0.977 + 0.057\frac{\phi_{c,3}}{\phi_0} - 0\right)$$

$$\therefore \quad \frac{\phi_{c,3}}{\phi_0} = 0.400$$

The solute concentration in each stream can now be calculated by substituting this value of $\phi_{c,3}/\phi_0$ into the above expressions; the results are:

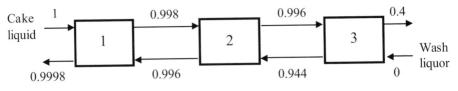

This example illustrates the effective utilisation of wash liquor that is achieved in a countercurrent washing system. If the three wash zones had been combined into a single zone and $3\times0.6 = 1.8$ wash ratios applied, then the output wash liquor concentration ($\phi_{1,av}/\phi_0$) would have been 0.559 (from Table 6.4) and the solute concentration in the cake liquid ($\phi_{c,3}/\phi_0$) would have been effectively zero (since $F = 1$ on Table 6.3). This points to the potential compromise that exists between high utilisation of the wash liquor and high cake purity; the objectives of washing must be identified carefully, as this will affect the number of wash zones and the wash liquor volume to be used.

The following example illustrates how the effectiveness of washing, as well as the cake formation parameters, are affected by the operation of a continuous filter where the total time available to carry out the cake formation and washing is limited by the total time available, using a horizontal belt filter as the illustrative filter.

Example 6.9 Washing filter calculations for a belt filter.

Phosphoric acid is obtained by reaction of sulphuric acid on a phosphate rock. Phosphoric acid and gypsum are then separated by vacuum filtration. To increase the recovery of the acid contained in the filter cake a thorough washing is necessary.

The acid-gypsum slurry is filtered at 60°C, and has the following properties: solids mass fraction $s = 0.3$ kg gypsum per kg slurry; density of the solids $\rho_s = 2350$ kg m^{-3}; density (ρ_l) and viscosity (μ_l) of the acid are 1390 kg m^{-3} and 1.31×10^{-3} Pa s.

Leaf tests were carried out on a filter with an area of 0.008 m^2 at a pressure difference of 53.4 kPa; a cake with a thickness of 45 mm and a porosity of 0.65 was formed. (The leaf test data are from Jemaa et al, 1974). The measured t/V vs. V data are:

t/V s cm^{-3}	0.054	0.076	0.08	0.086	0.093	0.109	0.115
V cm^3	18	54	71	81	96	128	151

Leaf test washing was carried out at 53.4 kPa with water at 40°C (viscosity $\mu_w = 0.656$×10^{-3} Pa s). Washing tests indicated that the washing process is composed of two consecutive steps:

(1) 49% of the cake acid is displaced by "plug flow", hence assume that the wash effluent viscosity is the same as that for pure acid at 60°C; the washing curve during this stage of the process can be represented by

$R = 1-W$ for $R \leq 0.49$

(2) the remainder of the wash corresponds to flow of dilute acid at 40°C, hence assume the wash effluent viscosity is the same as for water at 40°C. The washing curve during this stage can be represented by

$$R = 0.51\exp[-1.45(W-0.49)] \text{ for } R > 0.49$$

The proposed filter characteristics are a belt filter with a width (h_B) of 1 m, total belt length (z) of 10 m, and operated at a belt velocity (v_B) of 7 m min^{-1}.

Determine the filtrate flow rate, the acid concentration, cake thickness and acid loss for any given wash ratio.

Solution

1. Using equation (4.39), calculate the wet to dry cake mass ratio:

$$m = 1 + \frac{\rho_l \varepsilon_{av}}{\rho_s (1-\varepsilon_{av})} = 1 + \frac{1390 \times 0.65}{2350 \times 0.35} = 2.10$$

2. Using equation (4.13), calculate the mass of gypsum produced by unit filtrate volume:

$$c = \frac{\rho_l s}{1-ms} = \frac{1390 \times 0.3}{1-2.1 \times 0.3} = 1127 \text{ kg m}^{-3}$$

3. Using the t/V vs. V data and following the procedures in Example 4.1, calculate the cake specific resistance (the values below have been verified by plotting the data as in Example 4.1):

For $V_1 = 50$ cm^3 $(t/V)_1 = 0.074$ s cm^{-3}

For $V_2 = 150$ cm^3 $(t/V)_2 = 0.117$ s cm^{-3}

The slope of the t/V vs V plot is:

$$\frac{(t/V)_1 - (t/V)_2}{V_1 - V_2} = 4.3 \times 10^{-4} \text{ s cm}^{-6}$$

Hence

$$\alpha_{av} = \frac{2A^2 \Delta p \times slope}{\mu_l c} = \frac{2 \times (8 \times 10^{-3})^2 \times 5.34 \times 10^4 \times 4.3 \times 10^8}{1.31 \times 10^{-3} \times 1127}$$
$$= 1.99 \times 10^9 \text{ m kg}^{-1}$$

4. Noting equation (4.39), the cake liquid hold-up to filtrate volume ratio is:

$$\psi = \frac{c}{\rho_l}(m-1) = \frac{s(m-1)}{1-ms} = \frac{0.3(2.1-1)}{1-2.1 \times 0.3} = 0.89$$

At the end of cake formation the acid volume retained in the cake is of the same order as the acid volume that has been filtered, that is, cake washing is essential.

If Q_w is the wash liquid flow rate and Q_f is the filtrate flow rate, the wash ratio is

$$W = \frac{Q_w}{\psi Q_f}$$

5. Express the filtration and washing times as functions of the wash ratio. The filter cloth resistance is assumed to be negligible. Using equation (4.43), putting and $t_i = V_i = 0$ and $K_2 = 0$ (negligible medium resistance):

Filtration time, $t_f = \dfrac{\alpha_{av} c \mu_f}{2 A^2 \Delta p} V_f^2$

The washing time can be divided into two terms:

$$t_w = t_{w1} + t_{w2}$$

t_{w1} corresponds to the time necessary to displace 49% of the acid volume V_{w1}, whose viscosity is assumed to be equal to the pure acid viscosity:

$$\mu_{w1} = \mu_l = 1.31 \times 10^{-3} \quad \text{Pa s}$$

t_{w2} is the time necessary to remove volume V_{w2} of dilute acid whose viscosity is assumed to be the same as that of water at 40°C:

$$\mu_{w2} = 0.656 \times 10^{-3} \quad \text{Pa s}$$

Using equation (4.2) (Darcy's Law):

$$\Delta p = R_c \mu_{w1} \frac{1}{A} \frac{V_{w1}}{t_{w1}} = R_c \mu_{w2} \frac{1}{A} \frac{V_{w2}}{t_{w2}}$$

where the total resistance of the cake is obtained by combining equations (4.7) and (4.8):

$$R_c = \frac{\alpha_{av} c V_f}{A}$$

Hence

$$t_{w1} = \frac{\alpha_{av} c \mu_{w1}}{A^2 \Delta p} V_f V_{w1}$$

$$t_{w2} = \frac{\alpha_{av} c \mu_{w2}}{A^2 \Delta p} V_f V_{w2}$$

but

$$V_{w1} = 0.49 \psi V_f$$
$$V_{w2} = (W - 0.49) \psi V_f$$

Therefore

$$t_{w1} = \frac{\alpha_{av} c \mu_{w1}}{A^2 \Delta p} 0.49 \psi V_f^2$$

$$t_{w2} = \frac{\alpha_{av} c \mu_{w2}}{A^2 \Delta p} (W - 0.49) \psi V_f^2$$

6. Now apply these equations to the belt filter. For a continuous filtration and washing process we have the following relationships:

For the areas used in each stage of the process:

$$A_f = v_B h_B t_f \qquad A_{w1} = v_B h_B t_{w1} \qquad A_{w2} = v_B h_B t_{w2}$$

For the time required by each stage of the process:

$$t_f = \frac{z_f}{v_B} \qquad t_{w1} = \frac{z_{w1}}{v_B} \qquad t_{w2} = \frac{z_{w2}}{v_B}$$

The total time available is determined by:

$$t_{total} = \frac{z}{v_B} = t_f + t_{w1} + t_{w2}$$

For the volume of liquid collected from each stage of the process:

$$V_f = Q_f t_f \qquad V_{w1} = Q_{w1} t_{w1} \qquad V_{w2} = Q_{w2} t_{w2}$$

and hence the respective times required by each phase can be written in terms of the belt filter characteristics, the filtrate flow rate and cake formation time as follows:

$$t_f = \frac{\alpha_{av} c \mu_l}{2 v_B^2 h_B^2 \Delta p} Q_f^2$$

$$t_{w1} = \frac{\alpha_{av}c\mu_{w1}}{v_B^2 h_B^2 \Delta p} 0.49\psi Q_f^2 = 2\frac{\mu_{w1}}{\mu_f} 0.49\psi t_f$$

$$t_{w2} = \frac{\alpha_{av}c\mu_{w2}}{v_B^2 h_B^2 \Delta p}(W-0.49)\psi Q_f^2 = 2\frac{\mu_{w2}}{\mu_f}(W-0.49)\psi t_f$$

$$t_{total} = \frac{z}{v_B} = t_f\left[1 + 2\frac{\mu_{w1}}{\mu_f}0.49\psi + 2\frac{\mu_{w2}}{\mu_f}(W-0.49)\psi\right]$$

7. Establish the relationships between the wash ratio and the fraction of the belt filter length devoted to cake formation. The fraction of the belt filter length devoted to filtration is

$$\phi_f = \frac{z_f}{z} = \frac{t_f}{t_{total}} = \frac{1}{1 + 2\dfrac{\mu_{w1}}{\mu_f}0.49\psi + 2\dfrac{\mu_{w2}}{\mu_f}(W-0.49)\psi}$$

8. Express the filtrate flow rate, acid concentration, cake thickness and acid loss as a function of the wash ratio.

   (i) The filtrate flow rate can be expressed as a function of ϕ_f (using the above equations and $z_f = z\phi_f$):

   From 5.:   Filtration time, $t_f = \dfrac{\alpha_{av}c\mu_f}{2A^2\Delta p}V_f^2$

   Hence, using 6.:

   $$\frac{V_f}{t_f} = \frac{2A^2\Delta p}{\alpha_{av}c\mu_f}\frac{1}{V_f}$$

   $$= \frac{2v_B^2 h_B^2 \Delta p}{\alpha_{av}c\mu_f}\frac{t_f^2}{V_f}$$

   and

   $$Q_f = \frac{V_f}{t_f} = \sqrt{\frac{2v_B^2 h_B^2 \Delta p}{\alpha_{av}c\mu_f}}t_F$$

   $$= \sqrt{\frac{2v_B h_B^2 z_f \Delta p}{\alpha_{av}c\mu_f}}t_F$$

   $$= \sqrt{\frac{2v_B z h_B^2 \phi_F \Delta p}{\alpha_{av}c\mu_f}}$$

(ii) Referring to the diagram, the solute mass balance on the filter can be written as:

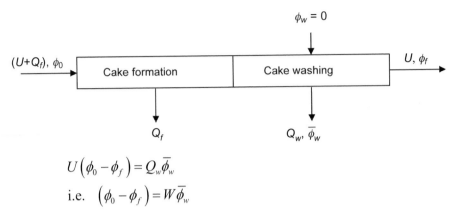

$$U\left(\phi_0 - \phi_f\right) = Q_w \overline{\phi}_w$$

i.e. $\left(\phi_0 - \phi_f\right) = W \overline{\phi}_w$

and the acid concentration in the wash effluent is:

$$\overline{\phi}^* = \frac{\overline{\phi}_w}{\phi_0} = \frac{1}{W}\left(1 - \frac{\phi_f}{\phi_0}\right) = \frac{1}{W}(1-R)$$

(iii) The cake thickness is calculated from:

$$L = \frac{cQ_f}{v_B h_B \left(1 - \varepsilon_{av}\right)\rho_s}$$

(iv) The acid lost in the cake hold-up is $\phi_f \psi Q_f$ (kg s^{-1}).

(v) The total acid fed to the filter (in kg s^{-1}) is

$$\phi_0\left(Q_f + U\right) = \phi_0 Q_f\left(1 + \frac{U}{Q_f}\right) = \phi_0 Q_f\left(1 + \psi\right)$$

and the acid loss fraction is

$$F_{AL} = \frac{\phi_f \psi Q_f}{\phi_0(1+\psi)Q_f} = R\frac{\psi}{1+\psi}$$

Numerical application of the equations:

For $W \le 0.49$ the acid concentration in the cake discharged from the filter is given by $R = 1 - W$, and the fraction of filter length devoted to the filtration is

$$\phi_f = \left(1 + 2\times1\times0.49\times0.89\right)^{-1} = 0.534$$

For $W > 0.49$ the acid concentration in the cake discharged from the filter is

$$R = 0.51 \exp\left[-1.45(W - 0.49)\right]$$

and the fraction of the filter length devoted to filtration is

$$\phi_f = \left(1 + 2 \times 1 \times 0.49 \times 0.89 + 2\frac{0.656}{1.31}(W - 0.49)0.89\right)^{-1}$$

$$= \left(1.872 + 0.891(W - 0.49)\right)^{-1}$$

The filtrate flow rate is

$$Q_f = \sqrt{\frac{2 \times 7 \times 10 \times 1^2 \times 5.34 \times 10^4}{60 \times 1.99 \times 10^9 \times 1127 \times 1.31 \times 10^{-3}}} \sqrt{\phi_f} = 6.51 \times 10^{-3} \sqrt{\phi_f}$$

The wash flow rate is

$$Q_w = W\psi Q_f = 0.89WQ_f$$

The acid loss fraction is

$$F_{AL} = R\frac{\psi}{1+\psi} = 0.471R$$

The cake thickness is

$$L = \frac{1127 \times 60 \times Q_f}{7 \times 1 \times 0.35 \times 2350} = 11.745Q_f$$

The overall results are summarised in the Table on page 246.

6.6 Conclusions

This chapter has sought to set out sequences of calculation procedures for cake washing, techniques which are fundamentally sound and which offer powerful tools that can be used in process calculations. The calculation methods can be elaborated further in respect of both their theoretical bases and their applications to practical problems, and in the interpretation of the effects of scale and equipment characteristics through the correction factors which are applied to the theory. For

Wash ratio	Fraction of acid recovered by washing	Fraction of belt length used for cake formation	Filtrate flow rate	Wash flow rate	Acid concentration in wash effluent	Acid loss fraction	Cake thickness
W	$1-R$	ϕ_f	Q_f $(\text{m}^3\,\text{s}^{-1})$	Q_w $(\text{m}^3\,\text{s}^{-1})$	$\bar{\phi}^*$	F_{AL}	L (m)
0	0	1	0.00652	0	1	0.47	0.077
0.49	0.49	0.53	0.00475	0.00207	1	0.24	0.056
0.6	0.57	0.51	0.00466	0.00249	0.95	0.20	0.055
0.8	0.67	0.47	0.00447	0.00318	0.84	0.16	0.052
1	0.76	0.43	0.00428	0.00381	0.76	0.11	0.050
1.5	0.88	0.36	0.00391	0.00522	0.59	0.056	0.046
2	0.943	0.31	0.00363	0.00646	0.47	0.027	0.043
2.5	0.972	0.27	0.00339	0.00754	0.39	0.013	0.040
3	0.987	0.24	0.00319	0.00853	0.33	0.006	0.038
3.5	0.994	0.22	0.00306	0.00953	0.28	0.003	0.036
4	0.997	0.20	0.00292	0.01040	0.25	0.001	0.034

many equipment types these factors have not been explored, and so it is necessary to use the best information currently available that is set out in the above paragraphs. However, the calculation methods presented here do allow preliminary material balances and equipment sizes to be quantified using the minimum of experimental data (if data on washing, for example, have been obtained, then this can always be used in place of the predicted characteristics and the ensuing calculation techniques remain similar). Changes of process conditions can be assessed and the effects of cake formation are implicitly written into the calculations through the cake specific resistance and porosity, so any effects of altering filtration conditions on washing are quantifiable.

7 Expression (compression deliquoring)

The separation of a liquid from a solid/liquid mixture by compression is often referred to as expression. The separation is caused by movement of a retaining wall or flexible diaphragm with which the suspension or filter cake is in contact, rather than pumping the solid/liquid mixture into a fixed volume chamber as in filtration. Expression of compressible filter cakes has become important in several diverse industries: in fruit pulp processing it is used to maximise the yields of liquid products; in sewage sludge treatments it is used to reduce the cake moisture content as far as possible prior to transportation or disposal by incineration or landfill; in the chemical industries thermal drying of wetter cakes incurs extra costs and it is desirable to eliminate as much liquor from the cake as possible by non-thermal processes.

Like filtration, expression can be classified into the three categories of constant pressure (Koo, 1942; Gurnham and Masson, 1946; Körmendy, 1964; Schwartzberg *et al*, 1977; Shirato *et al*, 1970–1974: Leclerc and Minery, 1980; and Wakeman *et al*, 1991), constant rate and variable pressure-variable rate (Shirato *et al*, 1971) operations depending on the variation of expression pressure and flow rate with time. The contents of this chapter are concerned only with expression under a constant applied pressure. The operation of a filter press may be divided into at least four stages (not considering cake washing or discharge), depending on the cake properties, the pump characteristics as well as the design of the press. The stages are: (i) Chamber filling, where the pressure is usually low, the flow rate from the pump is constant, but there is no filtrate discharged from the press; (ii) Cake formation at increasing pressure, possibly at a constant rate (if so the pressure will rise as the cake forms); (iii) Cake formation at constant pressure after the pump has reached its maximum pressure; and (iv) Cake compression, when the two cakes in the chamber have met and further pumping of the slurry into the press causes the cake to compress. This chapter deals with stage (iv) and those filters specifically designed to use compression as the means of filtering the feed suspension, for example the diaphragm filter press or the tube press. Figure 7.1 shows a tube press discharging its cake by lowering the inner filter tube, which is visible in the nearest tube shown; on this picture the dryness of the filter cake is apparent.

Figure 7.1 Tube presses discharging their cakes after lowering the filter candle which can be seen at the bottom of the nearest press casing. The friability of the cakes shown reflect their low moisture contents.

In general the original mixture to be expressed is likely to be concentrated or to be composed of fine particulates that do not settle appreciably under gravity, and may have the consistency of a slurry or a semi-solid. If the original mixture is able to flow as a slurry, applying mechanical pressure causes a sudden increase in hydraulic pressure uniformly throughout the slurry and the resulting hydraulic pressure is equal to the applied pressure, and expression initially proceeds following the principles of filtration (see Chapters 2 and 4). The hydraulic pressure remains equal to the applied pressure until the particles become networked or until a cake has been formed, whichever happens sooner during the expression process. Once networking of the solids or cake formation has taken place, or if mechanical pressure is applied to a pre-formed cake, further deliquoring is achieved by consolidation during which the hydraulic pressure in the pores of the cake decreases with time. This is accompanied by a decrease in the local porosity of the cake, and a transfer of the applied expression pressure from the pore liquid to the compressing solid structure. It does not follow that if a filter cake has a high porosity then it can be deliquored by using gas displacement (as discussed in Chapter 5), but it may be possible to deliquor the cake by compression (expression).

7.1 Experimental evaluation of expression

Experimental evaluation of the effects of compression on a filter cake or suspension can be measured using a laboratory piston press as illustrated in Figure 7.2. This comprises an upright cylinder mounted on a base plate, where both cylinder and plate are usually made from

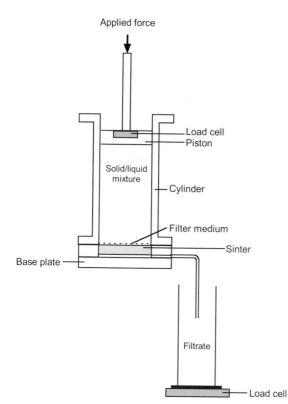

Figure 7.2 Compression filter (piston press) used to determine expression properties of slurries and filter cakes.

stainless steel. The base of the plate is recessed to accommodate a loosely fitting stainless steel sinter drainage plate which is used to support the filter medium. The drainage plate must also allow unrestricted passage of the filtrate. At the start of a test the cylinder is filled by suspension, and during the test liquid that is expelled from the cylinder is collected and its mass measured at various time intervals from the start of expresssion. The liquid is forced from the cylinder by the piston which is usually actuated either hydraulically or pneumatically; a force transducer is usually mounted in the piston shaft and used to measure the force applied to the suspension. The thickness of suspension (or cake towards the end of the test) in the cylinder can be calculated from a knowledge of the volume of liquid removed from the cylinder, or from measurements of the piston displacement from the beginning of the test.

As well as providing a good liquid seal between the piston and cylinder, the seals should be made from a low friction material. To simplify analysis of the resulting data, the experiment is usually carried out at a constant and controlled applied load. Under these conditions the characteristics of cakes formed in tube and diaphragm presses are represented well by the experiment, making the data particularly useful for design and scale-up calculations.

As ever with filtration testing, it is often preferable to carry out tests using pilot scale filters of a similar design to those expected to be used on plant. When this is done it is not possible to measure the cake thickness as in the laboratory piston press, but the filtrate volume must be monitored. The treatment of the data is then slightly different from what follows, since values of V rather than L are available.

7.1.1 General treatment of experimental expression data

Data are available from a piston press as the thickness of the solid/liquid mixture as a function of the pressing time, and/or as the volume of liquid expressed as a function of time. When both sets of data are available they can be converted one to the other, and as such serve to check that a material balance is preserved during the experiment. The thickness of the solid/liquid mixture in the cylinder at any time is:

$$L = \frac{V_0 - V}{A} \tag{7.1}$$

where V_0 is the volume of suspension in the piston press at the start of the test and L is the thickness of the solid/liquid mixture when a volume V of filtrate has been removed.

The mass fraction of solids in the initial suspension (s) should be ascertained before a test, and this can be converted into the corresponding solids volume fraction in the suspension using:

$$V_s = \frac{s\rho_l}{(1-s)\rho_s + s\rho_l} \qquad (7.2)$$

The thickness of the solid/liquid mixture and the solids volume fraction in the press at any time t_1 are related to the conditions at any other time t_2 by the solids mass balance:

$$(V_s L)_{t_1} = (V_s L)_{t_2} \qquad (7.3)$$

The ratio of the mass of wet cake to the mass of dry cake (m_{av}) (that is, after cake formation has occurred) is related to the volume fraction of solids in the press by:

$$m_{av} = 1 + \frac{\rho_l(1 - V_{s,av})}{\rho_s V_{s,av}} = 1 + \frac{\rho_l}{\rho_s} e_{av} \qquad (7.4)$$

where e is the voids ratio. Solids volume fraction and porosity are related by $V_s = 1 - \varepsilon$.

Expression is made up from two basic processes, filtration and consolidation. The transition is associated with a unique thickness of the solid/liquid mixture L_{tr}, which can be calculated from:

$$L_{tr} = \left(\frac{(m_{av})_{tr} - 1}{\rho_l} + \frac{1}{\rho_s} \right) \rho_s w_0 \qquad (7.5)$$

where $(m_{av})_{tr}$ is the ratio of the mass of wet "cake" to mass of dry "cake" in the cylinder at the end of the filtration process and w_0 is the volume of solids in the press per unit area (this is constant during a test and is equal to $(V_s L)$ at $t = 0$).

The value of $(m_{av})_{tr}$ cannot be measured experimentally during normal filtration or expression tests, and so it is estimated from the known characteristics of constant pressure filtration. The transition from filtration to consolidation mechanisms occurs when the solid/liquid mixture has a thickness L_{tr}. On a plot of $-\Delta L / \Delta \sqrt{t}$ against t, constant pressure filtration data should be represented by a straight line parallel to the t-axis (Shirato et al, 1986). It is often preferable to use logarithmic scales for this plot so that data scatter does not hide the transition point. When the concentration of solids in the mixture exceeds a limiting value, the mixture passes into a semi-solid state. From thereon it is consolidated and the value of $-\Delta L/\Delta \sqrt{t}$ decreases; at the end of the consolidation it reaches a value of zero, but this true equilibrium state can take a long time to achieve. The transition point for the mixture under test, L_{tr}, is determined from the graph as the point where $-\Delta L/\Delta \sqrt{t}$ starts to decrease rapidly. These points are shown in Figures 7.3 and 7.4, which are typical plots of experimental data obtained using different types of solid-liquid mixtures.

If the filtrate volume is measured in a test rather than a cake thickness then, noting equation (4.38) or similar ones, $\Delta V \propto \Delta L$ and a similar approach to the above analysis is possible.

7.2 Interpretation of expression data

The analysis of expression data was largely developed by Shirato *et al* (1970–1974, 1986) and is based on the equations describing filtration and consolidation. Expression of a stable suspension is made up of two parts: an initial stage in which filtration to form a cake occurs, followed by a consolidation stage during which the bulk volume of the deposited cake is reduced. If the starting suspension is unstable, the combined processes of filtration and sedimentation are responsible for the initial stage.

7.2.1 Analysis of the first stage – cake formation

The conventional filtration equation describes the first stage of expression of a slurry and is shown in Chapter 4 (Section 4.2) to be:

$$\frac{1}{A}\frac{dV}{dt} = \frac{A\Delta p}{\mu\left(\alpha_{av}cV + AR\right)} \tag{7.6}$$

The integrated form of this equation is generally more convenient to use as the expelled liquid data are usually collected as volumes rather than flow rates; at constant pressure equation (7.6) becomes:

$$\frac{t-t_i}{V-V_i} = \frac{K_1}{2}\left(V+V_i\right)+K_2 \tag{7.7}$$

where values of the constants are given by $K_1 = \alpha_{av}c\mu/A^2\Delta p$ and $K_2 = \mu R/A\Delta p$. Plotting the data from the filtration stage as $(t-t_i)/(V-V_i)$ against $(V+V_i)$ leads to a straight line (see Example 4.1) from which the cake formation properties characterising the rate of the first stage of

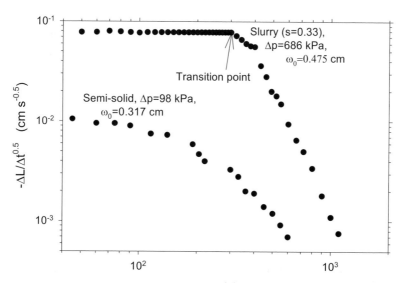

Figure 7.3 Identification of the filtration to consolidation transition point. The data are from Shirato *et al* (1987) for the expression of a mixture of Hara-Gairome clay and Solka Floc. When the starting mixture is a slurry the transition from filtration to consolidation is identified by the abrupt change in the slope of the plot, but no such change occurs when the initial mixture is at a concentration higher than the critical value and the entire process is a result of consolidation.

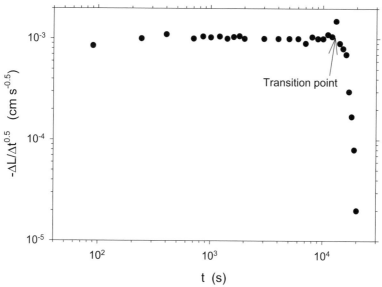

Figure 7.4 Identification of the filtration to consolidation transition point. The data are from Wakeman *et al* (1991) for the expression of a china clay suspension through a 0.1 μm medium (the mean particle size was 3.3 μm, pH 5.2 and applied pressure 6.4 MPa). The sharp transition point and steep slope of the consolidation data are indicative of a suspension that has formed a cake that is quite close to its maximum packing density, and undergoes minimal consolidation after the filtration stage.

expression may be evaluated; when the start of filtration coincides with the start of the integration $t_i = V_i = 0$. The effective concentration of solids in the feed to be used in equations (7.6) and (7.7) is:

$$c = \frac{\rho_l s}{1 - m_{av} s} \qquad (7.8)$$

where the ratio of the mass of wet cake to the mass of dry cake m_{av} is evaluated at the transition from slurry filtration to cake consolidation. Alternative relationships for the solids concentration are given in Section 4.2.6.

7.2.2 Analysis of the second stage – cake consolidation

When L decreases to a critical thickness L_{tr} (which corresponds to a critical solids concentration in the solid/liquid mixture) the filtration period ends and further expression takes place through a consolidation mechanism. During this latter stage the local (and hence also average) porosity in the filter cake decreases whilst there is a simultaneous transfer of the applied pressure from the pore liquid to the compressible solid structure. When the original concentration of the solid/liquid mixture to be expressed is greater than the critical value, the solids matrix possesses a structure (or the solids are networked). Application of a load to the mixture creates a more or less uniform hydraulic pressure distribution whose value is lower than the applied pressure.

The basic flow equations for consolidation were solved by Terzaghi and Peck (1948), assuming that the local voids ratio during the consolidation period was dependent only on the local compressive pressure. A so-called creep effect is often observed (also known as secondary consolidation), that is a rather slow movement of the particles before a final equilibrium between the compressive and frictional forces is obtained. This phenomenon is accounted for in the Terzaghi-Voigt combined model developed by Shirato et al (1974, 1978b).

For practical purposes the consolidation of a filter cake is most easily analysed through an empirical relationship relating the consolidation ratio:

$$U_c = \frac{L_{tr} - L}{L_{tr} - L_\infty} \qquad (7.9)$$

to a dimensionless consolidation time defined by:

$$T_c = \frac{i^2 C_e t_c}{\omega_0^2} \qquad (7.10)$$

where i is the number of drainage surfaces, C_e is a modified consolidation coefficient and t_c is the consolidation time ($t_c = 0$ at $U_c = 0$, i.e. at $L = L_{tr}$). Sivaram and Swamee (1977) developed a semi-empirical parametric equation which approximated the solution to the Terzaghi and Peck (1948) model for the consolidation of soils to within 3%, and Shirato et al (1986) modified the relationship so that it better described filter cake consolidation:

$$U_c = \frac{\sqrt{\dfrac{4T_c}{\pi}}}{\left[1+\left(\sqrt{\dfrac{4T_c}{\pi}}\right)^{2v}\right]^{1/2v}} \tag{7.11}$$

where v is the consolidation behaviour index which takes secondary consolidation effects into account (the Sivaram and Swamee (1977) equation took no account of the effects of creep and the behaviour index was assigned a constant value of 2.85). Values of $v > 2.85$ occur in filter cake consolidation (Wakeman et al, 1991). Numerical values of equation (7.11) are tabulated in Table 7.1, and plotted on Figure 7.5 together with some example experimental data.

Table 7.1 Variation of the consolidation ratio with (dimensionless consolidation time)$^{0.5}$ and the consolidation behaviour index according to equation (7.11). The data are also plotted on Figure 7.5

$\sqrt{T_c}$	$v=0.25$	0.375	0.5	0.625	0.75	1.0	1.25	1.5	2.0	3.0	4.0
0	0	0	0	0	0	0	0	0	0	0	0
0.2	0.104	0.155	0.184	0.201	0.211	0.220	0.223	0.225	0.226	0.226	0.226
0.4	0.161	0.251	0.311	0.351	0.378	0.411	0.429	0.438	0.451	0.451	0.451
0.6	0.204	0.322	0.404	0.462	0.504	0.561	0.596	0.619	0.645	0.667	0.673
0.8	0.237	0.377	0.474	0.545	0.597	0.670	0.718	0.751	0.795	0.840	0.862
1.0	0.265	0.421	0.530	0.609	0.667	0.748	0.801	0.839	0.887	0.936	0.960
1.2	0.289	0.456	0.575	0.659	0.720	0.804	0.857	0.894	0.937	0.975	0.989
1.4	0.310	0.489	0.612	0.699	0.762	0.845	0.895	0.928	0.963	0.990	0.997
1.6	0.329	0.516	0.644	0.732	0.794	0.875	0.921	0.950	0.978	0.995	0.999
1.8	0.345	0.540	0.670	0.759	0.820	0.897	0.939	0.964	0.986	0.998	1.0
2.0	0.360	0.561	0.693	0.781	0.841	0.914	0.952	0.973	0.991	0.999	
2.2	0.374	0.580	0.713	0.800	0.859	0.928	0.962	0.980	0.994	0.999	
2.4	0.387	0.596	0.730	0.817	0.873	0.938	0.969	0.985	0.995	1.0	
2.6	0.399	0.612	0.746	0.831	0.885	0.947	0.974	0.988	0.997		
2.8	0.410	0.626	0.760	0.843	0.896	0.953	0.978	0.991	0.997		
3.0	0.420	0.638	0.772	0.854	0.905	0.959	0.982	0.993	0.998		
3.2	0.429	0.650	0.783	0.864	0.913	0.964	0.984	0.994	0.999		
3.4	0.438	0.661	0.793	0.872	0.919	0.968	0.986	0.995	0.999		
3.6	0.447	0.671	0.802	0.880	0.925	0.971	0.988	0.996	0.999		
3.8	0.455	0.680	0.811	0.887	0.931	0.974	0.990	0.997	0.999		
4.0	0.462	0.689	0.819	0.893	0.935	0.976	0.991	0.998	0.999		

Consolidation ratio, U_c

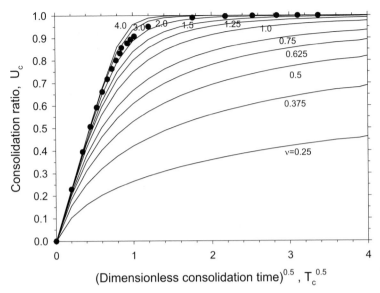

Figure 7.5 A plot of equation (7.11) showing the variation of the consolidation ratio with (consolidation time)$^{0.5}$ and the behaviour index ν. Experimental data can be superimposed onto this plot to estimate the value of ν (which may alternatively be evaluated using a curve fitting procedure); the points show the U_c versus $T_c^{0.5}$ data for consolidation of a hydromagnesite filter cake (pH 9.9) at a pressure of 1.65 MPa, from which $\nu \approx 2.5$ can be estimated.

According to equation (7.11), U_c plotted against $\sqrt{t_c}$ (or $\sqrt{T_c}$) is a straight line in the early stages of consolidation and C_e can be obtained from the slope of the line. For $T_c \ll 1$,

$$U_c = \sqrt{\frac{4T_c}{\pi}} = \sqrt{\frac{4i^2 C_e t_c}{\pi \omega_0^2}} \tag{7.12}$$

to give a line with a slope given by:

$$\text{slope} = \sqrt{\frac{4i^2 C_e}{\pi \omega_0^2}} \tag{7.13}$$

from which C_e is calculated.

Example 7.1 Calculation of expression parameters.

The time variation of the filtrate volume collected from expression of a suspension of peat in water is given in Table 7.2. The expression was carried out at an applied pressure of 10.8 MPa in a piston press with a height of 193 mm and an internal diameter of 43 mm. The mass fractions of solids in the initial and final solid/liquid mixtures were measured to be 0.06 and 0.47 respectively. The density of the solids was 1425 kg m^{-3}. Determine the values of $(m_{av})_{tr}$, the consolidation coefficient (C_e) and the behaviour index (v) for this mixture.

Table 7.2 Data measured for Example 7.1

Time t (s)	Filtrate volume V (cm^3)	Time t (s)	Filtrate volume V (cm^3)
0	0	5400	175
120	24	6600	188
240	45	7800	200
480	84	9000	209
720	93	10200	214
960	100.5	11400	217.5
1200	105.5	12600	219.3
1440	110	15000	220.5
1800	117.5	17400	223.5
3000	138	30600	232
4200	158	59400	235
		∞	250

Solution

The peat is a mixture of hydrocarbon and fibrous organic matter in water, with a wide range of particle sizes and shapes, and so scatter of the experimental data can be expected. The thickness of the solid/liquid mixture in the press at any instant can be calculated from the filtrate volume collected using equation (7.1); the volume in the press at the start of the test is $V_0 = \dfrac{\pi 4.3^2}{4} 19.37 = 281.3$ cm^3. The thickness is calculated and shown in Column 3 in Table 7.3.

The values of $-\Delta L/\Delta\sqrt{t}$ are then calculated, and are shown in Column 4. $-\Delta L/\Delta\sqrt{t}$ is then plotted against t on Figure 7.6, from which the transition from filtration to consolidation mechanisms is estimated to occur after about 8800 s.

For $t > 8800$ s the value of the consolidation time t_c is then obtained from $t_c = t - 8800$, ($t_c^{0.5}$ is shown in Column 6). Corresponding values for the consolidation ratio (U_c) are calculated using equation (7.9) (with $L_{tr} = 5.5$ cm and $L_\infty = 2.2$ cm) and shown in Column 7.

Table 7.3 Solution to Example 7.1

Time, t (s)	Filtrate volume, V (cm³)	Suspension thickness, L (cm)	−ΔL/Δ√t (cm s⁻⁰·⁵)		$t_c^{0.5}$ (s⁰·⁵)	U_c	$T_c^{0.5}$
Col. 1	Col. 2	Col. 3	Col. 4	Col. 5	Col. 6	Col. 7	Col. 8
0	0	19.3			–	–	–
			0.1095				
120	24	18.1			–	–	–
			0.3967				
240	45	16.3			–	–	–
			0.4208				
480	84	13.6			–	–	–
			0.1219				
720	93	13.0			–	–	–
			0.1205				
960	100.5	12.5			–	–	–
			0.1094				
1200	105.5	12.1			–	–	–
			0.0907				
1440	110	11.8			–	–	–
			0.1116				
1800	117.5	11.3			–	–	–
			0.1134				
3000	138	9.9			–	–	–
			0.1395				
4200	158	8.5			–	–	–
			0.1383				
5400	175	7.3			–	–	–
			0.1160				
6600	188	6.4			–	–	–
			0.1130	From graph,			
7800	200	5.6		$t_c = 0$ s	–	–	–
			0.0916	occurs	0	0	0
9000	209	5.0		at t ≈ 8800 s;	14.1	0.034	0.135
			0.0653	$L = L_{tr} =$			
10200	214	4.6		5.1cm.	37.4	0.172	0.357
			0.0346				
11400	217.5	4.4			51	0.241	0.487
			0.0183				
12600	219.3	4.3			61.6	0.276	0.589
			0.0098				
15000	220.5	4.2			78.7	0.310	0.752
			0.0212				
17400	223.5	4.0			92.7	0.379	0.886
			0.0140				
30600	232.0	3.4			148	0.586	1.414
			–				
59400	235.0	3.2			225	0.655	2.150
			–				
∞	250.0	2.2			–	1.0	–

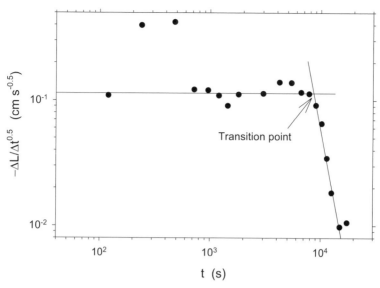

Figure 7.6 Identification of the transition point for the expression data in Example 7.1.

Converting the mass fraction of solids in the initial solid/liquid mixture to a volume fraction of solids gives (equation (4.34) is useful here):

$$V_s = \frac{s\rho_l}{s\rho_l + (1-s)\rho_s} = \frac{0.06 \times 1000}{0.06 \times 1000 + (1-0.06)1425} = 0.0429$$

Using the initial thickness (19.3 cm) in Column 3 of Table 7.3:

$$\omega_0 = (V_s L)_{t=0} = 0.0429 \times 19.3 = 0.828 \text{ cm} \quad (\text{i.e. } 0.00828 \text{ m})$$

Interpolating between the data in Column 2, the filtrate volume collected at the filtration/expression transition point is:

$$V_{tr} = 200 + (209 - 200)\frac{8800 - 7800}{9000 - 7800} = 207.5 \quad \text{cm}^3$$

and the corresponding suspension thickness is:

$$L_{tr} = \frac{281.3 - 207.5}{\pi\, 4.3^2 / 4} = 5.1 \quad \text{cm}$$

Rearranging equation (7.5) for $(m_{av})_{tr}$:

$$(m_{av})_{tr} = 1 + \frac{\rho_l}{\rho_s}\left(\frac{L_{tr}}{\omega_0} - 1\right) = 1 + \frac{1000}{1425}\left(\frac{0.051}{0.00828} - 1\right) = 4.62$$

The consolidation coefficient is calculated using the first two points in Columns 6 and 7 together with equation (7.12) as follows:

$$C_e = \frac{(\text{gradient})^2 \pi \omega_0^2}{4i^2}$$

where (gradient) is the initial slope of the U_c versus $\sqrt{t_c}$ plot, i is the number of drainage surfaces in the filter chamber ($i = 1$ in this case) and ω_0 is the total solids volume in the cake per unit filtration area.

$$(\text{gradient}) = \frac{0.152 - 0}{14.1 - 0} = 0.0108 \quad s^{0.5}$$

$$\text{Hence, } C_e = \frac{(0.0108)^2 \pi (0.00828)^2}{4} = 6.28 \times 10^{-9} \ m^2 \ s^{-1}$$

The values of $\sqrt{T_c}$ can now be calculated and are shown in Column 8 of Table 7.3. The consolidation behaviour index of the mixture can most easily be estimated by overlaying the ($\sqrt{T_c}$, U_c) values on Figure 7.5 – this is shown on Figure 7.7, from which it is seen that $v \approx 0.4$.

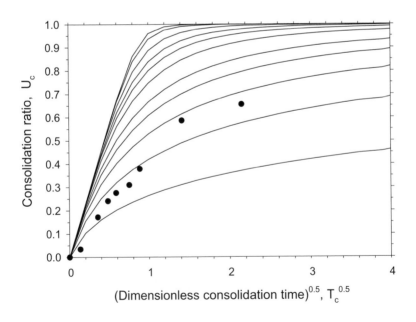

Figure 7.7 Estimation of the behaviour index v for the solid/liquid mixture in Example 7.1. The curves are for different values of v have the same labels as in Figure 7.5.

7.2.3 Expression on curved (2-dimensional) filter media

In a diaphragm type filter press the expression process is 1-dimensional, with the cake being compressed on a planar filter element and the expressed liquid flowing uniaxially and perpendicularly to the drainage surface.

In a tube press or a belt press, on the other hand, the solid/liquid mixture or cake is compressed on a cylindrical filter element and the expressed liquid flows radially (2-dimensionally). The rate of expression is then quite different from the rate associated with the 1-dimensional process due to the gradual change in the effective area on a cylindrical element as the filtration progresses.

Shirato *et al* (1986) have developed an effective consolidation area factor, j_{II}, defined by:

$$j_{II} = 0.297 n_D + 0.703 \qquad (7.14)$$

for $n_D < 3$, where n_D is ratio of the outer to inner diameters of the original suspension being filtered or expressed. Although creep may be present in some expression processes, it has no effect on the value of j_{II}. In most industrial presses, $n_D < 1.5$.

The 1-dimensional expression process is described fairly accurately by equation (7.11). The value of U_c for a 2-dimensional expression can be determined by substituting equation (7.14) into equation (7.11) to give:

$$U_c = \frac{\sqrt{\dfrac{4j_{II}^2 T_c}{\pi}}}{\left[1 + \left(\sqrt{\dfrac{4j_{II}^2 T_c}{\pi}}\right)^{2v}\right]^{1/2v}} \qquad (7.15)$$

This equation is used for analysis of data in the same way as equation (7.11).

7.3 Factors affecting expression characteristics

The properties of the suspension or solid/liquid mixture determine the extent to which it can be expressed. If a filter cake has a high porosity it does not mean that the liquid is easily removed by a compressive force, although it might indicate that the liquid can be removed by changing the structure of the cake. For example, many sewage sludges are difficult to dewater using compressive forces alone, but if these are combined with a shear force the liquid will exude from the sludge.

Particle size has a major effect on expression which becomes slower as particle sizes are reduced, all other factors remaining equal. As the particle size is reduced the magnitude of interparticle forces become more appreciable compared with gravitational or mechanically imposed forces, and the permeability of the solid/liquid mixture is lower. In general, provided the magnitude of the applied force is considerably greater than the net interparticle forces the final moisture content of the consolidated cake is affected to only a small extent by particle size. With this in mind it is evident that higher applied pressures lead to more rapid formation of more densely packed cakes, although this must be qualified by the magnitude of the surface charge on the particles.

Effects of pH on expression vary according to the magnitude of the surface charge (as characterised by the ζ-potential) on the particles. At lower ζ-potentials the particles tend to coagulate due to the lower repulsive forces between them; resulting cake structures can then be relatively open when low filtration pressures are used and expression can be used to further reduce the cake moisture content. If a high filtration pressure is used, little additional deliquoring may be achieved by expression. On the other hand, when the ζ-potential is high the repulsive forces between the particles are higher and open structure cakes are formed, in which cases expression may be necessary to further deliquor the solid/liquid mixture. The presence of macromolecules or other dissolved substances that cause a large osmotic pressure may also require that an expression process is used to deliquor a mixture.

The parameters α_{av} (the specific resistance of the cake) and e_{av} (the voids ratio) characterise the filtration period of expression (values and relationships for α_{av} and e_{av} are discussed in detail in Chapter 4). Similarly, the modified consolidation coefficient (C_e) and the mixture behaviour index (v) in equations (7.10) and (7.11) characterise the consolidation stage; C_e may be correlated by the equations:

$$C_e = C_{e0}p^\gamma \tag{7.16}$$

where C_{e0} and γ may be regarded as additional scale-up constants. Values of the coefficients arising from application of equation (7.16) to experimental data are summarised in Table 7.4 for a range of suspensions.

Table 7.4 Consolidation data that correspond to the constitutive equation coefficients shown in Table 3.3. For the above data $v = 4$ in all cases except for hydromagnesite where $v = 4p_s/(1.76+p_s)$. *This unusual value of γ is thought to be due to the relatively high solubility of hydromagnesite in water. (Wakeman, 1993)

Particle type	pH	ζ (mV)	x (μm)	C_{e0} (m² s⁻¹ MPa⁻$^\gamma$)	γ	p_s (MPa)
Anatase	4.0	−5	0.5	3.55×10^{-11}	0.915	0.3–11
	7.0	−39	0.5	1.66×10^{-12}	1.093	0.3–11
	9.1	−47	0.5	9.35×10^{-12}	0.912	0.3–11
Aragonite	11.0		1.7	5.45×10^{-8}	0.502	0.2–1.0
	11.0		3.3	3.17×10^{-8}	0.546	0.2–1.0
	11.0		8.2	3.92×10^{-9}	0.752	0.2–1.0
	11.0		10.0	1.87×10^{-10}	1.098	0.2–1.0
Calcite	10.0	~−20	2.3	7.61×10^{-10}	0.822	0.13–2.1
	10.0	~−20	26.5	7.98×10^{-10}	1.181	0.13–2.1
China clay	2.9	−15	5.0	2.87×10^{-8}	0.168	0.3–20
	5.2	−29	1.5	8.19×10^{-9}	0.263	0.3–20
	5.2	−29	5.0	1.29×10^{-9}	0.533	0.3–20
	5.2	−29	8.1	7.15×10^{-8}	0.116	0.3–20
	10.2	−57	5.0	2.75×10^{-10}	0.499	0.3–20
Hydromagnesite	9.9		16.4	4.11×10^{-7}	−0.08*	0.3–11
Wollastonite	9.2		12.7	2.95×10^{-9}	1.271	0.13–2.1

7.3.1 Expression rate increase by inserted strings

The optimum thickness of the original solid-liquid mixture to give maximum capacity by a batchwise expression process is dictated by the time required to attain a specified consolidation ratio; the expression time to achieve any specified consolidation ratio is proportional to the square of the cake loading in the filter (ω_0^2) (equation (7.12)). It has been reported that the rate of expression can be increased still further by placing several permeable strings perpendicular to the drainage surfaces (Shirato et al, 1978a). The strings act as drainage channels that allow easier passage of liquid out of the bulk volume of the mixture; average consolidation coefficients are increased by 10% or more by the addition of strings. This technique might be particularly useful when the original mixtures to be expressed are paste-like or semi-solid.

7.3.2 "Torsional" expression

Although expression is normally carried out with solely compressive forces, in a suitably designed filter simultaneous shear and compressive forces can be used – a process referred to here as torsional filtration. In fact, these combined forces act on cakes in pressure belt filters where a differential speed between the belts holding the cake in compression also creates a shear force in the cake. The imposition of shear into a consolidating solid-liquid mixture has little effect on the rate of expression but leads to a higher packing density in the final filter cake. The variation of solids content in the cake with rotational speed is shown in Figure 7.8 for cakes composed of particles of various shapes. Imposition of a small rotational speed during expression causes quite a substantial increase in the solids content of the cake; for the solid/liquid mixtures shown in the Figure, imposing a rotational speed of only 4 rpm increased the solids content of the titanium dioxide cake by 11%, the mica cake by 18% and the china clay cake by 23%. The titanium dioxide particles possess a bulkier shape whilst the mica and china clay particles have a flakey shape, probably accounting for the greater solids content increases. Koenders *et al* (2001) developed a qualitative analysis to describe filtration experiments with torsion shear flow of suspensions of interacting and non-interacting particles. The dominating effects of torsion shear relate to the compaction stage of filtration (expression).

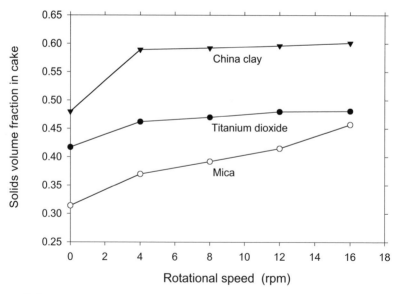

Figure 7.8 The effect of shear on the solids content of filter cakes formed from slurries of china clay (pH 11), titanium dioxide (pH 6.9) and mica (pH 3) expressed at 0.3 MPa.

7.4 Conclusions

Expression is useful for reducing the liquid content of compressible solid/liquid mixtures, and involves the use of a mechanical compressive force to reduce the bulk volume of the mixture. If the solids volume fraction in the slurry is low, the concentration is increased by filtration until a critical concentration is reached at which a consolidation process completes the deliquoring. The transition between the mechanisms is readily identified graphically. When the slurry has a high solids loading the entire expression process may be achieved through consolidation. The expression phase is interpreted using the usual filtration equations set out in Chapter 4, and consolidation is analysed using the equations in this chapter.

8 Membrane filtration

The three membrane filtration processes that are relevant here are microfiltration (MF), ultrafiltration (UF) and nanofiltration (NF). Reverse osmosis (RO) is a related process that utilises a different type of membrane and its successful operation relies on a different transport mechanism. Although all these use a membrane as the filter medium, this is not essential in the case of MF where media types other than membranes (for example, woven fabrics or wire cloths) can be used. RO is not considered in detail in this chapter as it is not strictly a filtration process, although it has characteristics closely related to NF. The reason for separating MF, UF and NF from other filtration processes is that they are often operated in a different mode, known as crossflow, and are also often subject to limitations that are little associated with cake filtration. Some general aspects of membrane media are considered in Section 1.1.2 and some examples are shown in Figure 1.6. In this text we shall refer to the media as membranes, although we must be aware that other media may be used. Membrane plant varies considerably is size from a single unit to several hundreds, mainly dependent on the volume throughput to be filtered. An example is shown in Figure 8.1.

Because of the way membrane filtration processes have evolved an alternate terminology is often associated with them. In MF, UF and NF, certain feed stream components are transmitted through the pores in the membrane into the permeate (the filtrate in cake filtration terminology), while other, usually larger, feed components are retained (rejected) by the membrane. These components accumulate in the retentate steam – see Figure 8.2.

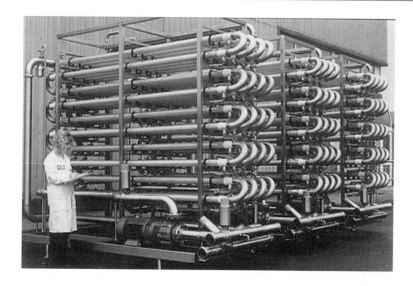

Figure 8.1 A 3-stage in series tubular UF system for wool scouring liquor (Memtech).

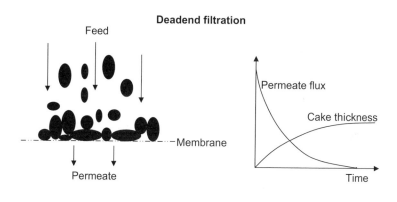

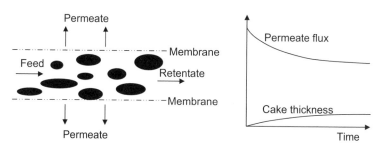

Figure 8.2 Diagrams of deadend and crossflow filtration showing the variations of cake growth and permeate flux with filtration time.

8.1 Membrane filtration processes

MF, UF, NF and RO are all pressure driven processes; MF and UF rely primarily on size-exclusion as the means of separation, although when the size of the species being separated is smaller the role of surface charge becomes more prominent. In NF and RO a further separation mechanism comes into play, the affinity of the membrane for specific species in the feed. The MF/UF pore size range spans approximately four orders of magnitude, from about 1 nm (0.001 μm) to about 20 μm, with UF generally taken to be in the range 0.001 to 0.1 μm and MF between 0.01 and 20 μm. UF and MF, and NF and UF, differ by degree and there is not a sharp demarcation between the each of the two processes based on pore size or on the size of components separated.

Historically RO has been called hyperfiltration, but the term reverse osmosis reflects the fact that there is a significant thermodynamic component to be overcome during the separation. Whereas MF is concerned with the separation of particles at the boundary of visibility, RO deals with the separation of ions and small molecules from a liquid. In between, NF is concerned with the separation of small low molecular weight organic species, and UF is used for the separation of larger molecules and macromolecules.

A guideline to the primary differences between the membrane filtration processes is given in Table 8.1. RO is included in the table because it completes the range of pressure driven membrane separation processes, and information relating to it is given for comparative purposes. Although NF and UF, and UF and MF, differ only by degree, the terminology is more than just one of semantics and from the point of view of the separation to be achieved these differences can be very important. The classification exists to convey differences in the primary features of the processes that have significance in the use, design and applications of the technology. The boundaries of each technology are not rigidly defined, but indicate differences between the technologies. Reflecting the different uses of the technologies, they are characterised in different ways; for example, MF membranes are characterised in terms of pore size, but UF and NF membranes are usually described by a molecular weight cut-off (MWCO).

Along with the size of the solute or particle that can be separated goes a change in the basic characteristic features of the separation devices. In going from RO to MF, the characteristic dimensions of the equipment become larger, the crossflow velocities used are higher, and the resultant fluxes are greater.

Table 8.1 Primary differences between the operational characteristics of membrane filtration processes

Operational parameter	Reverse osmosis	Nanofiltration	Ultrafiltration	Microfiltration
Separation principle	Size, charge, affinity	Size, charge, affinity	Size, charge	Size
Size of species separated (μm)	< 0.001	~ 0.001	0.001 – 0.1	0.1 – 20
Molecular weight cut-off value	< 200	200 – 1,000	500 – 500,000	–
Typical recovery (%)	30 – 90	50 – 95	80 – 98	90 – 99.99
Operating pressure (bar)	10 – 60	3 – 15	1 – 5	0.2 – 1
Typical flux ($m^3\ m^{-2}\ day^{-1}\ bar^{-1}$)	0.02 – 0.2	0.05 – 0.5	0.2 – 3	> 2
Crossflow velocity ($m\ s^{-1}$)	0.1 – 0.25	0.2 – 0.5	0.2 – 1	1 – 3
Typical rejected species	Salts, sugars	Sugars, pesticides	Proteins, viruses, endotoxins, pyrogens	Silts, bacteria, cysts, spores

There has been a tendency by many to regard membrane separations as a panacea that would solve the problems associated with most difficult to achieve separations; this is not the case and, like all other filtration processes, membrane filtrations are subject to economic and operational requirements. The separation method must be feasible technically; that is, it must be capable of accomplishing the desired separation, and must achieve a specified quality of product without product loss or damage. The separation must also be feasible economically; this depends strongly on the value of the products and is often related to the concentration of the feed, and energy and investment costs become more important when separating lower value products.

The basic nature of RO membranes is different from the others in that they are dense films through which there is no convective flow, the main transport mechanism of species through the membrane is by diffusion. In contrast, MF membranes have clear pores passing through them and are known as microporous and they conduct fluid under an imposed pressure gradient. NF and UF membrane characteristics fall between these two extremes, and although the transport mechanism is generally thought to be diffusive this may not necessarily be the case.

8.1.1 Crossflow filtration vs. normal (cake) filtration

MF, and to a lesser extent UF, are carried out in one of two types of configuration: crossflow (also referred to as tangential) and deadend (also referred to as normal) – see Figure 8.2. However, at the industrial scale both are more usually crossflow processes.

In deadend filtration the feed flow is towards the filter medium. A cake accumulates on the filter medium; its thickness increases during the course of the filtration, and as it does so the pressure loss across the cake increases in almost direct proportion. The permeate flow rate decreases simultaneously. Eventually, the pressure loss becomes equal to the pump delivery pressure, at which point further cake growth ceases and the permeate flow rate becomes zero.

In contrast, during crossflow filtration the feed flow is tangential to the filter medium. The thickness of the deposit (often referred to as the cake) formed is limited by the shear forces imparted by the feed flow, and so the pressure loss across the cake is also limited. The permeate flow rate decreases during the period of cake growth, but a steady state condition is reached whereby the cake thickness is stabilised and the permeate flow rate is maintained at a finite value.

In principle, therefore, the aim of crossflow filtration is to be able to establish an economical permeate flux that can be maintained over a long period. It should be clear that, under these conditions, cake formation does not occur for much of the filter cycle, but feed is actually being concentrated as liquid is being removed in the permeate.

8.1.2 How much permeate could be produced?

The permeate flow rate through the membrane, or the productivity, depends on the structure and design of the membrane as well as on the fluid. Consider a microfiltration membrane with cylindrical pores and noting that flow is invariably laminar, the Hagen-Poiseuille equation gives a simple relationship between the flux, J, and the pressure drop across the membrane, Δp:

$$J = \frac{\varepsilon r^2 \Delta p}{8 \mu L_m} \tag{8.1}$$

where ε is the membrane porosity, r the radius of the pores, L_m the length of the pores in a membrane (assumed to be the same as the membrane thickness), and μ the viscosity of the permeate. The theoretical permeate fluxes for different membrane thicknesses and pore sizes are shown on Table 8.2; in reality these fluxes cannot be achieved for several reasons. The primary reasons are that the pore structure is not a simple cylindrical one but is a complex tortuous channel, and during operation the membrane becomes fouled by components in the feed.

Table 8.2 The variation of the maximum permeate flux possible for different membrane properties (with $\varepsilon = 0.4$, $\mu = 10^{-3}$ Pa s)

Pore diameter (μm)	Membrane thickness (μm)	Flux/Δp (m^3 m^{-2} h^{-1} bar^{-1})
0.1	100	0.45
1	100	45
10	100	4500
1	100	45
1	10	450
1	1	4500

Equation (8.1) suggests that increasing the pressure will increase the flow rate through the pore, as is also suggested by the Darcy equation (see, for example, Chapters 2 and 4). This behaviour is seen during filtration as long as no consolidating layer is forming on the filter medium, which is not an uncommon occurrence.

8.1.3 Membrane fouling and concentration polarisation

For membrane separations to be economical, high flux of permeate is required. However, permeate flux through a new membrane declines quite rapidly and reaches a steady state value that is determined by the extent of fouling. Fouling, or deposition of matter onto or into the membrane, alters the flux performance of the membrane system and can also change the rejection. Fouling is a complex process, but two general classes of fouling are recognised:

(a) *Internal fouling* – the entrapment of solutes and particles within the internal pore structure of the membrane, or adsorption of components onto the internal surfaces of the membrane;

(b) *External fouling* – the formation of a filter cake layer on the membrane surface, or of a concentration polarisation layer close to the surface, caused by components being carried to the membrane by the feed flow which are then rejected by the membrane.

These represent serial resistances that increase during filtration until a steady state is achieved. Decreasing either one leads to an increased rate of productivity from the membrane system. In general terms, whereas external fouling is reversible and more readily controlled by operational parameters, internal fouling is often at least semi-permanent and can lead to an irretrievable loss of permeate flux.

Concentration polarisation arises during the filtration of fine dispersoids such as hydrocolloids or macromolecules. These larger components in the feed are rejected by the membrane, notably during UF, and form a concentrated layer that is fairly viscous and, depending on the solute, gelatinous. Concentration polarisation can have a major impact on UF performance, with a rapid reduction of permeate flux being observed in very short periods of filtration. The reduction in flux is thought to occur due to either:

(a) The increase of solute concentration at the membrane surface results in an osmotic pressure $\Delta\pi$, causing a decrease in the driving force ($\Delta p - \Delta\pi$) and a consequent reduction in flux; or

(b) The build-up of a steep concentration gradient of the solute within the boundary layer caused by convective transport of the solute towards the membrane, thereby increasing the hydrodynamic resistance of the boundary layer. The concentration gradient causes a back-diffusion of the solute, and after a while the two processes balance each other and a pseudo steady state, or pseudo equilibrium, is reached. The solute concentration close to the membrane surface reaches a maximum to form a consolidated "gel" or "cake" layer.

The consolidated layer causes the permeate flux to become independent of the transmembrane pressure (TMP); increasing the TMP results merely in a thicker or more consolidated layer. An increase of TMP will cause a momentary rise in flux, whose level then drops back to its previous or lower value.

The "gel", "cake" or "polarised" layer should not be confused with fouling; if a membrane filter is operated at a given set of conditions then the operating conditions are changed (for example, the TMP or the feed concentration are increased), lowering the pressure or feed concentration should enable the system to recover its previous operating state and yield its previous permeate flux.

With some process streams the permeate flux can be stable for weeks or even months. In most applications, however, there is a gradual flux decline with operating time, against which preventive measures are often taken. Prefilters or screens are used to remove large particles which might otherwise block thin channels or accumulate in stagnant areas of the module; although protection of the module is provided, fouling or cake formation on the membrane caused by the fine particles in the feed is not prevented. High crossflow velocities tend to reduce deposition, and low pressures help to avoid compaction of gels at the membrane surface. Some polymer membranes have a higher susceptibility to fouling, and chemical modification of an ultrafiltration membrane surface can have a great effect on its propensity to foul. If fouling does occur, the membrane can sometimes be cleaned by aggressive cleaning agents. However, even with repeated or periodic cleaning the flux cannot always be restored to its initial value.

The precise mechanisms of fouling are several, but can be related to the type of dispersoid in the feed (Table 8.3). There is no distinct boundary that separates one dispersoid from its neighbours in the Table, and it follows that the mechanism of fouling is not unique to any one fluid type nor to any one dispersoid. For example, surfactant dispersions at higher concentrations may well contain both micelles and monomer; the micelles may behave as colloids or in some cases as quite large particles and be separated primarily by a sieving process, whereas the monomer will have molecular characteristics and be separated by adsorptive processes.

Table 8.3 The main fouling mechanisms related to the type of dispersoid in the feed (Wakeman, 1996)

Feed type	Dispersoid	General fouling mechanism
Solutions ↑ Dispersions ↑ ↓ Suspensions ↑ ↓ ↓	Molecules Macromolecules Colloids Particles	Adsorptive ↑ Sieving ↑ ↓ ↓

Adsorptive effects can be important in UF. Of the contributing factors, charged and polar interactions should be considered at the fundamental level. At the practical level, the most important factors are membrane hydrophobicity and hydrophilicity and solution pH. The latter affects the charge and solubility of many solutes. At the isoelectric point (IEP) the net solute-solute repulsion is zero, aggregation occurs, and deposition of solute at the membrane surface becomes more dependent on module hydrodynamics, although structural rearrangements of the molecule may also occur. In UF it is more common to find molecular species in solution, and the rate and extent of fouling is then probably determined by molecular attractions between the membrane and the foulant rather than by physical mechanisms.

The effects of the main parameters during the crossflow filtration of solutions and dispersions are summarised in Table 8.4. The magnitude of the change of permeate flux or solute rejection brought about by altering any one of the parameters is dependent on and specific to the feed-membrane system.

Table 8.4 Influence of parameters during crossflow filtration of solutions and dispersions (Tarleton and Wakeman, 1994a and Williams and Wakeman, 2000)

Property	Comment
Crossflow velocity	An increase in crossflow velocity usually causes an increase in permeate flux, but within the normal range of crossflow velocities its effect on rejection is minimal.
Solute concentration in feed	Increasing the solute concentration in the feed causes more rapid flux decline, and lower steady state fluxes.
Filtration pressure	Effect of increasing transmembrane pressure (TMP) depends on properties of any fouling layer that is present: the cake or gel is often compressible and any increase in flux is not in proportion to the increase in TMP; if the cake or gel is highly compressible the flux may be reduced to zero.
pH	More fouling occurs, and hence a lower flux, when the feed is close to the isoelectric point. Structural as well as electrostatic changes may occur in the feed as the pH is varied, which are likely to be system specific.

8.1.4 Fouling by proteins and other biological materials

As soon as the upper surface of a membrane comes into contact with a feed solution, solute molecules adsorb at the membrane surface due to physico-chemical interactions (Belfort *et al*, 1994). Although protein molecules are much smaller than the pores of microfiltration membranes, severe fouling can occur under dynamic conditions (Güell *et al*, 1999). Protein aggregates play an important role in membrane fouling which can be significantly reduced by prefiltration to remove the aggregates (Chandevakar, 1990; Kelly *et al*, 1993). Experiments and scanning electron microscopy have supported the proposed mechanism of fouling and the role of protein aggregates, and also showed that membrane fouling by proteins is comprised of two separate steps. The first step involves protein adsorption or deposition, mainly on the pore walls and mouths (internal fouling), and the second step is characterised by the accumulation of a cake on the membrane surface due to the deposition and growth of aggregates (external fouling) (Bühler *et al*, 1993; Akay and Wakeman, 1994; Burrell and Reed, 1994; Burrell *et al*, 1994; Tracey and Davis, 1994).

Fouling of MF membranes by Bovine Serum Albumin (BSA) begins with the protein aggregates depositing on the membrane surface and thereby blocking some of the pores. As the filtration proceeds non-aggregated protein molecules chemically attach to the protein aggregates through disulphide interchange reactions and other interactions including van der Waals forces, electrostatic interactions, hydrophobic interactions and hydrogen bonding (Kelly and Zydney, 1997; Marshall *et al*, 1997).

The physical characteristics of the accumulated protein deposits is interesting. On polycarbonate membranes BSA and β-lactoglobulin form sheet-like deposits while immunoglobulin G forms granules which stack into layers creating a porous matrix (Lee and Merson, 1974). BSA deposits on membranes consist of an amorphous protein matrix that fully covers many of the pores, and the permeability of these protein deposits is a function of the solution pH and ionic strength. The solution environment can affect the properties of the protein deposit by altering the protein charge, shielding the electrostatic repulsion between adjacent proteins within the deposit, modifying the electro-osmotic counterflow that is generated by the solvent flow through the charged 'pores' in the protein deposit and finally by altering the protein conformation (Tracey and Davis, 1994; Palecek and Zydney, 1994a, b).

In addition to proteins and dissolved macromolecules there are a range of other colloidal and particulate matter which can foul MF membranes. Fouling by these materials is mediated by their role in pore narrowing and plugging and their deposition on to the upstream face of the membrane. The formation and effects of layers close to the membrane surface that act as secondary membranes in biological and food applications of membrane filters is affected by the larger components of the feed stream, such as microorganisms, γ–globulin and other polymers bound to form lattice-like structures on the surface of the membrane which are capable of entrapping small proteins (Arora and Davis, 1994; Lee and Merson, 1974).

The unwanted accumulation of microorganisms, cell debris and accumulated macromolecules on membrane surfaces is generally referred to as biofouling. In the case of the accumulation of micro-organisms, the extracellular polysaccharides excreted by the cells have been found to contribute to the fouling layer (Baker and Dudley, 1998). Experiments carried out by Schluep and Wilmer (1996) demonstrated that yeast cells irreversibly fouled a MF membrane operated in the crossflow mode, and the filtration performance was not able to be improved by creating transient pressure conditions through alternation of the flow direction in the module. McDonogh *et al* (1994) found that removal of an accumulated biofilm on the surface of an MF membrane is rarely achieved by standard cleaning techniques.

8.1.5 Fouling by inorganic particulates – effects on MF performance

The filtration of suspensions suggests that primarily particulate matter are to be removed from the liquid; the main process to effect the separation would be microfiltration (MF) and most of the discussion in this section is particularly relevant to MF. The extent of particulate fouling in crossflow MF is related to a matrix of feed stream, membrane and process parameters. Among the more important of these are the particle size distribution of the feed and the membrane pore size. The selection of a membrane for a given duty must be performed with care, as it is prudent to attempt to exclude particles from the internal pores of the membrane to enhance flux performance. The general effects of particle and pore size are given in Table 8.5.

Table 8.5 Summary of the influence of particle and pore sizes on crossflow MF (Tarleton and Wakeman, 1993)

Property	*Comment*
Particle size	At smaller particle sizes filtrate fluxes are lower and an equilibrium permeate flux is established more rapidly. The presence of a small percentage of fines significantly lowers fluxes. At higher crossflow velocities and longer filtration times similar fluxes are often observed for suspensions with differing median sizes.
Particle size distribution	Influence most pronounced at low crossflow velocity and low concentration where feeds containing the greater proportion of fines give lower filtration rates. At higher crossflows and concentrations, where the number of particles challenging each pore in unit time is increased, the effects on flux performance are negligible.
Membrane pore size	Little influence on flux or rejection when the majority of the particles in the feed are significantly larger than the pores in membrane. Filtrate quality and flux level often worse when a significant proportion of the particles in the feed are close to or smaller than the membrane pore size. If the pore sizes in the membrane are much larger than the particles in the feed, fluxes are better but solids rejection may be poor.

When filtering a relatively "large" particle size feed suspension an increased TMP results in an improved filtration rate. In crossflow filtration the permeate flux is rarely proportional to the applied hydraulic pressure gradient; frequently only small increases in flux are observed for quite substantial increases in pressure, particularly when feeds contain higher proportions of particle fines.

When the particle size of the suspension is reduced, there is a general tendency for the equilibrium flux to be established more rapidly at lower filtration pressures. When the suspension pH is such that the zeta potential around the particles is high (i.e. the suspension is well dispersed), an increased TMP often produces a reduced flux; the reduction can be to such an extent that filtration is essentially stopped after a short period, and raising the pressure further will not realise a flux. When the particles have a platelet or flakey shape, for example china clay, an increased pressure does not always increase the filtration rate. In contrast, for particles with a bulky shape an increase in flux is likely to be obtained by raising the filtration pressure.

When filtering fine particle suspensions an increased crossflow velocity produces a higher filtration flux. However, when the feed stream contains a greater proportion of larger particles the filtration rate generally falls with increasing crossflow velocity, despite a substantial thinning of the fouling layer at the higher crossflows. For a given suspension there may exist a critical particle size (and size distribution) where crossflow velocity has little or no effect on the flux decline curve.

Many of the effects of changing the crossflow velocity on filtration flux are directly attributable to the particle size and size distribution of the dispersed phase. When finer suspensions are filtered an equilibrium flux is often established more rapidly at lower crossflow velocities. If the crossflow velocity is raised the filtration flux can under some process conditions be seen to continually decline over a very long period. At lower crossflow velocities the shear caused by the crossflowing stream is insufficient to overcome the forces which cause particles to accumulate at the membrane. The increased permeation through the membrane at higher crossflows (for finer dispersions) tends to influence the particulates in the flowing suspension to a greater extent than at low crossflows, although other factors such as different combinations of pore plugging and blocking mechanisms may affect the fouling processes.

The general effect of increasing the solids concentration of the feed suspension is to lower the permeate flux, and increase the speed of establishment of an equilibrium flux; the latter effect is more pronounced at smaller particle sizes. When the feed stream is more

Table 8.6 Summary of the influence of process parameters on crossflow MF (Tarleton and Wakeman, 1994a)

Property	Comment
Suspension pH	When the particle sizes in the feed are sufficiently large hydrodynamic forces dominate and suspension pH/ionic strength has negligible effect on flux decline. With smaller particles surface forces are more dominant, and at high zeta potentials fluxes are lower than those at or near the IEP. Differences in flux levels are accentuated at higher feed concentrations. Filtrate quality (and solids retention) are generally improved near the isoelectric pH.
Suspension concentration	At greater suspension concentrations fluxes are often lower and equilibrium is established more rapidly. The fluxes at longer times are often similar over a range of concentrations, particularly when the feed sizes are smaller and when the feed particle shape is platelike.
Crossflow velocity	When the proportion of particle fines in the feed is high, an increased crossflow leads to thinner cakes and higher overall fluxes. Flux improvements are greater near the isoelectric pH of the feed, and less at pHs closer to the maximum particle surface charge. A steady state flux is established more rapidly at lower crossflows; filtrate clarity generally improves at higher crossflows. With reduced proportions of fines and larger particle size feeds an increased crossflow produces more fouling and lower flux levels. The size transition between flux "increase or decrease" with crossflow velocity depends on feed concentration.
Filtration pressure	For larger particle size suspensions there is a significant improvement in flux with increased pressure, particularly at lower suspension concentrations. The influence of pressure on flux levels is reduced for feeds with smaller median sizes and higher concentrations and some feeds containing particles of irregular shape.
Particle shape	The influences of irregular particle shape are difficult to quantify or predict, however significant effects on flux decline can result with an adverse shape.

concentrated there is a preference for filtration to occur with particles bridging membrane pores rather than plugging them. If a more dilute suspension is considered, however, there is a greater likelihood of pore plugging occurring. Crossflow MF should be performed with the particles in the feed suspension at or near the isoelectric point (IEP). Such an effect can be produced by altering the pH and/or ionic strength of the solution environment surrounding the particles. The main effects of the process parameters and membrane properties are summarised in Tables 8.6 and 8.7.

Table 8.7 Summary of the effects of membrane properties on crossflow MF (Tarleton and Wakeman, 1994b)

Property	Comment
Membrane morphology	With small fractions of fines present in a low concentration suspension containing a majority of particles larger than the membrane pore size, significant differences in flux can be observed with different membranes. At higher crossflows and concentrations and smaller particle sizes there is a reduced influence of membrane type on flux. When the majority of particles in the feed have sizes close to the pore sizes in the membrane, little influence of morphology on flux levels is observed.
Membrane wettability	Membranes exhibiting a higher contact angle and greater hydrophobicity produce rising fluxes during the initial periods of MF. At longer filtration times fluxes are similar to those for more hydrophilic membranes.
Membrane surface charge	Only minor differences in flux performance for suspension filtration.
Membrane pore size	Little influence on flux or rejection when the majority of the particles in the feed are significantly larger than the pores in the membrane. Filtrate quality and flux levels often worse when a significant proportion of the particles in the feed are close to or smaller than the membrane pores. If the pore sizes in the membrane are much larger than the particles in the feed stream, fluxes improve to higher levels although solids rejection is poor. Refer to Table 8.5 for other pore/particle size effects.

8.2 Ultrafiltration

UF processes are generally used in the dairy, food and beverage industries, for effluent treatment, and for biotechnology and medical applications. It has become important as a means of achieving disinfection or sterilisation of liquids. Typical species rejected by UF membranes include sugars, proteins and biomolecules, polymers, and colloidal particles. UF membranes are commonly characterised by their nominal molecular weight cut-off (MWCO), which is usually defined as the smallest molecular weight species for which the membrane has more than 90% rejection. However, the view that separation by UF is based on relative molecular size is true only to a first approximation; the chemistry of the solute-membrane interaction can be very important.

8.2.1 Rejection

In UF, total separation of the species of interest dispersed in the feed may not occur. This effect is generally characterised by the "rejection", and is defined in terms of the fraction of the species in the feed that appears in the permeate:

$$R = 1 - \frac{C_p}{C_f} \tag{8.2}$$

where C_p is the concentration in the permeate and C_f is the concentration in the feed. Sometimes the term "transmittance" is used, which is given by:

$$T = 1 - R = \frac{C_p}{C_f} \tag{8.3}$$

8.2.2 Mass transfer analysis

To provide a practical and usable approach to solute mass transfer in UF systems, the concept of a solute mass transfer coefficient is employed. The flux reduction has been analysed by a boundary layer model (Michaels, 1968; Kozinski and Lightfoot, 1972), shown in Figure 8.3. δ is the distance over which the solute concentration changes from C_f, the bulk feed concentration, to C_m, the solute concentration at the membrane surface. The distance δ is controlled by the flow regime in the module and the conditions of mass transfer. Solute is transported to the membrane surface by the convective flow of permeant, which is balanced by back diffusion of solute (Belfort and Nagata, 1985):

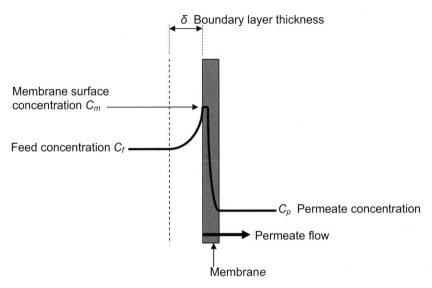

Figure 8.3 The boundary layer model for crossflow filtration.

$$J\left(C - C_p\right) = -D\frac{dC}{dx} \tag{8.4}$$

where J is the permeate flux and C_p is the solute concentration in the permeate. Integrating this equation across the boundary layer (noting that at $x = 0$, $C = C_m$, and at $x = \delta$, $C = C_f$), the permeate flux can be expressed as:

$$J = k \ln\frac{C_m - C_p}{C_f - C_p} \tag{8.5}$$

The mass transfer coefficient, k, which is generally assumed to be independent of the solute concentrations and of the permeate flux is then given by:

$$k = \frac{D}{\delta} \tag{8.6}$$

If the flux increases with the applied pressure, the value of C_m relative to C_f also increases. As C_m approaches the solubility limit for the solute, further concentration increases will cause a precipitate or thixotropic gel to be deposited onto the membrane surface. Once this has happened, further increases in the applied pressure do not lead to any further increase in permeate flux. Under these conditions the value of C_m becomes the concentration of the gel, C_g. Values of C_g have been found to be around 25% by weight for macromolecular solutes (with a range from 5 to 50%) and an average of 65% for colloidal dispersions (range 50 to 75%) (Kulkarni et al, 1992).

8.2.3 Determining the mass transfer coefficient

Experimental correlations for the solute mass transfer coefficient are conveniently expressed in dimensionless form as

$$\text{Sh} = a \, \text{Re}^{\alpha} \, \text{Sc}^{\beta} \left(\frac{D_h}{L} \right)^{\gamma} \tag{8.7}$$

where a, α, β and γ are constants whose values depend on the system geometry and flow conditions, and

$$\text{Sh} = \frac{kD_h}{D} = \text{the Sherwood Number}$$

$$\text{Re} = \frac{\rho u D_h}{\mu} = \text{the Reynolds Number} \tag{8.8}$$

$$\text{Sc} = \frac{\mu}{\rho D} = \text{the Schmidt Number}$$

L = the length of the channel on the feed side of the membrane

The values of the constants a, α, β and γ are summarised in Table 8.8.

Table 8.8 Values of the coefficients in the mass transfer correlation (equation (8.7))

Flow condition	a	α	β	γ	Reference
Laminar tube	1.62	0.33	0.33	0.33	Leveque, 1928
Laminar slit	1.86	0.33	0.33	0.33	Leveque, 1928
Developing	0.664	0.50	0.33	0.50	Grober et al, 1961
Turbulent	0.023	0.80	0.33	0	Gekas and Hallstrom, 1987
Spacer	0.664	0.50	0.33	0.50	Da Costa et al, 1994
Stirred cell	0.23	0.567	0.33	0	Smith et al, 1968
Rotating	0.75	0.50	0.33	0.42	Holeschovsky and Cooney, 1991

The hydraulic diameter, D_h, is defined as

$$D_h = 4 \times \frac{\text{cross-sectional area available for flow}}{\text{wetted perimeter of the channel}} \tag{8.9}$$

For tubes of circular cross-section, D_h is equal to the internal diameter of the tube. For slit shaped channels of width w and height h formed between two parallel plates,

$$D_h = \frac{2wh}{w+h} \tag{8.10}$$

$$\approx 2h \quad \text{when } w \gg h$$

The hydraulic diameter for a channel with a feed-side spacer is a function of the spacer porosity, ε_{sp}, and specific surface area, S_{sp}, defined as the surface area per unit solid-occupied volume of the spacer (Zeman and Zydney, 1996):

$$D_h = \frac{4\varepsilon_{sp}}{\dfrac{2}{h} + \left(1 - \varepsilon_{sp}\right) S_{sp}} \tag{8.11}$$

8.2.4 The diffusion coefficient

Diffusivity data for macromolecules and hydrocolloids are necessary in order to use the above equations. Ionic strength and pH of the solution affect the hydration and conformation of macromolecules, and hence affect the diffusion coefficient values. Theoretical equations to estimate diffusion coefficients depend on how well the structure of the liquid is described by the theory. A semi-theoretical relationship is the Wilke-Chang equation:

$$D = 117.3 \times 10^{-18} \frac{\left(\phi M\right)^{0.5} T}{\mu v^{0.6}} \tag{8.12}$$

where M is the molecular weight of the solvent in kg mol^{-1} (18 for water), T is the absolute temperature in K, μ is the solution viscosity in Pa s, v is the solute molal volume at normal boiling in m^3 kmol^{-1}, and ϕ is the association constant for the solvent (2.26 for water). This equation tends to predict diffusion coefficients that are very low in comparison to the shear enhanced diffusion in crossflow UF and MF.

Young et al (1980) suggest that the diffusion coefficient for a wide range of proteins can be estimated from:

$$D = 8.34 \times 10^{-12} \left(\frac{T}{\mu M^{1/3}}\right) \tag{8.13}$$

where D is the diffusion coefficient (m^2 s^{-1}), T the absolute temperature, μ the solution viscosity (cp) and M the protein molecular weight (g mol^{-1}). Experimental data for the diffusion coefficients of different molecular weight dextrans are correlated by (Granath, 1968):

$$D = 7.667 \times 10^{-9} M^{-0.4775} \tag{8.14}$$

with D in units of $m^2\ s^{-1}$ and M is the molecular weight in $g\ mol^{-1}$. Diffusivity data for polyethylene glycols (PEG) at 25°C have been correlated by (Pradanos et al, 1995):

$$D = 9.82 \times 10^{-9}\, M^{-0.52} \tag{8.15}$$

where D is in $m^2\ s^{-1}$ and M is the molecular weight of the PEG. Measured diffusion coefficients a selection of solutes that are filtered using membrane filtration are given in Table 8.9.

Table 8.9 Some values for diffusion coefficients

Solute	Molecular weight	Temper-ature (°C)	Diffusivity ×10¹¹ (m² s⁻¹)	Reference
Albumin, human serum	69,000	20	6.43	Walters et al, 1984
Albumin, bovine serum	66,500	20	6.3	Walters et al, 1984
	67,000	20	5.9	Smith, 1970
α-Amylase	96,920	20	5.7	Smith, 1970
α-1-Antitrypsin, human	45,000	20	5.2	Walters et al, 1984
Casein	24,000	25	1.9	Delaney and Donnelly, 1977
Collagen	345,000	20	1.2	Elias, 1977
DNA	6,000,000	20	0.13	Elias, 1977
Fibrinogen, human	339,700	20	3.0	Walters et al, 1984
Immunoglobulin G, human	153,000	20	4.1	Walters et al, 1984
Insulin, porcine (Na)	5,800	20	11.7	Walters et al, 1984
α-Lactalbumin	16,000	25	7.4	Delaney and Donnelly, 1977
β-Lactoglobulin	18,000	25	6.4	Delaney and Donnelly, 1977
Soy protein 7S	180,000	20	3.8	Koshiyama, 1968
Soy protein 11S	350,000	20	2.9	Wolf and Briggs, 1959
Sucrose	342.3	20	46.0	Polson, 1950
	342.3	37	69.7	Colton et al, 1970
Urea	60.1	20	120.0	Int. Crit. Tables, 1929
Urease	482,700	25	4.0	Cameron, 1973

Example 8.1 Prediction of the diffusivity of albumin.

Predict the diffusivity of bovine serum albumin at 293K in water as a dilute solution.

Solution

The molecular weight of bovine serum albumin is 66,500, the viscosity of water at 20°C is 1.005 cp. Using equation (8.13):

$$D = \frac{8.34 \times 10^{-12} \times 293}{1.005 \times (66500)^{1/3}} = 6.0 \times 10^{-11} \text{ m}^2 \text{ s}^{-1}$$

This value is 4.8% lower than the experimental value of 6.3×10^{-11} m^2 s^{-1} in Table 8.9.

It is generally known that higher mass transfer rates can be obtained at higher temperatures. The available equations for the diffusion coefficient show that it is an increasing function of temperature, with D generally increasing by about 3 to 3.4% per °C increase. However, product degradation can occur at higher temperatures so care must be taken to avoid, for example, protein denaturation, aggregation, or precipitation. The effect of temperature can be estimated through the Wilke modification of the Stokes-Einstein equation:

$$D_1 = D_2 \frac{\mu_2 T_1}{\mu_1 T_2} \tag{8.16}$$

where the subscripts refer to the two different temperatures.

Example 8.2 Estimation of permeate flux from the mass transfer equations.

Consider the ultrafiltration of milk at 50°C. Calculate the flux expected from a tubular membrane unit with the specification:

Inside diameter of tubes	12.5 mm
Length	2.5 m
Number of tubes	24
Feed flow rate	280 litres min^{-1}

The properties of the milk are:

Viscosity	8×10^{-4} Pa s
Diffusivity	7×10^{-11} m^2 s^{-1}
Protein concentration (C_f)	3.5% w/v

The gel concentration of the protein (C_g) is 22% w/v.

Solution

The velocity of milk in the tubes is:

$$u = \frac{0.280/60}{\pi \times (0.0125^2/4) \times 24} = 1.58 \text{ m s}^{-1}$$

The channel Reynolds Number is:

$$Re = \frac{\rho u D_h}{\mu} = \frac{1030 \times 1.58 \times 0.0125}{0.8 \times 10^{-3}} = 25428$$

The flow is turbulent (that is, Re > 3000), so using equation (8.7) with the relevant coefficients in Table 8.8:

$$Sh = 0.023(25428)^{0.8} \left(1.11 \times 10^4\right)^{0.33} = 1663$$

$$k = \frac{1663 \times 7 \times 10^{-11}}{0.0125} = 9.3 \times 10^{-6} \text{ m}^2 \text{ s}^{-1}$$

Using equation (8.5), the permeate flux is:

$$J = 9.3 \times 10^{-6} \ln\left(\frac{22}{3.5}\right) = 1.71 \times 10^{-5} \text{ m}^3 \text{ m}^{-2} \text{ s}^{-1}$$

or, in commonly used units the permeate flux is 61.6 litres m^{-2} h^{-1} (i.e. 61.6 LMH).

8.2.5 Limiting flux

In UF, a plot of the permeate flux against the transmembrane pressure (TMP), as shown in Figure 8.4, often indicates that the flux increases proportionately with TMP at low TMP's, but as the TMP is increased a lower proportionate increase of flux is experienced. The flux approaches a plateau value – the limiting flux, which increases with increasing crossflow velocity and decreasing feed concentration.

The increase in the limiting flux with crossflow velocity is due to the influence of the velocity on mass transfer, with a stronger effect occurring if the flow is turbulent. There is some evidence that for non-fouling systems the plateau is not actually a true plateau; at high pressures and low crossflow velocities the flux decreases slightly with increasing pressure (Field, 1996). The important features to note are the near linear increase of flux with TMP at low TMP and then a pressure independent flux at high TMP (for fixed values of u and C_f). Earlier works attributed this effect to the formation of a gel at the membrane surface, with a thicker gel being formed when the TMP is increased; thus, an increase in the TMP caused only a temporary increase in permeate flux as the flux was thought to fall away as the gel thickness grew, causing the steady state flux to be almost invariant with respect to TMP.

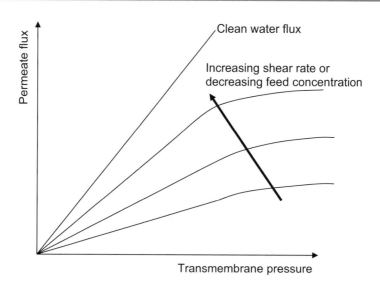

Figure 8.4 Permeate flux vs TMP, showing the limiting flux.

8.2.6 Limiting concentration

It has been observed many times with different filtering systems that there is an apparent linear semi-log relationship for steady state (or limiting flux) versus the bulk feed concentration. By extrapolating the line on such a plot to zero volumetric flux, it is tempting to say that a gel concentration is defined where the line meets the feed concentration axis. Indeed, models exist in the literature to explain this but it should be noted that such an exercise is misleading. Many experiments exist to show that by increasing the feed concentration the flux does not in fact fall to zero, but tends to level off at a low value – the limiting flux. This flux can be very low and often uneconomic as an operational flux.

8.3 Microfiltration

8.3.1 Filtration rate equations

Flux models are used to interpret crossflow MF data, with the aim of obtaining values for the permeate (filtrate) flux at the pseudo equilibrium condition. Correlating mass transfer data as in equation (8.7) for UF (or NF or RO) is more reliable than it is for MF, as solute motion behaviour in flowing fluids is better defined than is the motion of particles; molecular species have reliable values of diffusion coefficient associated with them. Particles in suspensions usually have a size distribution with no single size representative of the entire distribution; during crossflow filtration the size distribution of the deposit on the membrane is a function of the crossflow velocity (Tarleton and Wakeman, 1994a; 1994b), with coarser particles being scoured from the membrane surface.

8.3.2 Deadend filtration

For deadend MF the classical cake filtration models generally offer a reasonable way of interpreting experimental data. The form of Darcy's law that is usually applied in membrane filtration is:

$$J = \frac{1}{A}\frac{dV}{dt} = \frac{\Delta p}{\mu(R_c + R)} \tag{8.17}$$

where J is the permeate or volumetric flux, V is the volume of permeate, A is the membrane area, t is the filtration time, Δp is the pressure drop across the cake and membrane, μ is the permeate viscosity, and R and R_c are the resistances of the membrane and the cake layer respectively. It should be noted that this is identical to equation (4.6) for cake filtration, and that most of the discussion around equation (4.6) *et seq* is equally as relevant to deadend membrane filtration. The membrane resistance includes the effects of any penetrants into the membrane, and the cake resistance includes any effects of concentration polarisation or layers deposited onto the membrane surface.

The membrane resistance clearly depends on the membrane thickness, its pore size, and various morphological features such as its tortuosity and porosity. This is evident from consideration of equation (8.1).

The cake resistance is proportional to its thickness and the packing density of the particles. The factors that affect the cake resistance are considered in detail in Chapter 4.

By analogy with equation (8.17), the flux can be considered to be controlled by several resistances in series; hence, taking account of the boundary layer resistance (before true cake formation has occurred) R_{bl}, and of the fouling layer resistance R_f, the flux is:

$$J = \frac{\Delta p}{\mu(R_{bl} + R_f + R)} \tag{8.18}$$

R_{bl} represents the increase in flux that would be observed if the feed stream were replaced by a pure solvent, assuming the fouling layer remained intact. The values of the resistances depend on the mode of fouling, which may include adsorption or chemical interactions (which are more prevalent in UF), or cake formation or pore blocking (which may be more dominant in MF). The primary factors that contribute to cake formation are feed concentration and particle size and size distribution, whilst the particle shape and the ratio of particle size to pore size play an important role in determining the extent of pore blocking (Tarleton and Wakeman, 1993).

The modes of pore blocking are (see Section 1.1.4):

- Complete blocking (blocking filtration) – a particle seals the pore and prevents fluid flow through it;

- Pore bridging (bridging filtration) – particle bridge across the pore entrance, causing only a partial obstruction to fluid flow;

- Internal pore blinding (Standard blocking filtration) – material not rejected by the pore entrance is adsorbed or trapped at the pore wall inside the membrane or its support media.

The analysis proposed by Hermia (1982) for deadend cake filtration gives an understanding for the mechanisms that are likely to occur during the flux decline period of filtration. For constant pressure deadend filtration and the initial stages of crossflow filtration, Hermia's equations can be rewritten in terms of permeate flux (Field, 1996):

for standard blocking

$$J = J_0 \left(1 + 0.5 K_s A J_0 t\right)^{-2} \tag{8.19}$$

for intermediate blocking

$$J = J_0 \left(1 + K_i A J_0 t\right)^{-1} \tag{8.20}$$

for cake filtration

$$J = J_0 \left(1 + 2 K_c \left(A J_0\right)^2 t\right)^{-0.5} \tag{8.21}$$

Koltuniewicz *et al* (1995) grouped these together into the general equation:

$$J = J_0 \left(1 + k(2 - n)\left(A J_0\right)^{(2-n)} t\right)^{1/(n-2)} \tag{8.22}$$

where $n = 1.5$ for standard blocking filtration, $n = 1.0$ for intermediate blocking filtration, and $n = 0$ for cake filtration.

For constant pressure filtration with a driving pressure difference Δp, $R_0 = \Delta p / \mu J_0$. Also, $R = \Delta p / \mu J$ and hence

$$R = R_0 \frac{J_0}{J} = R_0 \left(1 + k(2-n)(AJ_0)^{(2-n)} t\right)^{1/(2-n)} \tag{8.23}$$

and

$$\frac{dR}{dt} = R_0 k (AJ_0)^{(2-n)} \left(1 + k(2-n)(AJ_0)^{(2-n)} t\right)^{(n-1)/(2-n)} \tag{8.24}$$

$$\frac{d^2 R}{dt^2} = R_0 k^2 (n-1)(AJ_0)^{2(2-n)} \left(1 + k(2-n)(AJ_0)^{(2-n)} t\right)^{(2n-3)/(2-n)} \tag{8.25}$$

Thus, the sign of d^2R/dt^2 is dependent on the $(n-1)$ term. A plot of dR/dt against time will reveal the resistance mechanism: a positive gradient indicates $n > 1$ and therefore pore blocking, whilst a negative gradient indicates cake filtration. In such a plot there are at least two unknown parameters (n and k) and sometimes more, and when experimental data are available curve fitting routines need to be used to evaluate their values.

8.3.3 Crossflow filtration

A robust model specifically for crossflow MF is still not available, but in models that are available the deposit is assumed to control the rate of filtration, and there is an implied assumption that particles do not penetrate the membrane surface. When this does occur, the membrane has a very high apparent resistance and controls the rate of filtration.

Hermia's approach can be extended to crossflow filtration: for complete pore blocking without any allowance for crossflow.

$$\frac{dJ}{dt} = K_b J \tag{8.26}$$

or

$$J = J_0 e^{-K_b t} \tag{8.27}$$

If a back flux mechanism occurs, noting that the limiting flux $J_{\lim}$ is the steady state flux, then:

$$\frac{dJ}{dt} = -K_b \left(J - J_{\lim}\right) \tag{8.28}$$

and

$$J = J_{\lim} + \left(J_0 - J_{\lim} \right) e^{-K_b t} \tag{8.29}$$

Following Arnot *et al* (2000), Field (1996) gives the equivalent equations for intermediate blocking and cake filtration with allowances for crossflow as, respectively:

$$\sigma t = \frac{1}{j_i} \left[\ln \frac{J_0 - j_i}{J_0} \cdot \frac{J}{J - j_i} \right] \tag{8.30}$$

$$Gt = \frac{1}{j_s^2} \left[\ln \left(\frac{J}{J_0} \cdot \frac{J_0 - j_s}{J - j_s} \right) - j_s \left(\frac{1}{J} - \frac{1}{J_0} \right) \right] \tag{8.31}$$

The equations can be used as a tool to interpret flux-time behaviour data and to investigate why one membrane may be working differently from another. In these equations there are unknown parameters that are used to match them to experimental data, requiring the use of curve fitting routines to evaluate values of the parameters.

8.4 Critical flux operation

Membrane filters, particularly microfilters, may sometimes be operated at a constant flux with no increase in transmembrane pressure (TMP); if that pressure is low or negligible, then fouling effects are slight. On the other hand, if the TMP is increased fouling occurs and the permeate flux declines. Once the lower TMP is restored, the flux is lower because of the additional fouling layer on the membrane surface.

The critical flux hypothesis for MF is that on start-up there exists a flux below which a decline of flux with time does not occur; above it fouling is observed (Field *et al*, 1995). This flux is the critical flux and its value depends on hydrodynamic and other variables.

The concept of critical flux suggests that there are advantages in operating below it so as to minimise the need for cleaning, but clearly the constant flux obtained should not be so low as to be uneconomic. Low pressure MF can be more effective than high pressure MF, and it would seem desirable to start-up the filtration operation at a low flux and hence reduce the effects of fouling.

The critical flux can be estimated using an experimental step-by-step technique by incrementally increasing the TMP, as shown in Figure 8.5. The figure indicates that at low TMP values the flux against time curve

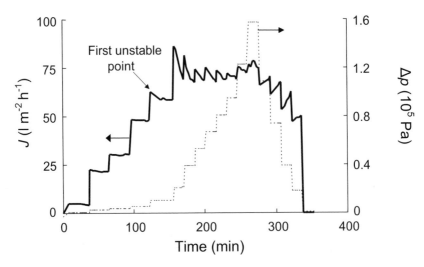

Figure 8.5 Determination of the critical permeation flux when filtering a 1.74 g kg^{-1} suspension of latex in a ceramic tube membrane with a 300 kg mol^{-1} cut off (Gésan-Guiziou *et al*, 2002).

shows a more or less constant value for *J*; the first unstable permeation flux is when the flux against time data becomes non-linear – that is, when the flux first shows a decline from the value obtained immediately after the TMP is increased. At higher values of the TMP the flux decline is far more severe, shown by the saw-tooth shape of the graph. After the flux at a maximum value of the TMP has been measured, as in Figure 8.5, the TMP can be successively decremented and the corresponding fluxes measured. When the flux obtained during decreasing the TMP is lower than that obtained at the same TMP whilst increasing it (as shown in these data), some permanent fouling of the membrane has occurred.

8.4.1 Deposit characteristics

The characteristics of a cake or deposit formed in a crossflow filter is affected to some extent by the wall shear stress on the retentate side of the membrane (which is largely controlled by the magnitude of the crossflow velocity in any one membrane filter) and by the transmembrane pressure. Figure 8.6 is illustrative of the effects of transmembrane pressure (Δp) on the deposit thickness, mass, porosity and specific resistance. As the TMP is increased, the mass deposited and thickness both increase in a similar manner but approach constant values. The specific resistance also increases with TMP, whilst there is a simultaneous small decrease in the porosity of the deposit. Figure 8.7 shows that a lower deposit mass and thickness are associated with a higher wall shear stress (or higher crossflow velocity). The same work found that the specific resistance

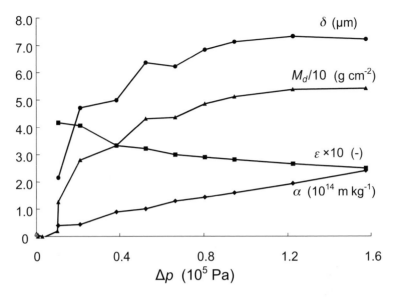

Figure 8.6 Variation of the deposit characteristics (M_d: deposited mass; δ: deposit thickness; ε: deposit porosity; α: deposit specific resistance) with transmembrane pressure (Δp) when filtering a 1.74 g kg^{-1} suspension of latex in a ceramic tube membrane with a 300 kg mol^{-1} cut off (Gésan-Guiziou *et al*, 2002).

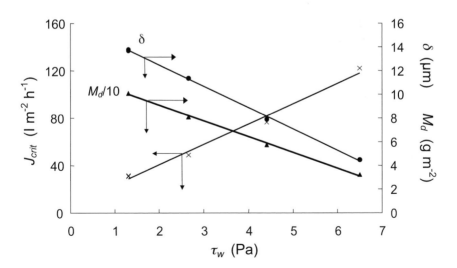

Figure 8.7 Variation of the critical flux and deposit characteristics (M_d: deposited mass; δ: deposit thickness) with wall shear stress (τ_w) when filtering a 4.9 g kg^{-1} suspension of latex at 50°C in a ceramic tube membrane with a 300 kg mol^{-1} cut off (Gésan-Guiziou *et al*, 2002).

and porosity of the deposit were independent of the wall shear stress. In that work the particle size distribution was very narrow; where wider size distributions are being filtered a size segregation effect occurs close to the membrane surface and it is then unlikely that the specific resistance would be independent of the wall shear stress. These deposit characteristics play a basic role is affecting the critical parameter J_{crit}/τ_w, which defines the operating conditions required for high stable filtration performance.

J_{crit}/τ_w is a critical parameter in crossflow filtration, and the critical permeation flux, J_{crit}, under which there is no deposition, increases almost linearly with wall stress. In microfiltration of dairy products (which contain particles like casein micelles, aggregates, or bacteria), J_{crit}/τ_w rules the permeability and selectivity performance of the separation (Gésan-Guiziou et al, 1999), but the deposit characteristics differ from one filtration application to another.

The parameter J_{crit}/τ_w is more appropriate than $\Delta p_{crit}/\tau_w$ which has also been proposed (Fischer and Raasch, 1986; Wakeman, 1994) because J_{crit} is independent of the initial hydraulic resistance of the clean membrane (or initial pore size of the membrane), in contrast with the transmembrane pressure. Wu et al (1999) observed a decrease in J_{crit} with increasing pore size when filtering bovine serum albumin solutions, silica dispersions and yeast suspensions, which they attributed to either differing charge effects and interactions as a result of different membrane materials or to changes in membrane porosity induced by internal fouling. Madaeni et al (1999) showed that the critical flux is insensitive to the pore size of the membrane.

For a given membrane separation, the independence of the critical parameter J_{crit}/τ_w from the membrane pore size indicates that there is not necessarily a need to work with the largest pore size membrane; larger pores will not induce higher critical flux and will not improve the area of the stability zone of the filtration. However, the risk of internal fouling would be greater. Smaller pores are therefore preferred, provided the transmembrane pressure required to obtain J_{crit} remains lower than the axial (transmodule) pressure drop in order not to induce an increase in energy consumption.

8.5 Conclusions

There have been numerous attempts to develop mathematical models to predict flux, but none are entirely satisfactory. Their main failing is their inability to describe completely all regions of filtration, from the pressure controlled to the mass transfer controlled regions (in the case of UF) or to the cake controlled region (in the case of MF). The value of the model used is to give an appreciation of the phenomena that may be controlling the process whilst simultaneously providing a practical approach to aspects of engineering design. In so doing, it must be capable of interpreting the effects of the operating variables or changing feed channel dimensions or configurations. The models described in this chapter are satisfactory for these purposes.

List of Notation

Unless otherwise defined in the text, the symbols used have the following meanings:

a — Effective fractional open area of a pore in a woven fabric, or scour coefficient (Section 2.4).

$a_0 \dots a_6$ — Constants.

a_p — Effective fractional open area of a plain pore, or precoat dosage (Chapter 4), (kg filter aid) m^{-2}.

a_T — Effective fractional open area of a twill pore.

A — Filtration area, m^2, or Hamaker constant, J, or plan area of a settling chamber, m^2.

A_{av} — Arithmetic average surface area of cake in centrifuge in direction of filtration, m^2.

A_{lm} — Log mean surface area of cake in centrifuge in direction of filtration, m^2.

A_0 — Effective area of an orifice or initial area of cylindrical filter element (candle), m^2.

A_p — Area of a single tilting pan, or projected area of a particle, m^2.

b — Radius of the spherical envelope enclosing a particle in the Happel model, m, or a constant (Chapter 4).

$b_0 \dots b_2$ — Constants.

c — Effective concentration of solids in the feed or suspension, kg m^{-3}.

c' — Solids concentration in the feed (refer to Chapter 4), kg m^{-3}.

c_L	Solids concentration in the capacity limiting layer (Chapter 3), kg m^{-3}.
C	Concentration of a species being separated by a membrane filter, kg m^{-3}.
C_e	Modified consolidation coefficient, m^2 s^{-1}.
C_D	Discharge coefficient, or drag coefficient (Section 2.1).
C_{e0}	Modified consolidation coefficient at unit applied pressure, m^2 s^{-1} kPa$^{-\gamma}$.
C_f	Concentration of a species in the feed to a membrane filter, kg m^{-3}.
C_m	Concentration of a species at the surface of a membrane in a filter, kg m^{-3}.
C_p	Concentration of a species in the permeate of a membrane filter, kg m^{-3}.
d_p	Effective diameter of an orifice (or pore), m.
d_y	Yarn diameter, m.
d_1	Diameter of the weft (fill) yarns, m.
d_2	Diameter of the warp yarns, m.
D	Diameter of tube, pipe, vessel or drum filter, m, or molecular diffusivity of solute, m^2 s^{-1}.
D_h	Hydraulic diameter, m.
D_L	Axial dispersion coefficient, m^2 s^{-1}.
D_n	Dispersion number ($= vL/D_L$).
D_s	Rotational speed of drum filter, s^{-1}.
e	Voids ratio.
e_0	Voids ratio at unit applied pressure.
e_c	Thickness of a cake formed on a filter candle, m.
e_p	Thickness of precoat cake on a filter candle, m.
(ec)	Warp yarns per cm of cloth, cm^{-1}.
E	Washing efficiency (defined as the fraction of solute removed by one wash ratio), or electric field strength, V m^{-1}.
$F, \bar{F}$	Fraction of solute removed from a filter cake by washing.
F_c	g–factor.

F_c	Drag force, N.
F_K	Filtrability coefficient, s m^{-2}.
F_s	Cumulative frictional drag, N.
$F*$	Fraction of feed solute removed by filtration and washing.
$F**$	Fraction of feed solute removed by washing in a multiple washing operation.
g	Acceleration due to gravity, m s^{-2}.
h	Candle length, or height of centrifuge basket, m.
h_B	Filter belt width, m.
h_D	Filter drum width, m.
h_L	Capillary drain height, m.
H	Interparticle distance, or depth of the clarification zone, or distance of the sampling point below the top of the suspension (Chapter 3), m.
i	Number of drainage surfaces.
I	Liquid flow rate associated with the input slurry to a filter, m^3 s^{-1}.
j	Area factor to account for a non-planar filter surface.
j_{II}	Consolidation area factor applied to a 2-dimensional expression.
J	Permeate flux from a membrane filter, m^3 m^{-2} h^{-1}.
J_{crit}	Critical flux of permeate from a membrane filter, m^3 m^{-2} h^{-1}.
k	Permeability, m^2, or the solute distribution coefficient, (kg solute/kg solid)/(kg solute/kg liquid), or mass transfer coefficient (in a membrane filter), m s^{-1}.
$k*$	Permeability term (Chapter 2), m^{-1}.
k_r	Rate constant, s^{-1}.
k_{ri}	Relative permeability of the cake to the ith phase flowing through its voids.
k_0	Constant in the Kozeny-Carman equation, or other constants.
k_1	Permeability of the cloth if the yarns were impermeable cylinders (Chapter 2), m^2.

k_1, k_2	Constants.
K	Fluid consistency index, Pa s^n, or proportionality constant.
K_b	Constant describing complete pore blocking, s^{-1}.
K_c, K_i, K_s	Constants during cake, intermediate and blocking filtration, m^{-3}.
K_0	Kozeny constant.
K_1, K_2	Constants.
L	Length of tube or pipe, or thickness of filter cake or bed, or the length of a membrane filter, m.
L_e	Effective distance travelled through a porous medium by a fluid element, m.
L_m	Membrane thickness, m.
L_{tr}	The thickness of the filter cake at the transition between filtration and consolidation stages of expression, m.
m	Ratio of the mass of wet cake to the mass of dry cake, or number of stages in a multiple reslurry system, or mass of a particle (Section 2.1)
m_{sol}	Mass of solute in a cake, kg.
M	Moisture content of a filter cake, expressed as the ratio of the mass of liquid in the cake to the total mass of the wet cake, or molecular weight of a solvent, kg mol^{-1}.
M_l	Mass of liquids (in a filter cake if not otherwise stated), kg.
M_l^*	Total mass of liquids, kg.
$M_l(t)$	Throughput of liquids in a filter cake at a point on a continuous filter, kg s^{-1}.
M_s	Mass of solids (in a filter cake if not otherwise stated), kg.
M_s^*	Total mass of solids, kg.
$M_s(t)$	Throughput of solids in a filter cake at a point on a continuous filter, kg s^{-1}.
M'	Ratio of the mass of liquid to the mass of solids in a filter cake.
n	Compressibility index, or the number of pores, or the number of particles in a suspension.
n_D	Ratio of the outer to inner diameters of the original suspension being expressed on a cylindrical filter element.

n_p	Number of tilting pans.
N	Rotational speed, revolutions per second (rps), or number of candles in a filter, or the flow behaviour index of a non-Newtonian fluid.
N_{cap}	Capillary number.
p	Pressure, usually the hydraulic pressure delivered by the pumping system, Pa.
p^*	Dimensionless pressure, defined in Table 5.1.
p_a	Pressure in the air phase, Pa.
p_{aei}	Pressure in air phase at air entry surface of cake on actual filter installation, Pa.
p_{aeo}	Pressure in air phase at air exit surface of cake on actual filter installation, Pa.
p_{ai}	Pressure in air phase at air entry surface of cake, Pa.
p_{ao}	Pressure in air phase at air exit surface of cake, Pa.
p_b	Threshold pressure, Pa.
p_B	Barometric pressure, Pa.
p_c	Capillary pressure, Pa.
p_i	Amount of solute adsorbed onto the cake solids, (kg solute)/(kg solid).
p_l	Pressure acting in the liquid phase, Pa.
p_r	Pressure due a centrifugal force acting at radius r, Pa.
p_s	Compressive drag pressure acting on the solids in a filter cake, Pa.
p_0	Pressure on the downstream side of a filter cloth, Pa.
p_1	Pressure at the cake-cloth interface, Pa.
Δp	Pressure difference, Pa.
Δp_m	$= p_1 - p_0$, the pressure difference across a filter cloth, Pa.
(pc)	Weft (fill) yarns per cm of cloth, cm^{-1}.
P	Filtrate flow rate from a continuous filter, m^3 s^{-1}.
Pe	Peclet number.
$P_y(\phi)$	Compressive yield stress, Pa.

q	$= dV/dt$, volume flow rate of filtrate, m^3 s^{-1}.
Q	Volumetric flow rate, m^3 s^{-1}.
Q_0	Overflow volume flow rate, m^3 s^{-1}.
r	Pore radius or radial co-ordinate, or half length/radius of a filter plate, or radius of rotation, m.
r_c	Radius of cake layer in centrifuge, m.
r_i	Inner radius of the cake in a centrifuge or a pan on a tilting pan filter, m.
r_l	Radius of liquid layer inside a centrifuge, m.
r_0	Inner radius of centrifuge drum or radius of a filter candle, m.
r', r''	Pore radius, m.
R	Resistance to fluid flow through a filter cloth, m^{-1}, or fraction of the solute originally in a filter cake which is retained after washing, or rejection by a membrane filter (equation (8.2)).
R_{bl}	Resistance of a boundary layer to fluid flow, m^{-1}.
R_c	Resistance to fluid flow through a filter cake, m^{-1}.
R_f	Resistance of a fouling layer to fluid flow, m^{-1}.
R_p	Radial length of a tilting pan, m
Re	Reynolds number.
s	Mass fraction of solids in a feed suspension.
S	Saturation, volume of liquid in a cake per unit volume of the voids.
S_R	Reduced saturation.
S_0	Specific surface of particles, m^2 m^{-3}.
S_{sp}	Specific surface area of a spacer in a membrane filter, m^2 m^{-3}.
S_∞	Irreducible saturation.
Sc	Schmidt number.
Sh	Sherwood number.
St	Stokes number.
t	Time, s.

t_d	Deliquoring time, or detention time, s.
t_d	Filter down time, s.
t_i	Time at which filtration is considered to commence, s.
T	Thickness of a filter press frame or chamber, m, or transmittance through a membrane (equation (8.3)), or temperature, K or C.
T_c	Dimensionless consolidation time.
ΔT	Distance moved by a diaphragm, m.
u	Superficial velocity, $m^3\,m^{-2}\,s^{-1}$, or particle-fluid relative velocity, $m\,s^{-1}$.
u'	Velocity of fluid flowing through a capillary or tube, $m\,s^{-1}$.
$\bar{u}^*$	Dimensionless flux during deliquoring averaged over the deliquoring time.
$u_a\vert_{des}$	Design air rate, $m^3\,m^{-2}\,s^{-1}$.
$u_a\vert_{tot}$	Design air rate on a total cycle basis, $m^3\,m^{-2}\,s^{-1}$.
u_0	Overflow volume flux, $m^3\,m^{-2}\,s^{-1}$.
u_h	Hindered settling velocity, $m\,s^{-1}$.
u_t	Terminal settling velocity, $m\,s^{-1}$.
U	Flow rate of liquid associated with cake solids, $m^3\,s^{-1}$.
v	Pore (or interstitial) velocity of a fluid, $m\,s^{-1}$, or molal volume of a solute, $m^3\,kmol^{-1}$.
v_B	Linear velocity of a filter belt, $m\,s^{-1}$.
v_L	Settling velocity of particles in the limiting layer, $m\,s^{-1}$.
$\bar{v}_L$	Upward propagation velocity of particles in the limiting layer, $m\,s^{-1}$.
V, V_f	Filtrate volume, or volume of the clarification zone (Chapter 3), m^3.
V_A	Potential energy of interaction due to van der Waals attraction, J.
V_{cz}	Volume of the compression zone, m^3.
V_i	Volume of filtrate at the start of filtration, m^3.
V_r	Residual volume of filtrate remaining in the filter cake, m^3.
V_R	Potential energy of interaction due to electrostatic repulsion, J.

V_T	Total potential energy of interaction, J.
V_0	Volume of suspension in a piston press at the start of a test, m^3.
V_s	Solids volume fraction in a suspension or solid/liquid mixture.
V_w	Wash liquor volume, m^3.
w	Mass of dry cake per unit area of filter, kg m^{-2}, or mass of dry solids, kg.
W	Wetted perimeter of an orifice (or pore), m, or the wash ratio (the amount of wash liquid passed through a cake per unit amount of liquid in the cake at the start of washing).
W_p	Wetted perimeter of a plain pore, m.
W_R	Solids throughput on a continuous filter, kg s^{-1}.
W_T	Wetted perimeter of a twill pore, m.
x	Particle diameter, m.
x_B	Distance along a belt filter, m.
x_f	Diameter of the fibres in a nonwoven fibre pad, m.
x_{sv}	Particle surface-volume mean diameter, m.
X	Normalised averaged solute concentration in the wash effluent from a continuous washing operation.
Y	Overall production rate from a filter, kg s^{-1} or m^3 s^{-1}.
z	Co-ordinate direction; or the length of filter devoted to a phase in a filter cycle, or the height of suspended solids (Chapter 3), m.
z_i	Height which a suspension would occupy if all solids were present at the concentration c_L, m.

Greek letters

α	Specific resistance of a filter cake, m kg^{-1}.
α_0	Specific resistance at unit applied pressure or at zero applied pressure, m kg^{-1} kPa^{-n}.
α_{av}	Specific resistance of the filter cake averaged over the compressive drag stress, m kg^{-1}.
$\alpha(r)$	Distribution of pore radii in a filter cake.

β	Compressibility index, or particle packing density of spheres, or bulking index (Section 2.4).
δ	$= 2/(r_0\rho_p)$, m² kg⁻¹, or length of the washing zone on a continuous filter, m, or thickness of the gel or cake layer in a membrane filter, m.
δ_c	$= 2/(r_0\rho_c)$, m² kg⁻¹.
ε	Porosity, volume of voids per unit volume of filter cake (or porous medium).
ε_0	Effective porosity at unit applied pressure, kPa⁻λ, or at zero applied pressure.
ε_p	Permittivity of a suspending fluid, C² J⁻¹ m⁻¹.
ε_{sp}	Porosity of a spacer in a membrane filter.
ϕ	$= ((ec)(d_1 + d_2))^{-1}$ (Chapter 2).
ϕ	Solute concentration in the liquid being discharged from a cake, kg m⁻³, or fraction of filter surface devoted to a phase in a filter cycle.
ϕ_w	Solute concentration in the feed wash liquor, kg m⁻³.
ϕ_0	Solute concentration in the liquid in cake voids prior to washing, kg m⁻³.
ϕ_{av}	Solute concentration in a wash effluent that is collected over a period and mixed, kg m⁻³.
φ	Included angle of a tilting pan, rad.
γ	Index in scale-up equation for consolidation.
$\dot{\gamma}$	Shear rate, s⁻¹.
κ	Reciprocal electrical double layer thickness, m⁻¹.
λ	Pore size distribution index, or defined in cake washing (Chapter 2), or compressibility index (Chapter 4), or filter coefficient (Section 2.4).
μ	Viscosity of the liquid in a feed or filtrate, Pa s.
θ	Dimensionless time, defined in Table 5.1.
ρ	Density of the liquid in a feed or filtrate, kg m⁻³.
ρ_c	Bulk density of a filter cake, kg m⁻³.
ρ_p	Bulk density of a precoat layer, kg m⁻³.

ρ_s	Density of solids or particles, kg m^{-3}.
σ	Surface tension, N m^{-1}, or specific deposit, (m^3 particles/m^2 filter).
τ	Shear stress, Pa.
τ_w	Shear stress at a wall or membrane surface, Pa.
υ	Consolidation index.
ω	Angular velocity, s^{-1}.
ω_0	Volume of solids per unit filter area, m^3 m^{-2}.
Ψ	Linear velocity of a filter cake, m s^{-1}, or cake liquid hold-up (Example 6.8).
Ψ_0	Voltage potential at a particle surface, V.
Ψ_1, Ψ_2	Voltage potential at the boundary of the Stern layer, V.
ζ	Zeta-potential, defined as the voltage potential at the plane of slip of a particle moving relative to a suspending fluid, V.

Subscripts

a	Referring to air.
av	Average value.
c	Referring to the consolidation phase in a filter cycle.
cake	Value for a filter cake.
d	Referring to the deliquoring phase in a filter cycle.
e	Referring to the end of a phase in a filter cycle.
eff	Effective value.
f	Referring to the filtration phase in a filter cycle, or to the filtrate, or to the feed to a thickener.
g	Referring to the gas phase (subscripts a and g are often used interchangeably).
i	Referring to a position in a table or sequence of calculations.
l	Referring to the liquid phase.
L	Refers to the capacity limiting layer in a thickening suspension.
opt	Optimum value.

pr	Referring to a value at the end of the previous phase in a filter cycle.
s	Referring to the solid phase.
susp	Value for a suspension.
t	Referring to the sum of parameter values over all phases of a filter cycle.
tr	Referring to the transition point between phases in a filter cycle.
u	In a thickener underflow.
w	Referring to the wash liquor or the washing phase in a filter cycle.
0	Initial value, unless otherwise stated.
∞	Equilibrium value.

Superscripts

*	Dimensionless value (unless otherwise stated).

Bibliography

Adorján L.A., 1975. A theory of sediment compression, *Proc. XIth Int. Min. Proc. Congress*, Cagliari, Paper 11.

Akay G. and Wakeman R.J., 1994. Mechanisms of permeate flux decay, solute rejection and concentration polarisation in crossflow filtration of a double chain ionic surfactant dispersion, *J. Memb. Sci.*, **88**, 177–195.

Akers R.J., 1974. Sedimentation techniques – a review, *Institution of Chemical Engineers Symposium Series No. 41*, H1–H14.

Alt C., 1975. *Practical problems in choosing filtration process and future developments*, in "The Scientific Basis of Filtration", Ed. K.J. Ives, pp. 411–444, Noordhoff, Leyden.

Ambler C.M., 1988. *Centrifugation*, in "Handbook of Separation Techniques", 2nd Edition, Ed. P.A. Schweitzer, pp. 4.59–4.88, McGraw-Hill, New York.

Antelmi D., Cabane B., Meireles M. and Aimar P., 2001. Cake collapse in pressure filtration, *Langmuir*, **17**, 7137–7144.

Andersen N.P.R., Christensen M.L. and Keiding K., 2004. New approach to determining consolidation coefficients using cake-filtration experiments, *Powder Technol.*, **142**, 98–102.

Arnot T.C., Field R.W. and Koltuniewicz A., 2000. Crossflow and deadend MF of oily water emulsions: Pt II Mechanism and modelling of flux decline, *J. Memb. Sci.*, **169**, 1–15.

Arora N and Davis R.H., 1994. Yeast cake layers as secondary membranes in deadend microfiltration of bovine serum albumin, *J. Membr. Sci.,* **92,** 247.

Atsumi K. and Akiyama T., 1975. A study of cake filtration – Formulation as a Stefan problem, *J. Chem. Eng. Japan*, **8**, 487–492.

Auzerais F.M., Jackson R. and Russel W.B., 1988. The resolution of shocks and the effects of compressible sediments in transient settling. *J. Fluid Mech.*, **195**, 437–462.

Auzerais F.M., Jackson R., Russel W.B. and Murphy W.F., 1990. The transient settling of stable and flocculated dispersions. *J. Fluid Mech.*, **221**, 613–639.

Baker J.S. and Dudley L.Y., 1998. Biofouling in membrane systems – A review, *Desalination,* **118**, 81.

Baluais G., Dodds J. and Tondeur D., 1983. The kinetics of displacement dewatering in filter cakes: An analytical approach, *IChemE Symposium Series No.* 69, 107–122.

Belfort G. and Nagata N., 1985. Fluid mechanics and crossflow filtration: some thoughts, *Desalination*, **53**, 57–59.

Bowen W.R. and Jenner F., 1995. Dynamic ultrafiltration model for charged colloidal dispersions: a Wigner-Seitz cell approach, *Chem. Eng. Sci.*, **50**(11), 1707–1736.

Bowen W.R., Mongruel A. and Williams P.W., 1996. Predictions of the rate of crossflow membrane ultrafiltration: a colloidal interaction approach. *Chem. Eng. Sci.*, **51**(18), 4321–4333.

Brenner H., 1961. Three-dimensional filtration on a circular leaf, *AIChEJ*, **7**, 666–671.

Brinkman H.C., 1948. On the permeability of media consisting of closely packed porous particles, *Appl. Sci. Res.*, **A1**, 81–86.

Brooks R.H. and Corey A.T., 1964. Hydraulic properties of porous media, *Hydrology Papers*, Colorado State University, Fort Collins, Colorado.

Brownell L.E. and Katz D.L., 1947. Flow of fluids through porous media, *Chem. Eng. Prog.*, **43**, 537–548; 601–612; 703–712.

Brownell L.E. and Gudz G.B., 1949. Blower requirements of rotary drum vacuum filters, *Chem. Engrg.*, **56**, 112–115.

Bühler T.M., Burrell K., Eggars H.U. and Reed R.J.R., 1993. The application of membranes for new approaches to brewery operations, *Proc. Eur. Brew. Conv.*, Oslo.

Burak N. and Storrow J.A., 1950. The flow relationships in a basket centrifuge, *J. Soc. Chem. Ind.*, **69**, 8–13.

Burdine N.T., 1953. Relative permeability calculations from pore-size distribution data, *Trans AIME*, **198**, 71–77.

Burrell K.J. and Reed R.J.R., 1994. Cross-flow microfiltration of beer: laboratory scale studies of the effect of pore size, *Filtration and Separation*, **31**, 399.

Burrell K.J., Gill C., McKechnie M. and Murray J., 1994. Advances in separations technology for the brewer: ceramic cross-flow microfiltration of rough beer, *Q. Master Brew. Assoc. Am.*, **31**, 42.

Burger R., Concha F. and Karlsen K.H., 2001. Phenomenological model of filtration processes: 1. Cake formation and expression, *Chem. Eng. Sci.*, **56**, 4537–4553.

Buscall R. and White L.R., 1987. The consolidation of concentrated suspensions – The theory of sedimentation, *J. Chem. Soc. Faraday Trans.*, **1**, 83, 873–891.

Cameron J.R., MS Thesis, Ohio State University, 1973.

Carleton A.J. and Mehta K.B., 1983. Leaf test predictions and full-scale performance of vacuum filters, *Proc. Filtech Conference*, pp. 120–127, The Filtration Society, Karlsruhe.

Carleton A.J. and Mackay D.J., 1988. Assessment of models for predicting the dewatering of filter cakes by gas blowing, *Filtration and Separation*, **25**, 187–191.

Carleton A.J. and Salway A.G., 1993. Dewatering of cakes, *Filtration and Separation*, **30**, 641–646.

Carman P.C., 1938. Fundamental principles of industrial filtration, *Trans IChemE*, **16**, 168–188.

Carman E.H.D. and Steyn D.P., 1965. Some observations on thickening, *VIIIth Comm. Min. Metall. Congress*, Sydney, Australia, 443–454.

Chan S.H., Kiang S. and Brown M.A., 2003. One-dimensional centrifugation model, *AIChEJ*, **49**, 925–938.

Chandavarkar A.S., 1990. *Dynamics of fouling of microporous membranes by proteins*, PhD thesis, Massachusetts Institute of Technology, U.S.A.

Channell G.M., Miller K.T. and Zukoski C.F., 2000. Effects of microstructure on the compressive yield stress, *AIChEJ*, **46**, 72–78.

Chase G.G. and Dachavijit P., 2003. Incompressible cake filtration of a yield stress fluid, *Separ. Sci. Technol.*, **38**, 745–766.

Chu C.P., Chang M.J. and Lee D.J., 2003. Cake structure of consolidated biological sludge, *Separ. Sci. Technol.*, **38**, 967–976.

Coe H.S. and Clevenger G.H., 1916. Methods for determining the capacities of slime thickening tanks, *Trans AIME*, **55**, 356–384.

Colton C.K., Smith K.A., Merrill E.W. and Reece J.M., 1970. *Chem. Eng. Progr. Symp.*, **66**, 85.

Condie D.J., Hinkel M. and Veal C.J., 1996. Modelling the vacuum filtration of fine coal, *Filtration and Separation*, **33**, 825–834.

Couturier S., Valat M., Vaxelaire J. and Puiggali J.R., 2003. Liquid pressure measurement in filtration-compression cell, *Separ. Sci. Technol.*, **38**, 1051–1068.

Cunningham C.E., Broughton G. and Kraybill R.R., 1954. Flow through textile filter media, *Ind. Eng. Chem.*, **46**, 1196–1200.

Da Costa A.R., Fane A.G. and Wiley D.E., 1994. Spacer characterization and pressure drop modelling in spacer-filled channels for ultrafiltration, *J. Membr. Sci.*, **87**, 79–98.

Dahlstrom D.A., 1978. Predicting performance of continuous filters, *Chem. Eng. Prog.*, **74**, 69–74.

Darcy H.P.G., 1856. *Les Fontaines Publiques de la Ville de Dijon*, Dalamont, Paris.

Davies C.N., 1952. The separation of airborne dust and particles, *Proc. Inst. Mech. Engrs.*, B, **1**, 185–198.

de Kretser R.G., Usher S.P., Scales P.J., Boger D.V. and Landman K.A., 2001, Rapid filtration measurement of dewatering design and optimization parameters, *AIChEJ*, **47**, 1758–1769.

Delaney R.A.M. and Donnelly J.K., 1977. In *Reverse Osmosis and Synthetic Membranes. Theory, Technology, Engineering*, Ed. S. Sourirajan, National Research Council, Ottawa, Canada.

Dick R.I. and Ewing B.B., 1967. Evaluation of activated sludge thickening theories, *J San. Engrg. Div., Proc. ASCE*, **95**, 333.

Dixon D.C., 1979. *Theory of gravity thickening*, in "Progress in Filtration and Separation 1", Ed. R.J. Wakeman, Elsevier, Amsterdam.

Dullien F.A.L., 1975. Single phase flow through porous media and pore structure, *Chem. Eng. J.*, **10**, 1–34.

Elias H.G., 1977. *Macromolecules Vol.1*, Plenum Press, New York.

Eriksson G., Rasmuson A. and Theliander H., 1996. Displacement washing of lime mud: Tailing effects, *Separations Technology*, **6**, 201–210.

Eriksson G. and Theliander H., 1994. Displacement washing of lime mud, *Nord. Pulp Pap. Res. J.*, **9**, 60–66.

Fasoli U. and Melli P., 1971. Lavaggio per filtrazione Parte II – Diffusione negli alveoli a regime Rigamonti, *Ing. Chim. Ital.*, **7**, 135–144.

Field R.W., 1996. In *Industrial Membrane Separation Technology*, Eds. K. Scott and R. Hughes, Blackie, Glasgow.

Field R.W., Wu D., Howell J.A. and Gupta B.B., 1995. Critical flux concept for microfiltration fouling, *J. Membrane Sci.*, **100**, 259–272.

Filter Design Software, 2005. Filtration Solutions, UK. (www.filtrationsolutions.co.uk).

Fischer E. and Raasch J., 1986. Model tests of the particulate deposition at the filter medium in crossflow filtration, *Proc. 4th World Filtration Congress*, pp. 11.11–11.17, Ostend.

Fitch B., 1966. Current theory and thickener design. *Ind. Eng. Chem.*, **58**(10), 18–28.

Fitch E.B., 1975. Current theory and thickener design: Part 3 – Design procedures, *Filtration and Separation*, **12**, 636–638.

Fitch E.B., 1986. Gravity Separation Equipment, in *Solid/Liquid Separation Equipment Scale-Up*, Eds. D.B. Purchas and R.J. Wakeman, Uplands Press and Filtration Specialists Ltd, London.

Garrido P., Concha F. and Burger R., 2003. Settling velocities of particulate systems: 14. Unified model of sedimentation, centrifugation and filtration of flocculated suspensions, *Int. J. Min. Proc.*, **72**, 57–74.

Gaudin A.M. and Fuerstenau M.C., 1962. Experimental and mathematical modelling of thickening, *Trans. AIME*, **223**, 122–129.

Gekas V. and Hallstrom B., 1987. Mass transfer in the membrane concentration layer under turbulent cross flow. 1. Critical literature review and adaptation of existing Sherwood correlations to membrane operation, *J. Membr. Sci.*, **30**, 153–170.

Gésan-Guiziou G., Boyaval E. and Daufin G., 1999. Critical stability conditions in crossflow microfiltration of skimmed milk: transition to irreversible deposition, *J. Membr. Sci.*, **158**, 211–222.

Gésan-Guiziou G., Wakeman R.J. and Daufin G., 2002. Stability of latex crossflow filtration: Cake properties and critical conditions of deposition, *Chemical Engineering Journal*, **85**, 27–34.

Gonsalves V.E., 1950. A critical investigation on the viscose filtration process, *Rec. Trav. Chim. des Pays-Bas*, **69**, 873.

Grace H.P., 1953. Resistance and compressibility of filter cakes, *Chem. Eng. Prog.*, **49**, 303–318; **49**, 367–376.

Grace H.P., 1956. Structure and performance of filter media, *AIChEJ*, **2**, 307–336.

Green M.D., Landman K.A., de Kretser R.G. and Boger D.V., 1998. Pressure filtration technique for complete characterization of consolidating suspensions, *Ind. Eng. Chem. Res.*, **37**, 4152–4156.

Granath K.A., 1958. Solution properties of branched dextrans, *J. Colloid Interface. Sci.*, **13**, 308–328.

Gray V.R., 1958. The dewatering of fine coal, *J. Inst. Fuel*, **31**, 96–108.

Gren U., 1972. Washing packed beds of fibres, *Filtration and Separation*, **9**, 265–270.

Grober H., Erk S. and Grigoli V., 1961. *Fundamentals of Heat Transfer*, McGraw-Hill, New York.

Güell C., Czekaj P. and Davis R.H., 1999. Microfiltration of protein mixtures and the effects of yeast on membrane fouling, *J. Membr. Sci.*, **155**, 113.

Gurnham C.F. and Masson H.J., 1946. Expression of liquids from fibrous materials, *Ind. Eng. Chem.*, **38**, 1309–1315.

Hallit J., 1975. Sugar and sugar centrifuges, *Filtration and Separation*, **12**, 675–680.

Han C.D., 1967. Washing theory of the porous structure of aggregated materials, *Chem. Eng. Sci.*, **22**, 837–846.

Han C.D. and Bixler H.J., 1967. Washing of the liquid retained by granular solids, *AIChEJ*, **13**, 1058–1066.

Happel J., 1958. Viscous flow in multiparticle systems: Slow motion of fluid relative to beds of spherical particles, *AIChEJ*, **4**, 197–201.

Happel J. and Brenner H., 1965. *Low Reynolds Number Hydrodynamics*, Prentice-Hall, Englewood Cliffs, N.J.

Harris C.C., Jowett A. and Morrow N.R., 1964. Effect of contact angle on the capillary properties of porous masses, *Trans IMM*, **73**, 335–351.

Haruni M.M. and Storrow J.A., 1952a. Hydroextraction: Techniques for permeability measurements, *Ind. Eng. Chem.*, **44**, 2751–2756.

Haruni M.M. and Storrow J.A., 1952b. Hydroextraction: Relationships between hydroextraction and filtration permeability, *Ind. Eng. Chem.*, **44**, 2756–2763.

Haruni M.M. and Storrow J.A., 1953a. Hydroextraction: Effects of cake shape, *Chem. Eng. Sci.*, **2**, 164–172.

Haruni M.M. and Storrow J.A., 1953b. Hydroextraction: Residual moisture in whizzed cakes, *Chem. Eng. Sci.*, **2**, 203–212.

Hasset N.J., 1965. Mechanisms of thickening and thickener design, *Trans IMM*, **74**, 627–656.

Heertjes P.M., 1957. Studies in filtration – the initial stages of the cake filtration, *Chem. Eng. Sci.*, **6**, 269–276.

Helfferich F.H., 1962. *Ion Exchange*, McGraw Hill, New York.

Hermans P.H. and Bredée H.L., 1935. Zur Kenntnis der Filtrationgesetze, *Rec. Trav. Chim. des Pays-Bas*, **54**, 680.

Hermia J., 1981. Calculation method for counter current washing on continuous vacuum filters, *Hydrometallurgy 81*, Soc. Chem. Ind., C5/1–10.

Hermia J., 1982. Constant pressure blocking filtration laws – Application to power law non-Newtonian fluids, *Trans IChemE*, **60**, 183–187.

Hermia J. and Letesson Ph., 1982. The universal washing curve and its application to multistaged counter current washing processes, *Proc. 3rd World Filtration Congress*, 426–435, Downingtown.

Hermia J. and Taeymans D., 1978. Considerations on the choice of a washing process, *Proc. Liquid-Solid Filtration Symposium*, pp. 61–74, Société Belge de Filtration, Antwerp.

Hermia J. and Brocheton S., 1993. Comparison of modern beer filters, *Proc. Filtech Conference*, 13pp., The Filtration Society, Karlsruhe.

Holdich R.G., 1993. Prediction of solid concentration and height in a compressible filter cake, *Int. J. Mineral Processing*, **39**, 157–171.

Holeschovsky U.B. and Cooney C.L., 1991. Quantitative description of ultrafiltration in a rotating filtration device, *AIChEJ*, **37**, 1219.

Hosseini M., 1977. PhD Thesis, Manchester University.

Hosten C. and San O., 1999. Role of the clogging phenomena in erroneous implications of conventional data analysis for constant pressure cake filtration, *Separ. Sci. Technol.*, **34**, 1759–1772.

Howells I., Landman K.A., Panjkov A., Sirakoff C. and White L.R., 1990. Time-dependent batch settling of flocculated suspensions, *Appl. Math Modelling*, **14**, 77.

International Critical Tables, Vol. V, National Research Council, McGraw Hill, New York, 1929.

Ives, K.J., 1973. *Mathematical models of deep bed filtration*, in "The Scientific Basis of Filtration", Ed. K.J. Ives, pp. 203–224, NATO Advanced Study Institute Series E2, Noordhoff International Publishing, Leyden.

Iwasaki, T., 1937. Some Notes on Sand Filtration, *J. Am. Water Works Association*, **29**, 1591.

Jemaa E., Krempff R. and Depyre D., 1974. Modélisation mathématique de la filtration et du lavage dans la fabrication de l'acide phosphorique, *Chem. Eng. J.*, **8**, 103–111.

Johansson C. and Theliander H., 2003. Measuring concentration and pressure profiles in deadend filtration, *Filtration*, **3**, 114–120.

Kapur P.C., Laha S., Usher S., de Kretser R.G. and Scales P., 2002, Modeling of the consolidation stage in pressure filtration of compressible cakes, *J. Coll. Interface Sci.*, **256**, 216–222.

Kelly S.T., Opong W.S and Zydney A.L., 1993. The influence of protein aggregates on the fouling of microfiltration membranes during stirred cell filtration, *J. Membr. Sci.*, **80**, 175.

Kelly S.T. and Zydney A.L., 1997. Protein fouling during microfiltration: comparative behaviour of different model proteins. *Biotechnology and Bioengineering*, **55**, 91.

Koenders M.A. and Wakeman R.J., 1996. The initial stages of compact formation from suspensions by filtration, *Chem. Eng. Sci.*, **51**, 3897–3908.

Koenders M.A. and Wakeman R.J., 1997a. Filter cake formation from structured suspensions, *Trans IChemE*, **75**, Part A, 309–320.

Koenders M.A. and Wakeman R.J., 1997b. Initial deposition of interacting particles by filtration of dilute suspensions, *AIChEJ*, **43**, 946–958.

Koenders M.A., Reymann S. and Wakeman R.J., 2000. The intermediate stages of the dead-end filtration process, *Chem. Eng. Sci.*, **55**, 3715–3728.

Koenders M.A., Liebhart E. and Wakeman R.J., 2001. Dead-end filtration with torsional shear: Experimental findings and theoretical analysis, *Trans IChemE*, **79**, Part A, 249–259.

Koltuniewicz A.B., Field R.W. and Arnot T.C., 1995. Crossflow and deadend MF of oily water emulsions: Pt II Experimental study and analysis of flux decline, *J. Memb. Sci.*, **102**, 193.

Koo E.C., 1942. Expression of vegetable oils, *Ind. Eng. Chem.*, **34**, 342–345.

Körmendy I., 1964. *J. Food Sci.*, **29**, 631.

Kos P., 1974. Gravity thickening of water-treatment plant sludges, *94th Annual Conference of AWWA*, Boston.

Kos P., 1975. Transport phenomena applied to sludge dewatering, *J. Environ. Eng. Div.*, ASCE, **101**, 947–965.

Koshiyama I., 1968. *Cereal Chem.*, **45**, 394.

Koudstaal J., 1976. PhD Thesis, University of Natal.

Kozeny J., 1927. Über kapillare Leitung des Wassers im Boden, *Sitzungsber. Akad. Wiss. Wien*, **136**, 271–306.

Kozicki W., 1990. Factors affecting cake resistance in non-Newtonian filtration, *Can. J. Chem. Eng.*, **68**, 69–80.

Kozinski A.A. and Lightfoot E.N., 1972. Protein ultrafiltration: a general example of boundary layer filtration, *AIChEJ*, **18**, 1030.

Kulkarni S.S., Funk E.W. and Li N.N., 1992. In *Membrane Handbook*, Eds. W.S. Ho and K.K. Sirkar, Van Nostrand Reinhold, New York.

Kuo K.T., 1960. Filter cake washing performance, *AIChEJ*, **6**, 566–568.

Kuo M.T. and Barrett E.C., 1970. Continuous filter cake washing performance, *AIChEJ*, **16**, 633–638.

Kynch G.J., 1952. A theory of sedimentation, *Trans. Faraday Society*, **48**, 166–176.

Landman K.A., White L.R. and Buscall R., 1988. The continuous-flow gravity thickener: steady state behaviour, *AIChEJ*, **34**, 239–252.

Landman K.A., Sirakoff C. and White L.R., 1991. Dewatering of flocculated suspensions by pressure filtration, *Phys. Fluids A*, **3**(6), 1495–1509.

Landman K.A. and White L.R., 1992. Determination of the hindered settling factor for flocculated suspensions, *AIChEJ*, **38**, 184–192.

Landman K.A. and Russel W.B., 1993. Filtration at large pressures for strongly flocculated suspensions, *Phys. Fluids A*, **5**(3), 550–560.

Landman K.A., White L.R. and Eberl M., 1995. Pressure filtration of flocculated suspensions, *AIChEJ*, **41**, 1687–1700.

Leclerc D. and Minery E., 1980. Comparison du comportment des boues floculées et non floculées lors de leur déshydratation par compression mécanique, *La Tribune Cebedeau*, **33**, 441–449.

Lee D.N. and Merson R.L., 1974. Examination of cottage cheese whey by scanning electron microscopy: relationship to membrane fouling during ultrafiltration, *J. Dairy Sci.*, **58**, 1423.

Leu W., 1986. *Principles of compressible cake filtration*, in "Encyclopedia of Fluid Mechanics", Vol. 5, Ed. N.P. Cheremisinoff, Gulf Publishing, Houston.

Leveque M.D., 1928. Les lois de la transmission de chaleur by convection, *Ann. Mines*, **13**, 201.

Li N.N. and Ho W.S., Membrane Processes, in *Perry's Chemical Engineers Handbook*, Eds. R.H. Perry and D. Green.

Lloyd P.J. and Dodds J., 1972. Liquid retention in filter cakes, *Filtration and Separation*, **9**, 91–96.

Lu W.M., Huang Y.P. and Hwang K.J., 1998a. Dynamic analysis of constant rate filtration data, *J. Chem. Eng. Japan*, **31**, 969–976.

Lu W.M., Huang Y.P. and Hwang K.J., 1998b. Methods to determine the relationship between cake properties and solid compressive pressure, *Separ. Purif. Technol.*, **13**, 9–23.

MacDonald I.F., El-Sayed M.S., Mow K. and Dullien F.A.L., 1979. Flow through porous media – the Ergun equation revisited, *Ind. Eng. Chem. Fundam.*, **18**, 199–208.

Madaeni S.S., Fane A.G. and Wiley D.A., 1999. Factors influencing critical flux in membrane filtration of activated sludge, *J. Chem. Technol. Biotechnol.*, **74**, 539–543.

Marshall A.D., Munro P.A. and Tragardh G., 1997. Influence of permeate flux on fouling during the microfiltration of lactoglobulin solutions under crossflow conditions, *J. Membr. Sci.*, **130**, 23.

McDonogh R.M., Schaule G. and Flemming H.C., 1994. The permeability of biofouling layers on membranes, *J. Membr. Sci.,* **87**, 199.

McGregor J., 1965. *J. Soc. Dyers Colour.,* **81**, 429.

Meeten G.H., 2000. Septum and filtration properties of rigid and deformable particle suspensions, *Chem. Eng. Sci.,* **55**, 1755–1767.

Michaels A.S., 1968. Ultrafiltration, in *Progress in Separation and Purification*, Ed. E.S. Perry, Vol.1, Wiley-Interscience, New York.

Michaels A.S. and Bolger J.C., 1962. Settling rates and sediment volumes of flocculated kaolin suspensions, *Ind. Eng. Chem. Fund.,* **1**, 24–33.

Michaels A.S., Baker W.E., Bixler H.J. and Vieth W.R., 1967. Permeability and washing characteristics of flocculated kaolinite filter cakes, *Ind. Eng. Chem. Fundam.,* **6**, 33–40.

Mints, D.M., 1966. Modern theory of filtration, *7th Congress of the International Water Supply Association*, Barcelona, Volume 1, I.W.S.A., London.

Moncrieff A.G., 1964. Theory of thickener design based on batch sedimentation tests, *Trans IMM*, **73**, 729–759.

Nenniger E. and Storrow J.A., 1958. Drainage of packed beds in gravitational and centrifugal force fields, *AIChEJ*, **4**, 305–316.

Oyama Y. and Sambuichi S., 1954. On the fundamental study of centrifugal filtration, *J. Chem. Eng. Japan*, **18**, 593–600.

Palecek S.P. and Zydney A.L., 1994a. Hydraulic permeability of protein deposits formed during microfiltration: effect of solution pH and ionic strength, *J. Membr. Sci.,* **95**, 71.

Palecek S.P. and Zydney A.L., 1994b. Intermolecular electrostatic interactions and their effect on flux and protein deposition during protein filtration. *Biotechnology Progress,* **10**, 207.

Pedersen G.C., 1969. Fluid flow through monofilament fabrics, *Paper presented at 64th National Meeting of AIChE*, New Orleans.

Philip J.R., 1968. *Aust. J. Soil Res.,* **6**, 249.

Polson A., 1950. *J. Phys. Colloid Chem.,* **54**, 649.

P. Prádanos, J.I. Arribas and A. Hernández, Mass transfer coefficient and retention of PEGs in low pressure cross-flow ultrafiltration through asymmetric membranes, *J. Membrane Sci.,* **99**, 1–20, 1995.

Purchas D.B. and Sutherland K., 2002. *Handbook of Filter Media*, 2nd Edition, Elsevier Advanced Technology, Oxford.

Purchas D.B. and Wakeman R.J. (Eds.), 1986. *Solid/liquid separation equipment scale-up*, 2nd Edition, Uplands Press & Filtration Specialists Ltd, London.

Puttock S.J., Fane A.G., Fell C.J.D., Robins R.G. and Wainwright M.S., 1986. Vacuum filtration and dewatering of alumina trihydrate – The role of cake porosity and interfacial phenomena, *Int. J. Miner. Process.,* **17**, 205–224.

Rasmuson A., 1985. The effects of particles of variable size, shape and properties on the dynamics of fixed beds, *Chem. Eng. Sci.,* **40**, 621–629.

Rasmuson A. and Neretnieks I., 1980. Exact solution of a model for diffusion in particles and longitudinal dispersion in packed beds, *AIChEJ*, **26**, 686–690.

Richardson J.F. and Zaki W.N., 1954. Sedimentation and fluidisation, Part 1. *Trans IChemE*, **32**, 35.

Roberts E.J., 1949. Thickening – Art or science?, *Trans. Am. Inst. Mining Met. Engrs.*, **184**, 61–64.

Robins W.H.M., 1964. The theory of the design and operation of settling tanks, *Trans IChemE*, **42**, 158–163.

Rushton A. and Griffiths P.V.R., 1971. Fluid flow in monofilament filter media, *Trans IChemE*, **49**, 49–59.

Rushton A. and Griffiths P.V.R., 1972. Role of the cloth in filtration, *Filtration and Separation*, **9**, 81–89.

Rushton A. and Spear M., 1975. Centrifugal filtration and permeation, *Filtration and Separation*, **12**, 254–256.

Rushton A. and Griffiths P.V.R., 1987. *Filter media*, in "Filtration", Eds. M.J. Matteson and C. Orr, Marcel Dekker.

Rushton A., Ward A.S. and Holdich R.G., 1996. *Solid-Liquid Filtration and Separation Technology*, VCH Verlagsgesellschaft, Weinheim.

Ruth B.F., Montillon G.H. and Montonna R.E., 1933. Studies in filtration: I. Critical analysis of filtration theory; II. Fundamentals of constant pressure filtration, *Ind. Eng Chem.*, **25**, 76–82 and 153–161.

Ruth B.F., 1946. Correlating filtration theory with industrial practice, *Ind. Eng. Chem.*, **38**, 564–571.

Scales P.J., Dixon D.R., Harbour P.J. and Stickland A.D., 2004. The fundamentals of wastewater sludge characterization and filtration, *Water Sci. Technol.*, **49**, 67–72.

Scarlett B. and Ward A.S., 1986. *Particle size analysis*, in "Solid/Liquid Separation Equipment Scale-Up", 2nd Edition, Eds. D.B. Purchas and R.J. Wakeman, Uplands Press & Filtration Specialists Ltd, London.

Schluep T. and Widmer F., 1996. Initial transient effects during crossflow microfiltration of yeast suspensions, *J. Membr. Sci.,* **115***, 133.

Schwartzberg H.G., Rsenau J.R. and Richardson G., 1977. The removal of water by expression, *AIChE Symp. Ser.* 163, **73**, 177–190.

Scott K.J., 1968. Theory of thickening: factors affecting settling rate of solids in flocculated pulps, *Trans IMM*, **77**, C85–C97.

Scott K.J., 1970. Continuous thickening of flocculated suspensions – comparison with batch settling tests and effects of floc compression using pyrophyllite pulp, *Ind. Eng. Chem. Fund.*, **9**, 422–427.

Sherman W.R., 1964. The movement of soluble material during the washing of a bed of packed solids, *AIChEJ*, **10**, 855–860.

Shirato M., Sambuichi M., Kato H. and Aragaki T., 1967. *J. Chem. Eng. Japan*, **31**, 359.

Shirato M., Murase T., Kato H. and Fukaya S., 1970. Fundamental analysis for expression under constant pressure, *Filtration and Separation*, **7**, 277–282.

Shirato M., Murase T., Negawa M. and Moridera H., 1971. Analysis of expression operations, *J. Chem. Eng. Japan*, **4**, 263–268.

Shirato M., Murase T., Tokunaga A. and Yamada O., 1974. Calculations of consolidation period in expression operations, *J. Chem. Eng. Japan*, **7**, 229–231.

Shirato M., Aragaki T., Iritani E., Wakimoto M., Fujiyashi S. and Nanda S., 1977. Constant pressure filtration of power law non-Newtonian fluids, *J. Chem. Eng. Japan*, **10**, 54–60.

Shirato M., Murase T., Hayashi N. and Noguchi T., 1978a. Effects on expression rate increase of solid-liquid mixture with inserted strings, *J. Chem. Eng. Japan*, **11**, 315–318.

Shirato M., Murase T., Atsumi K., Nagai T. and Suzuki H., 1978b. Creep constants in expression of compressible solid-liquid mixtures, *J. Chem. Eng. Japan*, **11**, 334–336.

Shirato M., Aragaki T. and Iritani E., 1980a. Analysis of constant pressure filtration of power law non-Newtonian fluids, *J. Chem. Eng. Japan*, **13**, 61–66.

Shirato M., Aragaki T., Iritani E. and Funahashi T., 1980b. Constant rate and variable rate filtration of power law non-Newtonian fluids, *J. Chem. Eng. Japan*, **13**, 473–478.

Shirato M., Murase T. and Mori H., 1983. Centrifugal dehydration of a packed particulate bed, *Int. Chem. Eng.*, **23**, 298–306.

Shirato M., Murase T. and Iwata M., 1986. *Deliquoring by expression – theory and practice*, in "Progress in Filtration and Separation 4", Ed. R.J. Wakeman, Elsevier, Amsterdam.

Shirato M., Murase T., Iwata M. and Kurita T., 1986. *Principles of expression and design of membrane compression-type filter press operation*, in "Encyclopedia of Fluid Dynamics", Vol. 5, Ed. N.P. Cheremisinoff, Gulf Publishing, Houston.

Shirato M., Murase T., Iritani E., Tiller F.M. and Alciatore A.F., 1987. *Filtration in the chemical process industry*, in "Filtration", Eds. M.J. Matteson and C. Orr, Marcel Dekker, New York.

Silverblatt C.E. and Dahlstrom, D.A., 1954. Moisture content of a fine-coal filter cake, *Ind. Eng. Chem.*, **46**, 1201–1207.

Simons C.S. and Dahlstrom, D.A., 1966. Steam dewatering of filter cakes, *Chem. Eng. Prog.*, **62**, 75–81.

Sivaram B. and Swamee P.K., 1977. *J. Jpn. Soc. Soil Mech. Found. Eng.*, **17**, 48.

Smiles D.E., 1970. A theory of constant pressure filtration. *Chem. Eng. Sci.*, **25**, 985–996.

Smiles, D.E., 1986. *Principles of constant pressure filtration*, in "Encyclopedia of Fluid Mechanics", Vol. 5, Ed. N.P. Cheremisinoff, Gulf Publishing, Houston.

Smith K.A., Colton C.K., Merrill E.W. and Evans L.B., 1968. Convective transport in a batch dialyzer: determination of the true membrane permeability from a single measurement, *AIChE Symp. Ser.*, **64**, 45.

Smith M.H., 1970. In *Handbook of Biochemistry*, 2nd Edition, Ed. H.A. Sober, CRC Press, Cleveland, Ohio.

Sørensen B.L. and Wakeman R.J., 1996. Filtration characterisation and specific surface area measurement of activated sludge by Rhodamine B adsorption, *Wat. Res.*, **30**, 115–121.

Stamatakis K. and Chi Tien, 1991. Cake formation and growth in cake filtration. *Chem. Eng. Sci.*, **46**, 1917.

Stokes G.G., 1851. On the effect of the internal friction of fluids on the motion of a pendulum. *Trans. Cam. Phil. Soc.*, **9**, 8.

Svarovsky L., Ed., 1990. *Solid-Liquid Separation*, 3rd Edition. Butterworths, London.

Talmage W.L. and Fitch E.B., 1955. Determining thickener unit areas, *Ind. Eng. Chem.*, **47**, 38–41.

Tarleton E.S., 1998. A new approach to variable pressure cake filtration, *Minerals Engineering*, **11**, 53–69.

Tarleton E.S., 1998. Predicting the performance of pressure filters, *Filtration and Separation*, **35**, 293–298.

Tarleton E.S. and Wakeman R.J., 1993. Understanding flux decline in crossflow microfiltration: Part I – Effects of particle and pore size, *Trans IChemE, Part A*, **71**, 399–410.

Tarleton E.S. and Wakeman R.J., 1994a. Understanding flux decline in crossflow microfiltration: Part II – Effects of process parameters, *Trans IChemE, Part A*, **72**, 431–440.

Tarleton E.S. and Wakeman R.J., 1994b. Understanding flux decline in crossflow microfiltration: Part III – Effects of membrane morphology, *Trans IChemE, Part A*, **72**, 431–440.

Tarleton E.S. and Hancock D.L., 1997a. Using mechatronics for the interpretation and modelling of the pressure filter cycle, *Trans IChemE*, **75**, Part A, 298–308.

Tarleton E.S. and Willmer S.A., 1997b. The effects of scale and process parameters in cake filtration, *Trans IChemE*, **75**, Part A, 497–507.

Tarleton E.S. and Morgan S.A., 2001. An experimental study of abrupt changes in cake structure during dead-end pressure filtration, *Filtration*, **1**, 93–100.

Tarleton E.S. and Hadley R.C., 2003. The application of mechatronic principles in pressure filtration and its impact on filter simulation, *Filtration*, **3**, 40–47.

Teoh S.K., Tan R.B.H., He D. and Tien C., 2001. A multifunctional test cell for cake filtration studies, *Filtration*, **1**, 81–90.

Terzaghi K. and Peck P.B., 1948. *Soil Mechanics in Engineering Practice*, Wiley, New York.

Tien C., 1989. *Granular Filtration of Aerosols and Hydrosols*, Butterworths, Stoneham.

Tien C., 2002. Cake filtration research – a personal view, *Powder Technol.*, **127**, 1–8.

Tien C., Teoh S.K. and Tan R.B.H., 2001. Cake filtration analysis – the effect of the relationship between the pore liquid pressure and the cake compressive stress, *Chem. Eng. Sci.*, **56**, 5361–5369.

Tien C. and Bai R.B., 2003. An assessment of the conventional cake filtration theory, *Chem. Eng. Sci.*, **58**, 1323–1336.

Tiller F.M., 1953. The role of porosity in filtration, *Chem. Eng. Prog.*, **49**, 467–479.

Tiller F.M., 1955. The role of porosity in filtration 2: Analytical formulas for constant rate filtration, *Chem. Eng. Prog.*, **51**, 282–290.

Tiller F.M., 1958. The role of porosity in filtration 3: Variable pressure-variable rate filtration, *AIChEJ*, **2**, 171–174.

Tiller F.M., 1974. Bench scale design of SLS systems, *Chem. Engng*, **81**, 117–119.

Tiller F.M., 1975. *Compressible cake filtration*, in "The Scientific Basis of Filtration", Ed. K.J. Ives, Noordhoff, Leyden.

Tiller F.M., 1975. "Solid-Liquid Separation", 2nd Edition, University of Houston.

Tiller F.M., 1981. Revision of Kynch sedimentation theory, *AIChEJ*, **27**, 823–829.

Tiller F.M. and Huang C.J., 1961. Filtration theory, *Ind. Eng. Chem.*, **53**, 529–537.

Tiller F.M. and Cooper H., 1962. The role of porosity in filtration, part V: Porosity variation in filter cakes, *AIChEJ*, **8**, 445–449.

Tiller F.M. and Shirato M., 1964. The role of porosity in filtration, part VI: New definition of filtration resistance, *AIChEJ*, **10**, 61–67.

Tiller F.M. and Khatib Z., 1984. Theory of sediment volumes of compressible particulate structures, *J. Colloid Interface Sci.*, **100**, 55–67.

Tiller F.M., Li W.P. and Lee J.B., 2001. Determination of the critical pressure drop for filtration of super-compactible cakes, *Water Sci. and Technol.*, **44**, 171–176.

Tiller F.M. and Li W.P., 2003. Radial flow filtration for super-compactible cakes, *Separ. Sci. Technol.*, **38**, 733–744.

Tosun I., 1986. Formulation of cake filtration, *Chem. Eng. Sci.*, **41**, 2563–2568.

Tomiak A., 1994. *Pulp washing calculation manual*, Canadian Pulp and Paper Association, Montreal.

Tracey E.M. and Davis R.H., 1994. Protein fouling of track-etched polycarbonate microfiltration membranes, *J. Colloid and Interface Sci.*, **167**, 104.

Usher S.P., de Kretser R.G. and Scales P.J., 2001. Validation of a new filtration technique for dewaterability characterisation, *AIChEJ*, **47**, 1562–1570.

van Brakel J., van Rooijen P.H. and Dosoudil M., 1984. Prediction of the air consumption when dewatering a filter cake obtained by pressure filtration, *Powder Technol.*, **40**, 235–246.

Valleroy V.V. and Maloney J.O., 1960. Comparison of the specific resistances of cakes formed in filters and centrifuges, *AIChEJ*, **6**, 382–390.

Underwood A.J.V., 1926. A critical review of published experiments on filtration, *Trans IChemE*, **4**, 19–41.

Wakeman R.J., 1972. Prediction of the washing performance of drained filter cakes, *Filtration and Separation*, **9**, 409–415.

Wakeman R.J., 1973. Theoretical analyses of filter cake washing, *The Chem. Engr. (London)*, **280**, 596–602.

Wakeman R.J., 1974. The role of internal stresses in filter cake cracking, *Filtration and Separation*, **11**, 357–360.

Wakeman R.J., 1975a. Packing densities of particles with log-normal size distributions, *Powder Technol.*, **11**, 297–299.

Wakeman R.J., 1975b. *Filtration Post-Treatment Processes*, Elsevier, Amsterdam.

Wakeman R.J., 1976a. Diffusional extraction from hydrodynamically stagnant regions in porous media, *Chem. Eng. J.*, **11**, 39–56.

Wakeman R.J., 1976b. Vacuum dewatering and residual saturation of incompressible filter cakes, *Int. J. Miner. Process.*, **5**, 193–206.

Wakeman R.J., 1978. A numerical integration of the differential equations describing the formation of and flow in compressible filter cakes, *Trans IChemE*, **56**, 258–265.

Wakeman R.J., 1979a. Low pressure dewatering kinetics of incompressible filter cakes: I Variable total pressure loss or low capacity systems, *Int. J. Miner. Process.*, **5**, 379–393.

Wakeman R.J., 1979b. Low pressure dewatering kinetics of incompressible filter cakes: II Constant total pressure loss or high capacity systems, *Int. J. Miner. Process.*, **5**, 395–405.

Wakeman R.J., 1979c. The performance of filtration post-treatment processes: 1. The prediction and calculation of cake dewatering characteristics, *Filtration and Separation*, **16**, 655–660.

Wakeman R.J., 1981a. *Cake washing*, in "Solid-Liquid Separation", 2nd Edition, Ed. L. Svarovsky, pp. 408–451, Butterworths, London.

Wakeman R.J., 1981b. Material balance calculations for multiple washing filter installations, *Proc. Symposium on "Economic Optimisation Strategy in Solid/Liquid Separation Processes"*, pp. 159–176, Société Belge de Filtration, Louvain-la-Neuve.

Wakeman R.J., 1981c. The analysis of continuous countercurrent washing systems, *Filtration and Separation*, **18**, 35–41.

Wakeman R.J., 1981d. The formation and properties of apparently incompressible filter cakes on downward facing surfaces, *Trans IChemE*, **59**, 260–270.

Wakeman R.J., 1982. An improved analysis for the forced gas deliquoring of filter cakes and porous media, *J. Separ. Proc. Technol.*, **3**, 32–38.

Wakeman R.J., 1986a. Transport equations for filter cake washing, *Trans IChemE*, **64**, 308–319.

Wakeman R.J., 1986b. *Theoretical approaches to thickening and filtration*, in "Encyclopedia of Fluid Mechanics", Vol. 5, Ed. N.P. Cheremisinoff, Gulf Publishing, Houston.

Wakeman R.J., 1993. Scale-up procedures and test methods in solid/liquid separation: 2. Filtration of binary particle mixtures and flocculated suspensions, *IFPRI Annual Research Report ARR 25–02*.

Wakeman R.J., 1994. Modelling slurry dewatering and cake growth in filtering centrifuges, *Filtration and Separation*, **31**, 75–81.

Wakeman R.J., 1996. Fouling in crossflow ultra- and micro-filtration, *Membrane Technology*, **70**, 5.

Wakeman R.J., 1997. Post filtration processes and their impact on drying, *Chemical Process Technology International*, **11**, 117–125.

Wakeman R.J., 1999. Visualisation of cake formation in crossflow microfiltration, *Trans IChemE*, Part A, **152**, 89–98.

Wakeman R.J. and Rushton A., 1974. A structural model for filter cake washing, *Chem. Eng. Sci.*, **29**, 1857–1865.

Wakeman R.J. and Rushton A., 1977. Dewatering properties of particulate beds, *J. Powder & Bulk Solids Technol.*, **1**, 64–69.

Wakeman R.J. and Mulhaupt B., 1985. Process design and scale-up of multi-stage washing pusher centrifuges, *Filtration and Separation*, **22**, 231–234.

Wakeman R.J. and Vince A., 1986a. Kinetics of gravity drainage from porous media, *Trans IChemE,* **64**, 94–103.

Wakeman R.J. and Vince A., 1986b. An engineering model for the kinetics of drainage from centrifuge cakes, *Trans IChemE*, **64**, 104–108.

Wakeman R.J. and Attwood G.J., 1988. Developments in the applications of cake washing theory, *Filtration and Separation*, **25**, 272–275.

Wakeman R.J., Thuraisingham S.T. and Tarleton E.S., 1989. Colloid science in solid/liquid separation technology – Is it important?, *Filtration and Separation*, **26**, 277–283.

Wakeman R.J. and Attwood G.J., 1990. Simulations of dispersion phenomena in filter cake washing, *Trans IChemE*, **68**, 161–171.

Wakeman R.J., Sabri M.N. and Tarleton E.S., 1991. Factors affecting the formation and properties of wet compacts, *Powder Technol.*, **65**, 283–292.

Walas S.M., 1941. PhD Thesis, University of Michigan, Ann Arbor, Michigan.

Walters R.R., Graham J.F., Moore R.M. and Anderson D.J., 1984. *Anal. Biochem.*, **140**, 190.

Williams C. and Wakeman R.J., 2000, Membrane fouling and alternative techniques for its alleviation, *Membrane Technology*, **124**, 4–10.

Willis M.S., 1983. *A multiphase theory of filtration*, in "Progress in Filtration and Separation 3", Ed. R.J. Wakeman, Elsevier, Amsterdam.

Wolf W.J. and Briggs D.R., 1959. *Arch. Biochem. Biophys.*, **85**, 186.

Wu D., Howell J.A. and Field R.W., 1999. Critical flux measurement for model colloids, *J. Membrane Sci.*, **152**, 89–98.

Wyllie M.R.J. and Gardner G.H.F., 1958. The generalised Kozeny-Carman equation II. A novel approach to problems of fluid flow, *World Oil Prod. Sect.*, 210–228.

Yim S.S., Song Y.M. and Kwon Y.D., 2003. The role of p(i), p(o), and p(f) in constitutive equations and new boundary conditions in cake filtration, *Korean J. Chem. Eng.*, **20**, 334–342.

Yoshioka N., Hotta Y., Tanaka S., Naito S. and Tongami S., 1957. Continuous thickening of homogeneous slurries, *J. Chemical Engineering Japan*, **21**, 66–74.

Young M.E., Carroad P.A. and Bell R.L., 1980. Estimation of diffusion coefficients of proteins, *Biotech. Bioeng.*, **22**, 947.

Zeitsch K., 1977. *Proc. International Symposium*, Société Belge de Filtration, Antwerp.

Zeman L.J. and Zydney A.L., 1996. *Microfiltration and Ultrafiltration*, Marcel Dekker, New York.

Additional bibliography

The following are useful additional sources of reference information that have been helpful during the writing of this book.

Akers R.J., 1972. *Flocculation*, IChemE, London.

Alles C.M. and Anlauf H., 2003. Efficient process strategies for compressible cake filtration, *Chemie Ingenieur Technik*, **75**, 1221–1230.

Alt C., 1985. *Centrifugal separation*, in "Mathematical Models and Design Methods in Solid-Liquid Separation", Ed. A. Rushton, Martinus Nijhoff, Dordrecht.

Ambler C.M., 1971. Centrifuge selections, *Chem. Engng*, **78**, 55–62.

Andersen N.P.R., Agerbaek M.L. and Keiding K., 2003. Measurement of electrokinetics in cake filtration, *Colloids and Surfaces A – Physicochemical and Engineering Aspects*, **213**, 27–36.

Besra L., Sengupta D.K., Roy S.K. and Ay P., 2004. Influence of polymer adsorption and conformation on flocculation and dewatering of kaolin suspension, *Separ. Purif. Technol.*, **37**, 231–246.

Biesheuvel P.M., 2000, Particle segregation during pressure filtration for cast formation, *Chem. Eng. Sci.*, **55**, 2595–2606.

Birss R.R. and Parker M.R., 1981. *High intensity magnetic separation*, in "Progress in Filtration and Separation 2", Ed. R.J. Wakeman, pp. 171–304, Elsevier, Amsterdam.

Blagden H.R., 1975. Separating solids from liquids: Filter selection I, *Chemical Processing*, **21**, 29–33.

Bollinger J.M. and Adams R.A., 1984. Electrofiltration of ultrafine aqueous dispersions, *Chem. Eng. Prog.*, **80**, 54–58.

Bratby J., *Coagulation and Flocculation*, Uplands Press, 1980.

Brewer P., 1981. Sheet filtration applied to the pharmaceutical and food industries, *Filtration and Separation*, **18**, 242–246.

Bruncher B., 1984. *Filtration and Wire Cloth*, Gantois, Saint-Dié.

Chen N.H., 1978. Liquid-solid filtration: Generalised design and optimisation equations, *Chem. Engng*, **85**, 97–101.

Civan F., 1998. Incompressive cake filtration: Mechanism, parameters, and modelling, *AIChEJ*, **44**, 2379–2387.

Civan F., 1998. Practical model for compressive cake filtration including fine particle invasion, *AIChEJ*, **44**, 2388–2398.

Dahlstrom D.A., 1978. *How to select and size filters*, in "Mineral Processing Plant Design", Eds. A.L. Mular and R.B. Bhappu, pp. 578–600, AIME., New York.

Das S. and Ramarao B.V., 2002. Inversion of lime mud and papermaking pulp filtration data to determine compressibility and permeability relationships, *Separ. Purif. Technol.*, **28**, 149–160.

Davies E., 1965. Selection of equipment for solid/liquid separations, *Trans IChemE*, **43**, 256–259.

Davies E., 1970. What is the right choice of filter or centrifuge, *Filtration and Separation*, **7**, 76–79.

Day R.W., 1974. Techniques for selecting centrifuges, *Chem. Engng*, **81**, 98–104.

Dickenson C., 1997. *Filters and Filtration Handbook*, 4th Edition, Elsevier, Oxford.

Ehlers S., 1961. The selection of filter fabrics re-examined, *Ind. Eng. Chem.*, **53**(7), 552–556.

Emmett R.C. and Silverblatt C.E., 1974. When to use continuous filtration hardware, *Chem. Eng. Prog.*, **70**, 38–42.

Emmett R.C. and Silverblatt C.E., 1975. When to use continuous filtration hardware, *Filtration and Separation*, **12**, 577–581.

Ernst M., Talcott R.M., Romans H.C. and Smith G.R.S., 1991. Tackle solid-liquid separation problems, *Chem. Eng. Prog.*, **87**, 22–28.

Ernst M., Talcott R.M., San Giovanni J. and Romans H.C., 1987. Methodology for selecting solid-liquid separation equipment, *Proc. AIChE Spring National Meeting*, pp. A1–A11, AIChE, Houston.

Fan D., Wakeman R.J. and Huang Z., 1991. An analysis of gyratory forces and wobble angles in tumbler centrifuges, *Trans IChemE*, **69**, 409–416.

Fan D., Wakeman R.J. and Huang Z., 1992. Effects of step drums on solids residence times in conical basket and tumbler centrifuges, *Filtration and Separation*, **29**, 147–154.

Fitch B., 1974. Choosing a separation technique, *Chem. Eng. Prog.*, **70**, 33–37.

Fitch B., 1977. When to use separation techniques other than filtration, *AIChE Symposium Series*, **73(171)**, 104–108.

Fitzgibbons D.P., 1976. Filtration rates of rotary vacuum filters, *Filtration and Separation*, **13**, 227–232.

Flood J.E., Porter H.F. and Rennie F.W., 1966. Filtration practice today, *Chem. Eng.*, **73**, 163–181.

Garg M.K., Douglas P.L. and Linders J.G., 1991. An expert system for identifying separation processes, *Canadian Journal Chemical Engineering*, **69**, 67–75.

Gaudfrin G. and Sabatier E., 1978. Tentative procedure to choose a filtration equipment, *Proc. Symposium on Liquid/Solid Filtration*, pp. 29–47, Soc. Belge de Filtration, Amsterdam.

Gale R.S., 1971. Recent research on sludge dewatering, *Filtration and Separation*, **8**, 531–538.

Glover S.M., Yan Y.D., Jameson G.J. and Biggs S., 2004. Dewatering properties of dual-polymer-flocculated systems, *Int. J. Min. Proc.*, **73**, 145–160.

Grace H.P., 1951. What type of filter and why, *Chem. Eng. Prog.*, **47**, 502–507.

Gregory J., 1973. Rates of flocculation of latex particles by cationic polymers, *J. Colloid Interface Sci.*, **42**, 448–456.

Gucbilmez Y., Tosun I. and Yilmaz L., 2000. Optimization of the locations of side streams in a filter cake washing process, *Chem. Eng. Comm.*, **182**, 49–67.

Hardman E., 1994. Some aspects of the design of filter fabrics for use in solid/liquid separation processes, *Filtration and Separation*, **31**, 813–818.

Hawkes R.O., 1970. Optimum utilisation of equipment characteristics, *Filtration and Separation*, **7**, 311–318.

Heertjes, P.M., 1957. Studies in filtration – blocking filtration, *Chem. Eng. Sci.*, **6**, 190–203.

Heertjes P.M. and Haas H., 1949. Studies in filtration, *Rec. Trav. Chim.*, **68**, 361–383.

Heertjes P.M. and Nijman J., 1957. On the instability and inhomogeneity of filter cakes, *Chem. Eng. Sci.*, **7**, 15–25.

Hermia, J., 1980. Pre- and post-treatment techniques for industrial separation, *Filtration and Separation*, **17**, 362–370.

Hicks C.P. and Hillgard A., 1970. Pressure filtration or centrifugal separation – complementary or competitive, *Filtration and Separation*, **7**, 456–460.

Hirata Y., Onoue K. and Tanaka Y., 2003. Effects of pH and concentration of aqueous alumina suspensions on pressure filtration rate and green microstructure of consolidated powder cake, *J. Ceramic Society Japan*, **111**, 93–99.

Hosten C. and San O., 2002, Reassessment of correlations for the dewatering characteristics of filter cakes, *Minerals Engineering*, **15**, 347–353.

Hunter R.J., 1995. *Foundations of Colloid Science*, Clarendon Press, Oxford.

Hwang K.J. and Lu W.M., 1997. A simple model for estimating surface porosity of cake in cake filtration of submicron particles, *J. Chinese Inst. Chem. Eng.*, **28**, 121–129.

Hwang K.J., Huang C.Y. and Lu W.M., 1998. Constant pressure filtration of suspension in viscoelastic fluid, *J. Chem. Eng. Japan*, **31**, 558–564.

Iritani E., Mukai Y. and Yorita H., 1999. Effect of sedimentation on properties of upward and downward cake filtration, *Kagaku Kogaku Ronbunshu*, **25**, 742–746.

Iritani E., Mukai Y. and Hagihara E., 2002. Measurements and evaluation of concentration distributions in filter cake formed in dead-end ultrafiltration of protein solutions, *Chem. Eng. Sci.*, **57**, 53–62.

Ives K.J. (Ed.), 1975. *The Scientific Basis of Filtration*, Noordhoff, Leyden.

Kelsey G.D., 1965. Some practical aspects of continuous rotary vacuum filters, *Trans IChemE*, **43**, T248–T255.

Kirk-Othmer Encyclopedia of Chemical Technology, 1980, 3rd Edition, Wiley, New York.

Kobayashi Y., Ohba S. and Shimuzu K., 1993. Analysis of dewatering performance of belt press filter, *Proc. 6th World Filtration Congress*, pp. 778–781, Japanese Filtration Society, Nagoya.

Komline T.R., 1980. Sludge dewatering equipment and performance, *AIChE Symposium Series*, **76**(197), 321–332.

Korhonen E., Lahdenperä E. and Nyström L., 1989. Selection of equipment for solid-liquid separation by expert systems, *Proc. Filtech Conference*, pp. 436–443, The Filtration Society, Karlsruhe.

Kosvintsev S., Holdich R.G., Cumming I.W. and Starov V.M., 2002. Modelling of dead-end microfiltration with pore blocking and cake formation, *J. Mem. Sci.*, **208**, 181–192.

Kukreja V.K., Ray A.K. and Singh V.P., 1998. Mathematical models for washing and dewatering zones of a rotary vacuum filter, *Indian J. Chem. Technol.*, **5**, 276–280.

La Mer, V.K. and Healy, T.W., 1966. *The nature of the flocculation reaction in solid-liquid separation*, in "Solid/Liquid Separation", Eds. J.B. Poole and D. Doyle, pp. 44–59, HMSO, London.

Landman K.A., Stankovich J.M. and White L.R., 1999. Measurement of the filtration diffusivity D(phi) of a flocculated suspension, *AIChEJ*, **45**, 1875–1882.

Lee D.J. and Wang C.H., 2000. Theories of cake filtration and consolidation and implications to sludge dewatering, *Water Research*, **34**, 1–20.

Lin I.J. and Benguigui L., 1983. *High-intensity, high-gradient dielectrophoretic filtration and separation processes*, in "Progress in Filtration and Separation 3", Ed. R.J. Wakeman, pp. 149–204, Elsevier, Amsterdam.

Lu W.M., Tung K.L. and Hwang K.J., 1997. Effect of woven structure on transient characteristics of cake filtration, *Chem. Eng. Sci.*, **52**, 1743–1756.

Lu W.M., Tung K.L., Hung S.M., Shiau J.S. and Hwang K.J., 2001. Constant pressure filtration of mono-dispersed deformable particle slurry, *Separ. Sci. Technol.*, **36**, 2355–2383.

Maloney G.F., 1972. Selecting and using pressure leaf filters, *Chem. Engng*, **79**(11), 88–94.

Matteson M.J. and Orr C. (Eds.), 1987. *Filtration*, Marcel Dekker, New York.

Matthews H.B. and Rawlings J.B., 1998. Batch crystallization of a photochemical: Modeling, control, and filtration, *AIChEJ*, **44**, 1119–1127.

Meyer E. and Lim H.S., 1989. New nonwoven microfiltration membrane material, *Fluid/Particle Separation Journal*, **2**, 17–21.

Moody G.M., 1995. Pre-treatment chemicals, *Filtration and Separation*, **3**, 329–336.

Moos S.M. and Dugger R.E., 1979. Vacuum filtration: Available equipment and recent innovations, *Minerals Engineering*, **31**, 1473–1486.

Moyers C.G., 1966. How to approach a centrifuge problem, *Chem. Engng*, **73**, 182–189.

Nelson P.A. and Dahlstrom D.A., 1957. Moisture-content correlation of rotary vacuum filter cakes, *Chem. Eng. Prog.*, **53**(7), 320–327.

Murkes J. and Carlsson C.G., 1988. *Crossflow Filtration*, Wiley, Chichester.

Nachinkin O.I., 1991. *Polymeric Microfilters*, Ellis Horwood, New York.

Nakakura H., Sambuichi M., Ishitoku H. and Osasa K., 2001. Filtration mechanism of gel particle slurry, *J. Chem. Eng. Japan*, **34**, 862–868.

Nyström, L.H.E., 1993. Simulation of disc filter for the pulp and paper industries, *Filtration and Separation*, **30**, 554–556

Perry R.H. and Green D., 1984. *Perry's Chemical Engineers' Handbook*, 6th Edition, McGraw-Hill, New York.

Pierson H.G.W., 1990. *The selection of solid/liquid separation equipment* in *Solid-Liquid Separation*, Ed. L. Svarovsky, pp. 525–538, Butterworths, London.

Porter M.C. (Ed.), 1990. *Handbook of Industrial Membrane Technology*, Noyes Publications, New Jersey.

Purchas D.B., 1970. A non-guide to filter selection, *Chemical Engineer*, **237**, 79–82.

Purchas D.B., 1972a. Guide to trouble-free plant operation, *Chem. Engng*, **79**, 88–96.

Purchas D.B., 1972b. Cake filter testing and sizing. A standardised procedure, *Filtration and Separation*, **9**, 161–171.

Purchas D.B., 1978. Solid/liquid separation equipment. A preliminary experimental selection programme, *Chemical Engineer*, **328**, 47–49.

Purchas D.B., 1980. Art, science & filter media, *Filtration and Separation*, **17**, 372–376.

Purchas D.B., 1981. *Solid/Liquid Separation Technology*, Filtration Specialists, UK.

Purchas D.B., 1996. *Handbook of Filter Media*, Elsevier Advanced Technology, Oxford.

Rivet P., 1981. *Guide de la Séparation Liquide-Solide*, Société Française de Filtration.

Rushton A., 1969. The effect of concentration in rotary vacuum filtration, *Filtration and Separation*, **6**, 136–139.

Rushton A., 1978. Pressure variation effects in rotary drum filtration with incompressible cakes, *Powder Technol.*, **20**, 39–46.

Rushton A., 1981. Centrifugal filtration, dewatering and washing, *Filtration and Separation*, **18**, 410–415.

Rushton A. and Wakeman R.J., 1978. Theory vs. practice in vacuum, pressure and centrifugal filtration, *J. Powder & Bulk Solids Tech.*, **1**, 58–65.

Sambuichi M., Nakakura H., Nishigaki F. and Osasa K., 1994. Dewatering of gels by constant pressure expression, *J. Chem. Eng. Japan*, **27**, 1994.

Sanstedt H.N., 1980. Non-wovens in filtration applications, *Filtration and Separation*, **17**, 358–361.

Schweitzer P.A. (Ed.), 1997. *Handbook of Separation Techniques for Chemical Engineers*, 3rd Edition, McGraw Hill, New York.

Scott K., 1997. *Handbook of Industrial Membranes*, Elsevier, Oxford.

Sedin P. and Theliander H., 2004. Filtration properties of green liquor sludge, *Nordic Pulp & Paper Research Journal*, **19**, 67–74.

Shaw D.J., 1992. *Introduction to Colloid and Surface Chemistry*, 4th Edition, Butterworths.

Sis H. and Chander S., 2000. Pressure filtration of dispersed and flocculated alumina slurries, *Minerals & Metallurgical Processing*, **17**, 41–48.

Smiles D.E., 1999. Centrifugal filtration of particulate systems, *Chem. Eng. Sci.*, **54**, 215–224.

Smiles D.E., 2000. Material coordinates and solute movement in consolidating clay, *Chem. Eng. Sci.*, **55**, 773–781.

Smith J.C., 1955, How to approach your separation problem, *Chem. Engng*, **62(6)**, 177–184.

Sørensen B.L. and Sørensen PB, 1997. Structure compression in cake filtration, *J. Environmental Engineering-ASCE*, **123**, 345–353.

Sperry D.R., 1924. What is the most suitable filter, *Chem. Metall. Engng*, **31**, 422–428.

Stahl W. and Nicolaou I., 1990. Calculation of rotary vacuum plant, *Proc. 5th World Filtration Congress*, pp. 37–44, Société Française de Filtration, Nice.

Suh C.W., Kim S.E. and Lee E.K., 1997. Effects of filter additives on cake filtration performance, *Korean J. Chem. Eng.*, **14**, 241–244.

Sullivan M.S. and Johnson M., 1997. The use of high filtration pressure in the dewatering of industrial mineral effluents for disposal to landfill, *Proc. Filtech Conference*, pp. 63–73, The Filtration Society, Düsseldorf.

Svarovsky L., 1984. *Hydrocyclones*, Holt, Rinehart & Winston, Eastbourne.

Tan W., Lu S.Q., Wu Y.T. and Zhu Q.X., 2003. Theoretical study and analysis on properties of filter aids, *Chinese J. Chem. Eng.*, **11**, 249–252.

Tarleton E.S., 1992. The role of field assisted techniques in solid/liquid separation, *Filtration and Separation*, **29**, 246–252.

Tarleton E.S., 1996. A mechatronics approach to solid/liquid separation, *Proc. 7th World Filtration Congress*, pp. 311–315, Hungarian Chemical Society, Budapest.

Tarleton E.S., 1998. The control of pressure in constant rate cake filtration, *Proc. International Symposium Filtration and Separation*, pp. 87–94, Ibérica de Filtración y Separación, Las Palmas.

Tarleton E.S. and Wakeman R.J., 1991. *Solid/Liquid Separation Equipment Simulation & Design: p^C-SELECT – Personal computer software for the analysis of filtration and sedimentation test data and the selection of solid/liquid separation equipment*, Separations Technology Associates, Loughborough.

Tarleton E.S. and Wakeman R.J., 1994. The simulation, modelling and sizing of pressure filters, *Filtration and Separation*, **31**, 393–397.

Tarleton E.S. and Hancock D.L., 1996. The imaging of filter cakes through electrical impedance tomography, *Filtration and Separation*, **33**, 491–494.

Trawinski H.F., 1980. Current solid/liquid separation technology, *Filtration and Separation*, **17**, 326–335.

Wakeman R.J., 1976. The influence of entrapped gas and resaturation on diffusional extraction in porous media, *Chem. Eng. J.*, **16**, 73–78.

Wakeman R.J., 1980. The estimation of cake washing characteristics, *Filtration and Separation*, **17**, 67–73.

Wakeman R.J., 1981. The application of expression in variable volume filters, *J. Separ. Process Technol.*, **2**, 1–8.

Wakeman R.J., 1982. Effects of solids concentration and pH on electrofiltration, *Filtration and Separation*, **19**, 316–319.

Wakeman R.J., 1983. Some effects of pretreatment on the optimal design of washing filters, *Filtration and Separation*, **20**, 195–199.

Wakeman R.J., 1984. Filtration and washing on vacuum filters – process optimisation for economy, *Filtration and Separation*, **21**, 201–205.

Wakeman R.J., 1984. Residual saturation and dewatering of fine coals and filter cakes, *Powder Technol.*, **40**, 53–63.

Wakeman R.J., 1985. *Filtration Dictionary and Glossary*, 295pp., The Filtration Society.

Wakeman R.J., 1995. Selection of equipment for solid/liquid separation processes, *Filtration and Separation*, **32**, 337–341.

Wakeman R.J., 1998. Washing thin and non-uniform filter cakes: Effects of wash liquor maldistribution, *Filtration and Separation*, **35**, 185–190.

Wakeman R.J. and Rushton A., 1976. The removal of filtrate and soluble material from compressible filter cakes, *Filtration and Separation*, **13**, 450–454.

Wakeman R.J., Rushton A. and Brewis L.N., 1976. Residual saturation of dewatered filter cakes, *The Chem. Engr. (London)*, **314**, 668–670.

Wakeman R.J., Mehrotra V.P. and Sastry K.V.S., 1981. Mechanical dewatering of fine coal and refuse slurries, *Bulk Solids Handling*, **1**, 281–293.

Wakeman R.J. and Jimenez-Novoa C.E., 1985. Lavado de tortas de filtrado, *Ingeneria Quimica*, **17**, 147–156.

Wakeman R.J., Davies T.P. and Manning C.J., 1988. The HW filter – a new concept for clarification filtration, *Filtration and Separation*, **25**, 407–410.

Wakeman R.J. and Tarleton E.S., 1990. Modelling, simulation and process design of the filter cycle, *Filtration and Separation*, **27**, 412–419.

Wakeman R.J. and Fan D., 1991. The control ring centrifuge – a new type of conical basket centrifuge, *Trans IChemE*, **69**, Part A, 403–408.

Wakeman R.J. and Tarleton E.S., 1991. An experimental study of electroacoustic crossflow microfiltration, *Trans IChemE*, **69**, Part A, 386–397.

Wakeman R.J. and Tarleton E.S., 1991. Solid/liquid separation equipment simulation and design – an expert systems approach, *Filtration and Separation*, **28**, 268–274.

Wakeman R.J. and Tarleton E.S., 1993. Sensitivity analysis for solid/liquid separation equipment selection using an expert system, *Proc. Filtech Conference*, pp. 43–57, The Filtration Society, Karlsruhe.

Wakeman R.J. and Tarleton E.S., 1994a. A framework methodology for the simulation and sizing of diaphragm filter presses, *Minerals Engineering*, **7**, 1411–1425.

Wakeman R.J. and Tarleton E.S., 1994b. Scale-up procedures and test methods in solid-liquid separation: methodology for the simulation and sizing of pressure filters, *Proc. 15th annual IFPRI meeting*, Goslar.

Wakeman R.J. and Wei X., 1995. Simulating the performance of tilting pan filters, *Filtration and Separation*, **32**, 979–984.

Wakeman R.J., Burgess D.R. and Stark R.J., 1994. The Howden-Wakeman filter in wastewater treatment, *Filtration and Separation*, **31**, 183–187.

Wakeman R.J. and Akay G., 1997. Membrane-solute and liquid-particle interaction effects in filtration, *Filtration and Separation*, **34**, 511–519.

Watson J.H.P., 1990. *High gradient magnetic separation*, in "Solid-Liquid Separation", 3rd Edition, Ed. L. Svarovsky, pp. 661–684, Butterworths, London.

Wills B.A., 1992. *Mineral Processing Technology*, 5th Edition, Pergamon, Oxford.

Wu Y.X. and Wang B.Y., 2004. Analysis of the medium resistance for constant pressure filtration, *Chinese J. Chem. Eng.*, **12**, 33–36.

Yelshin A. and Tiller F.M., 1989. Optimising candle filters for incompressible cakes, *Filtration and Separation*, **26**, 436–437.

Yim S.S., 1999. A theoretical and experimental study on cake filtration with sedimentation, *Korean J. Chem. Eng.*, **16**, 308–315.

Yim S.S. and Kim J.H., 2000. An experimental and theoretical study on the initial period of cake filtration, *Korean J. Chem. Eng.*, **17**, 393–400.

Yim S.S. and Kwon Y.D., 1997. A unified theory on solid-liquid separation: Filtration, expression, sedimentation, filtration by centrifugal force, and cross flow filtration, *Korean J. Chem. Eng.*, **14**, 354–358.

Yim S.S., Kwon Y.D. and Kim H.I., 2001. Effects of pore size, suspension concentration, and pre-sedimentation on the measurement of filter medium resistance in cake filtration, *Korean J. Chem. Eng.*, **18**, 741–749.

Index